General
Applied
Statistics

General Applied Statistics

THIRD EDITION

Fadil H. Zuwaylif

California State University, Northridge

ADDISON-WESLEY PUBLISHING COMPANY

Reading, Massachusetts

Menlo Park, California

London • Amsterdam

Don Mills, Ontario • Sydney

Second printing, July 1980

ISBN 0-201-08994-7
ABCDEFGHIJ-DO-89876543210

To Suzanne and Samara

Preface to the Third Edition

This edition includes new material and a greater emphasis on some subjects in several areas. The distinction between objective and subjective probabilities is greatly emphasized in Chapter 4. The treatment of type I and type II errors in Chapter 7 is expanded by adding new illustrative problems and a new set of exercises.

This edition also contains a completely new chapter dealing with inferences in regression. Although this subject is somewhat difficult, it is treated with the same clarity and simplicity that characterize the text. Finally, a rich variety of new exercises have been added throughout the text.

I would like to take this opportunity to thank Professors Thomas Wedel and Ernest Scheuer for their helpful comments on the new material. In addition, I wish to express my appreciation to Rita Hary, Debbie Kitch, and Lisa Boultinghouse for their typing efforts; and to George Abbott, my editor, for his help and guidance.

Northridge, California   F. H. Z.
March, 1979

Preface to the Second Edition

This edition includes new material in several areas. Chapter 4, which deals with probability distributions, is greatly expanded. The cumulative binomial distribution, binomial probabilities table, the Poisson distribution, and Poisson probabilities table are added to this chapter. Type I and Type II errors in hypothesis testing are now treated in Chapter 7.

This edition also contains a complete new chapter on nonparametric statistics. The two chapters dealing with time series in the first edition of this text are entirely omitted. In addition to many new, illustrative problems throughout the chapters, three sets of review exercises are provided in this edition.

I wish to take this opportunity to thank all who used my book for their suggestions for its improvement. I am also indebted to D. Van Nostrand Company, Inc., for permission to reproduce, in Table K, material from E. C. Molina's *Poisson Exponential Limit*, and to Richard D. Irwin, Inc., for permission to reproduce Figs. 6–3, 6–4, 6–5, and 6–6 from *Statistics for Business Decisions*, E. Kurnow, G. J. Glasser, and F. Ottman.

Northridge, California   F.H.Z.
December 1973

Preface to the First Edition

This book is written to serve as a text for a first course in statistics. The main features of the book are its simplicity of exposition and its extensive use of illustrative problems and exercises. No mathematical knowledge beyond high school algebra is needed as a prerequisite.

Another important feature of this book is its organization. A clear distinction between a population and a sample is made at the outset in Chapter 1, and this distinction is maintained throughout all subsequent chapters. Chapter 2 is devoted exclusively to the treatment of population parameters. A general survey of probability is presented in Chapters 3 and

4, and the subject of sampling is treated in Chapter 5. Since the book deals primarily with statistical inference concerning the arithmetic mean and the proportion, Chapters 6, 7, and 8 are devoted to the problems of estimating the mean and testing hypotheses concerning the mean. Statistical inference concerning the proportion is the subject matter of Chapters 9 and 10. An elementary treatment of regression and correlation is presented in Chapter 11.

Chapters 12 and 13 are written primarily for those who desire to use this book as a text for business statistics. These two chapters deal with the analysis of time series. After a concise treatment of linear and semiloga-rithmic trend measurement and seasonal index construction, the use of these tools in forecasting is thoroughly discussed.

I wish to express my appreciation to all those who made the preparation of this book possible. I am particularly indebted to Professor Robert Ferber for stimulating my interest in statistics and for making my training in that field possible. I am also grateful to Professors David Hitchcock and Max Astrachan for their valuable criticism and suggestions; to my wife Suzanne, to Sally Gary, and to Masayo Kubota for typing the manuscript.

I am indebted to the Literary Executor of the Late Sir Ronald A. Fisher, F. R. S., and to Oliver & Boyd Ltd., Edinburgh, for their permission to reprint Tables III and IV from *Statistical Methods for Research Workers;* to Professors E. S. Pearson and H. O. Hartley, and Cambridge University Press for permission to reprint in abridged form, as Tables B and C, Table 18 from *Biometrika Tables for Statisticians*, Vol. I; to Professors G. W. Snedecor and W. G. Cochran, and Iowa State University Press, Ames, Iowa, for permission to reprint Table A11 from *Statistical Methods*, 6th ed.

Northridge, California   F. H. Z.
November 1969

Contents

1
Introduction

Introduction

Statistics

As purchasing manager for the *Pocket Radio Corporation* you must decide between two brands of 9-volt batteries which are available locally in ample supply. The wholesale prices of the two brands, *A* and *B*, are identical. The choice must therefore be based on the comparative quality of the two products.

There are various measures for quality, but you have chosen the average life of a battery as the most important measure. You wish to purchase the brand with the highest average life.

To make your choice, you test 100 batteries of each brand. Brand *A* shows an average life of 48 hours and brand *B*, 47 hours. Concluding that *A* has a longer life than *B*, you purchase *A*.

Are you *absolutely* sure that *A* is superior to *B*? The answer clearly is "no." Most probably brand *A* is superior to *B*. However, the possibility still exists that the two brands are of the same quality. It is also possible that brand *B* is superior to *A*, although this possibility is less probable.

How can we then compare the average lives of the two brands if we desire our conclusion to be absolutely correct? We must examine not only 100 batteries, not even 1000 batteries, but all the batteries that are made and will be made of the two brands.

If the life of every battery of brand *A* is measured (this is an impossible task indeed, since more and more batteries are produced every day), we obtain what is called the *statistical population* of brand *A*. The statistical population of brand *A*, therefore, consists of thousands and thousands of numbers, where each number measures the life of an individual battery. A statistical population is complete data. If such complete data are available, the average life of a brand *A* battery can be determined with absolute certainty.

In the same manner, the statistical population of brand B consists of very many numbers, each describing the life of a particular battery of brand B.

The statistical populations of the two brands are infinite, and complete data are impossible to obtain. Therefore, the average life of each brand must be determined from a given number of batteries, say 100 batteries. These 100 batteries represent a part or a *sample* of the complete population. A sample is a part of a population.

When we compare the average lives of the two types of batteries, our conclusion will be based only on a sample selected from each brand. But any conclusion based on a sample rather than on complete population data may be erroneous because the sample may not be representative of the population from which it was selected.

It is true that the risk of incorrect conclusions is always present unless one obtains data on the entire population. However, conclusions based on sample data can be evaluated by various statistical methods and techniques. Knowledge of statistics enables us to attach certain confidence to our conclusions. We may be able to state, for example, that there is .95 or .99 confidence that brand A is superior to brand B. It must be emphasized again that absolute certainty is obtained only when complete population data are examined.

Statistics, therefore, is a field of knowledge that enables the investigator to derive and evaluate conclusions about a population from sample data. Specifically, statistics deals with theorems, tools, methods, and techniques to be used in

1. gathering, selecting, and classifying data;
2. interpretation and analysis of data;
3. deriving and evaluating the reliability of conclusions based on sample data.

Statistical Populations

The distinction between a sample and a population must be made clear at the outset. A set of data is a population if decisions and conclusions based on these data can be made with absolute certainty. In other words, if population data are available, the risk of arriving at incorrect decisions is completely eliminated. In contrast, a sample is only a part of a population, and conclusions based on any part of a population could be erroneous indeed.

The data of which a statistical population consists are either measured or described characteristics of certain objects. These objects are called the elementary units of the population. We shall give some examples of statistical populations to clarify these concepts.

1. Let us say that we wish to determine the average annual family income in a community of 10,000 families. Then our population consists of 10,000 numbers ($5273, $4000, $10,500, ...), each number measuring the income of a particular family. The elementary unit is a family living in that community, and the characteristic to be measured is its annual income.

2. We wish to determine the average age of a student attending San Francisco State University, where total enrollment is 15,000 students. Then our statistical population consists of 15,000 numbers (18, 23, 22, ...), each indicating the age of a particular student. The elementary unit is a student attending San Francisco State University, and the characteristic to be measured is the student age.

3. Not all characteristics can be measured; some can only be described. We want to determine the vacancy rate in Northridge, California, where there are 20,000 dwelling units. Our population of interest consists of 20,000 terms (occupied, occupied, occupied, vacant, occupied, ...), each term describing the occupancy status of a particular dwelling unit in Northridge.

Thus, a statistical population is a set of measured or described observations made on each and every elementary unit. Measurable observations are called quantitative observations. The annual income of a family, the age of a student, the life of an electric bulb, and the shearing strength of a weld are all measurable and are, therefore, *quantitative* observations. In comparison, the vacancy status of a dwelling unit, the marital status of a person, the make of the car we drive, and the brand of cigarettes we smoke cannot be measured; they can only be described. Such nonmeasurable observations are called *qualitative* observations.

The values assumed by quantitative observations are called *variates*. Thus, if the annual income of a family is $15,320.17, the number $15,320.17 is called a variate. Similarly, if the life of an electric bulb is 852 hours, the number 852 is also called a variate.

Quantitative observations are further classified as either discrete or continuous. A discrete quantitative observation can assume only *a limited* number of values on a measuring scale. The number of students in a statistics class, for example, can be 25 or 26, but it certainly cannot be 25.3 or 25.712. Similarly, the size of a family can be either 1, 2, 3, 4, ..., but it cannot be 2.37 or 4.15. The number of students in a statistics class and the family size are thus considered as discrete, since each can assume only a limited number of values on a measuring scale.

Some quantitative observations, on the other hand, can assume an infinite number of values on a measuring scale. These quantitative observations are called continuous.   The age of a student in this class, for example, can be 18 or 19 years; but it can also be, at least theoretically, one of an infinite number of values between 18 and 19 such as 18.31 or 18.813 years. In the same manner, a pineapple can weigh 3 or 4 pounds as well as an infinite number of other values between 3 and 4 pounds such as 3.178 pounds. The age of a student and the weight of a pineapple are thus considered as continuous quantitative observations since each can assume an infinite number of values on a measuring scale.

Qualitative observations cannot be measured; they can only be described. The vacancy status of a dwelling unit, for example, can be described by the terms occupied or vacant. These terms are called *attributes*. Similarly, the marital status of a person can be described by such terms as single, married, divorced, or widowed. These terms, again, are called attributes.

The distinction between a population of variates (a set of *measured observations*) and a population of attributes (a set of *described observations*) is important. Since variates are measurable, their arithmetic mean or simple average can be calculated. The average age of a student, the average family income, the average life of an electric bulb, and the average shearing strength of a weld can be and most often are calculated.

Ratios or percentages, on the other hand, are calculated for attributes. The percentage of vacant dwelling units, the percentage of defective articles in a given lot, the percentage of senior students in a college are usually calculated from data that consist of attributes.

In conclusion, it must be emphasized that although the decision in a given problem depends on such information as the value of the average variate in the population, such average value is seldom calculated using the entire population data. The value of the average variate is usually estimated from a carefully selected sample.

In many problems, the statistical population is infinite, and therefore complete data are impossible to obtain. Even when population data can be obtained, the cost and time involved in collecting and analyzing such data would be prohibitive. Information is therefore obtained from sample data, and decisions based on sample information comprise the hard core of statistics.

Frequency Distribution

Statistical data may consist of a very large number of observations.   The larger the number of observations, the greater the need to present the data in a summarized form which may omit some details, but reveals the general nature of a mass of data. A frequency distribution is a summary of a large number of observations. Table 1–1 below summarizes the ages of 1,763,000 unemployed males in the United States in a certain month.

Table 1–1

Unemployed males looking for full or part-time work,
by age

Age	Number of unemployed males
14 to 19 years	218,000
20 to 24 years	313,000
25 to 55 years	977,000
55 years and over	255,000
TOTAL	1,763,000

Although the above frequency distribution does not list the age of every unemployed male, it certainly presents a better general picture of the ages of unemployed males than a list of 1,763,000 individual numbers occupying hundreds of pages.

Table 1–2 summarizes the major fields of study of 1301 students attending the School of Business in California State University, Northridge during the Fall of 1980.

Table 1–1 shows a frequency distribution of variates since the age of an unemployed person can be measured in years. Table 1–2, on the other hand, is a frequency distribution of attributes. A student's major field cannot be measured but can only be described as accounting, marketing, finance, etc.

Table 1–2

Major fields of study for students at School of Business,
California State University, Northridge, 1980.

Major field	Number of students
Accounting	472
Business education	42
Economics	88
Finance	224
Marketing	163
Office administration	50
Personnel	73
Production	39
Quantitative methods	34
Undecided	116
TOTAL	1301

Frequency Distribution of Attributes

The construction of a frequency distribution of attributes is a simple task indeed. One lists the various attributes, together with the frequency of their occurrence.

Example 1

Twenty students in business statistics class reported their major fields as follows:

> Accounting, Accounting, Marketing, Management,
> Management, Marketing, Accounting, Management,
> Finance, Accounting, Marketing, Finance,
> Accounting, Finance, Finance, Accounting,
> Marketing, Marketing, Accounting, Accounting.

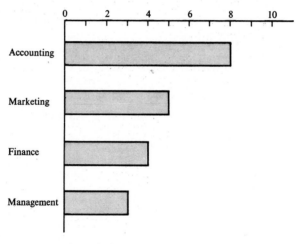

Fig. 1–1 Major fields of 20 students in business statistics class.

The above set of attributes can be summarized in the frequency distribution shown in Table 1–3.

Table 1–3

Major field	Number of students
Accounting	8
Finance	4
Management	3
Marketing	5
TOTAL	20

A frequency distribution of attributes may be represented graphically by a bar chart, in which the length of each bar is determined by the frequency of the attribute it represents. The bars should have the same width, and the space between them should be uniform. See Fig. 1–1.

Frequency Distribution of Variates

Summarizing a large number of variates by a frequency distribution is in general more difficult than summarizing a set of attributes, because of the very many different values of the variates. A ledger of accounts receivable, for example, may list amounts of $20.24, $83.17, $9.92, ... Thus a list of all the values of the accounts receivable in the ledger, together with the frequency of their occurrence, may lead exactly nowhere if no two accounts in the ledger have the same value. Even when some accounts are of the same value, the resulting summary may be as detailed as, and no more useful than, the original unsummarized data.

The difficulty of summarizing a set of variates, however, can be overcome by grouping the various variates into a limited number of classes called *class intervals*. A concrete example will best explain the construction of these class intervals.

Example 2

The following are the balances (in dollars) of 100 accounts receivable taken from the ledger of the XYZ Store.

31, 38, 41, 52, 59, 46, 74, 69, 39, 60
69, 83, 78, 74, 77, 35, 79, 80, 71, 65
56, 69, 34, 33, 92, 37, 60, 43, 51, 61
74, 68, 83, 49, 34, 71, 58, 83, 94, 66
78, 48, 34, 50, 68, 65, 64, 95, 92, 81
77, 84, 41, 40, 38, 38, 60, 67, 50, 86
76, 99, 38, 94, 48, 70, 80, 95, 98, 42
55, 49, 54, 60, 62, 70, 88, 94, 85, 51
59, 68, 51, 87, 53, 57, 54, 46, 46, 76
69, 64, 61, 63, 78, 55, 66, 73, 75, 64

These accounts are summarized in the distribution given in Table 1–4.

In constructing the frequency distribution in Table 1–4, we group the 100 accounts into seven class intervals: 30–39, 40–49, 50–59, 60–69, 70–79, 80–89, 90–99. Each of the seven classes has a lower and an upper limit. The lower limit of a given class lists the smallest number that falls into this class, and the upper limit is the largest number that falls into the same class. The first class interval, 30–39, has a lower limit of 30 and an upper limit of 39. The lower limit of the second class is 40.

Table 1–4

Distribution of accounts
receivable, XYZ Store

Class, dollars	Number of accounts
30–39	11
40–49	12
50–59	16
60–69	23
70–79	17
80–89	11
90–99	10
TOTAL	100

Overlapping class limits (for example, first class interval 30–40, second class interval 40–50) should be avoided to eliminate confusion as to which class each individual amount belongs to.

Once again, the upper limit of the first class interval is 39 and the lower limit of the second class interval is 40. The midpoint between these two successive class limits is 39.5. This midpoint is called the *upper boundary* for the first class interval and the *lower boundary* for the second class interval. Similarly, the point 49.5 is the upper boundary for the second class interval and the lower boundary for the third class interval. All variates in a given class must be greater than its lower boundary and smaller than its upper boundary. Class boundaries must be carried to *one more decimal place* than the variates to be grouped, to avoid ambiguous assignments of variates.

To continue our definitions, the difference between the upper and lower boundaries of a class is called *class width*. The width of the first class is $10.0 ($39.5 − $29.5), and the width of the second class is also $10.0 ($49.5 − $39.5). It is mandatory for all class intervals to have the same width.

The midpoint between the two limits (or boundaries) of a given class is called the *class mark*. The class mark for the first class interval is

$$\left(\frac{30 + 39}{2} \text{ or } \frac{29.5 + 39.5}{2}\right) = 34.5.$$

Finally, the number of variates falling in a given class interval is called the *frequency* of that class. Table 1–5 shows the class intervals, class boundaries, class mark, and frequency of each class interval in the frequency distribution of our 100 accounts receivable.

Table 1–5

Class interval	Lower boundary	Upper boundary	Class mark	Frequency
30–39	29.5	39.5	34.5	11
40–49	39.5	49.5	44.5	12
50–59	49.5	59.5	54.5	16
60–69	59.5	69.5	64.5	23
70–79	69.5	79.5	74.5	17
80–89	79.5	89.5	84.5	11
90–99	89.5	99.5	94.5	10

How to Construct a Frequency Distribution of Variates

In constructing a frequency distribution of variates, we must first decide how many classes there are to be. In general, the number of classes to be used in constructing a frequency distribution of variates depends to a large extent on the nature of the data to be summarized as well as on the purpose for summarizing the data. However, some guidelines may prove to be helpful in determining the number of classes. First, the number of classes should not be too small or too large. A small number of classes may conceal the general nature of the data, and a large number of classes may be too detailed to reveal any useful information. As a general rule, it is recommended that the number of classes should be between 5 and 20. Sturges' Rule, furthermore, may provide a convenient approximation to the number of classes. According to this rule,

$$\text{Number of classes} = 1 + 3.3 \log N,$$

where N is the number of observations to be summarized. Applying Sturges' Rule to the 100 accounts receivable of the XYZ Store, we find that

$$\begin{aligned} \text{Number of classes} &= 1 + 3.3 \log 100 \\ &= 1 + (3.3)2 \\ &= 7.6 \text{ classes.} \end{aligned}$$

Thus, according to Sturges' Rule, we should have about 7 or 8 classes summarizing our accounts receivable. In constructing our frequency of these accounts we have used 7 classes.

Once the number of classes has been determined, we must decide on the class width. All class intervals should have the same class width. In such a case, the class width can be approximated as follows:

$$\text{Class width} = \frac{\text{highest variate} - \text{lowest variate}}{\text{number of classes}} .$$

Referring to our 100 accounts receivable, we have

$$\text{Class width} = \frac{\$99 - \$31}{7} = \$9.7.$$

For ease of computing, we shall use $10 rather than $9.7 as a class width.

Having determined the approximate number of classes (7) and the width of each class ($10), we shall now choose the bottom limit (the lower limit of the first class). Since the smallest account receivable is $31, the bottom limit should be $31 or less. Here again, because of convenience, a bottom limit of $30 is chosen.

Having determined the lower class limit for the first class interval as 30, our next step is to establish the lower class *boundary* for this first class interval. Since our accounts receivable are rounded to the nearest dollar, and since class boundaries must be carried out to one more decimal place than the accounts receivable, the lower class boundary for the first class interval must be established as 29.5. (This lower boundary would be 29.995 had the accounts receivable been given to the nearest cent.)

The upper class *boundary* for the first class interval can now be obtained by adding the class width, 10, to the lower boundary.

$$\begin{aligned}\text{Upper boundary} &= \text{lower boundary} + \text{class width}\\ &= \quad 29.5 \quad + \quad 10 \quad = 39.5;\end{aligned}$$

consequently, the upper *limit* for the first class interval is 39.

Class interval	Lower boundary	Upper boundary	Class width
30–39	29.5	39.5	10

The successive class limits and boundaries can now be obtained by adding the class width, 10, to the preceding limits and boundaries (see Table 1–6).

Table 1–6

Class interval	Lower boundary	Upper boundary	Class width
30–39	29.5	39.5	10
40–49	39.5	49.5	10
50–59	49.5	59.5	10
60–69	59.5	69.5	10
70–79	69.5	79.5	10
80–89	79.5	89.5	10
90–99	89.5	99.5	10

Once the appropriate class intervals have been constructed, the number of accounts receivable falling in any given class is counted and regarded as the frequency of that class. The resulting frequency distribution is shown in Table 1–7.

Table 1–7

Class interval	Frequency
30–39	11
40–49	12
50–59	16
60–69	23
70–79	17
80–89	11
90–99	10
TOTAL	100

Finally, a frequency distribution of variates may be represented graphically by a *histogram*. In constructing such a diagram, class boundaries are marked off along the *horizontal* axis of the graph, and a rectangle is drawn to represent each class. The base of the rectangle corresponds to the class width and the height of the rectangle corresponds to the frequency of the class (see histogram, Fig. 1–2).

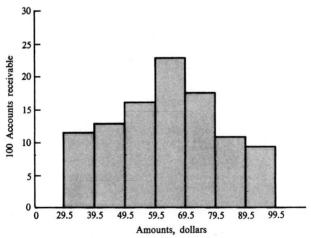

Fig. 1–2 Distribution of 100 accounts receivable.

EXERCISES

1. Determine the class boundaries, the class width, and the class mark for the class intervals

    10–14, 15–19, 20–24, 25–29.

2. The lifetimes in hours of sixty 100-watt electric bulbs are

807, 811, 620, 650, 817, 732, 747, 823, 844, 907
660, 753, 1050, 918, 857, 867, 675, 880, 878, 890
881, 872, 869, 841, 847, 833, 829, 827, 822, 811
766, 787, 923, 792, 803, 933, 947, 717, 817, 753
1056, 1076, 958, 970, 776, 828, 831, 781, 1088, 1082
832, 863, 852, 788, 980, 889, 1030, 897, 755, 891

a) Construct a frequency distribution of the lifetimes of these 60 electric bulbs (use equal class widths).

b) Draw a histogram of the frequency distribution.

3. Draw a bar chart to represent the following frequency distribution of attributes:

Surgical operations performed in Dillingham
Memorial Hospital, 1973.

Type of operation	Number of cases
General surgery	98
Thoracic surgery	20
Abdominal surgery	112
Proctologic surgery	67
Urologic surgery	79
Eye, ear, nose, and throat surgery	56
Neurosurgery	25
Operations on bones and joints	44

4. The annual sales of 80 branches of McCarthy Department Stores (millions of dollars) are

38 30 36 32 33 27 38 45
43 21 24 45 33 49 41 36
28 32 43 24 36 47 33 24
35 38 36 44 24 39 21 26
20 42 30 42 43 46 41 46
27 34 32 23 27 34 32 23
25 23 21 44 35 34 33 28
21 24 44 44 23 28 36 30
24 27 25 29 36 47 42 40
27 31 46 48 20 23 48 21

a) Construct a frequency distribution of the annual sales for these 80 branches.

b) Draw a histogram of the frequency distribution.

5. Draw a bar chart to represent the following frequency distribution of attributes:

Wholesale trade by type of operation, 1978	
Type of operation	Establishments (1000)
Merchant wholesalers	200
Manufacturers' sale branches and offices	50
Petroleum bulk plants and terminals	60
Merchandise agents and brokers	55
Assemblers of farm products	130

2
Population
Parameters

Population
Parameters

Introduction

Although statistics deals with the analysis and evaluation of sample data, we shall for the moment set aside the problem of sampling and pretend that decisions and conclusions are to be based on complete population data. In other words, we assume that population data are available and that decisions, therefore, are made with complete certainty. Although this assumption is unrealistic, it is nevertheless a useful and convenient step in the development of our topic.

But if population data are available, how can such data be utilized in solving specific problems? This chapter deals with the organization and summarization of population data into a form that enables us to make decisions and derive conclusions.

A Parameter

Decisions and conclusions can often be made with absolute certainty if a single value (or values) that describes a certain aspect of a population is determined. For example, to determine whether to accept or reject a lot of 1000 articles received from a supplier, one must know the percentage of defective articles in the entire lot. To determine whether Los Angeles or Chicago is economically more prosperous, one must know the average family income in each city. A single value such as the percentage of defective articles in a lot or the average family income in Los Angeles is called a *parameter*. Thus a parameter is a single number that describes one aspect of an entire population; to determine its value, one must utilize complete population data.

For any given population, a wide variety of parameters can be computed. This chapter, however, deals with parameters that are most frequently used in statistical analysis. The meaning, significance, properties, and method of computation of each parameter are thoroughly examined. The usefulness of individual parameters in solving specific problems is also discussed here.

The Aggregate

The aggregate is a parameter that can be computed only for a population of variates. Denoted by A, the aggregate is the sum of the values of all variates in the population.

Example 1

Let our population of interest consist of the ages of the following five houses:

$$X_1 = 2 \text{ years}, \qquad X_2 = 2 \text{ years},$$
$$X_3 = 4 \text{ years}, \qquad X_4 = 5 \text{ years},$$
$$X_5 = 2 \text{ years},$$

where X_1 is the age of the first house in the population, X_2 is the age of the second house in the population, etc. The aggregate, A, is the sum of the ages of the five houses:

$$A = 2 + 2 + 4 + 5 + 2, \qquad A = 15 \text{ years}.$$

In a more general form,

$$A = X_1 + X_2 + X_3 + X_4 + X_5,$$

or, in short-hand notation,

$$A = \sum_{i=1}^{5} X_i.$$

The Greek letter sigma, Σ, indicates the summation of variates. Since we have only five variates, i goes from 1 to 5. For the sake of completing the notion of notation, the aggregate of 7 terms would be written as

$$\sum_{i=1}^{7} X_i = X_1 + X_2 + X_3 + X_4 + X_5 + X_6 + X_7,$$

and that of 89 terms as

$$\sum_{i=1}^{89} X_i = X_1 + X_2 + \cdots + X_{89}.$$

Finally,

$$\sum_{i=1}^{N} X_i = X_1 + X_2 + \cdots + X_N,$$

where there are N variates in the population.

It must be emphasized here that an aggregate can be computed for a population of variates since the population consists of measurable observations whose values can be added. A population of attributes, on the other hand, consists of *nonmeasurable* observations that cannot be summed.   The aggregate as the sum of the values of all variates in a population must not be confused with the population size, N, which refers to the number of observations in the population.

The above discussion dealt with the computation of an aggregate for a population of variates. The question now arises: For what type of problem do we compute the aggregate and utilize it as an end in itself? The aggregate is calculated whenever the decision in a given problem depends on the aggregate value. Example 2 illustrates a situation in which the aggregate serves as a basis for decision.

Example 2

In 1980 city **X** needs an additional revenue of 50 million dollars; the existing sales tax rate of 4% therefore must be increased. The increase depends on the aggregate retail sales in 1980. Thus, if the aggregate sales are 10,000 million dollars, the increase in tax rate would be

$$\frac{50}{10,000} \times 100 = \tfrac{1}{2}\%.$$

If, on the other hand, aggregate sales are 5000 million dollars, the increase in tax rate is

$$\frac{50}{5000} \times 100 = 1\%.$$

Proportion

Proportion, denoted by π, refers to a fraction of the population that possesses a certain property. In Example 1 (population of five houses), the proportion of houses more than two years of age is .40.

$$\pi = \tfrac{2}{5}, \qquad \pi = .40.$$

In a population of 1000 articles, 82 of which are defectives, the proportion of defective items in the population is .082.

$$\pi = \tfrac{82}{1000}, \qquad \pi = .082.$$

The proportion is the parameter used most often in describing a population of attributes. Its use as a basis for decision can be illustrated as follows.

Example 3

In California, Thomas Rees, a Democrat, and Dr. John Marshall, a Republican, are the two contestants in a run-off election for the U.S. 27th Congressional seat. The decision as to which candidate will be the future California Congressman certainly depends on the proportion of total votes received by each candidate.

MEAN, MEDIAN, AND MODE

When the population consists of observations of variates, the investigator is often interested in finding a value that can best describe the average or typical variate in the population. He may wish to know, for example, the average family income in Los Angeles, the average life of a certain brand of tire, or the average tensile strength of a particular fiber. The arithmetic mean, the median, and the mode are parameters, designed to represent the average variate.* But why three different measures?

Each of these three parameters is calculated in a different way, and this difference in their calculation may render one of the parameters a better representative of the average variate *in a given problem* than either of the other two. The three measures are discussed separately.

The Arithmetic Mean

The arithmetic mean, denoted by μ, is simply the familiar average. It is obtained by dividing the sum of all variates in the population by their number. Symbolically,

$$\mu = \frac{\sum\limits_{i=1}^{N} X_i}{N}, \quad \text{or} \quad \mu = \frac{A}{N}.$$

The arithmetic mean of the ages of five houses is

$$\mu = \frac{X_1 + X_2 + X_3 + X_4 + X_5}{5},$$

$$\mu = \frac{2 + 2 + 4 + 5 + 2}{5}, \quad \mu = \frac{15}{5}, \quad \mu = 3 \text{ years.}$$

---

* Other parameters that describe the average variates are the geometric and harmonic means. Because of their limited use, however, these two measures are not treated here.

The arithmetic mean has an interesting property. The sum of the deviations of the individual variates from their arithmetic mean is always equal to zero. Thus, if we subtract the mean, which is 3, from the age of each house, the resulting five deviations will add up to zero:

X_i	$(X_i - \mu)$ $(X_i - 3)$
2	-1
2	-1
4	1
5	2
2	-1
$\Sigma X = 15$	$\Sigma (X - \mu) = 0$
$\mu = 3$	

The arithmetic mean is used much more often than the median or the mode to describe the average variate in the population. It best describes the values such as the average life of an electric bulb, the average nicotine content of a certain brand of cigarettes, the average tensile strength of a fiber, the average carbohydrate content of a grain of wheat, and the average corn yield per acre of a given farm. It often serves as a basis for decision making in a great variety of problems. A few examples will illustrate this point.

Example 4

An agronomist wishes to compare the effectiveness of two different types of fertilizer mixes used in wheat production. He decides to use the average yield per acre as a measure of effectiveness. The average yield is the arithmetic mean.

Example 5

A manufacturer of a certain brand of cake mix packages his product in boxes of one pound net weight. As a representative of the Federal Fair Trade Commission you must decide whether or not the manufacturer's operation is satisfactory. The average weight per package as determined by the arithmetic mean can be used as a basis for your decision.

The Median

Like the arithmetic mean, the median is designed to describe the average variate in the population. The method of computing the median, however, is different from that used in computing the mean. To compute the median for a population of variates, the variates must be arranged first in an increasing or decreasing order. The median is the middle variate if the number of the variates, N, is odd. If N is even, the median is the arithmetic mean of the two middle variates. The median of the five variates (3, 3, 7, 11, 19), for example, is 7, while the median of the six variates (3, 3, 7, 11, 19, 22) is

$$\frac{7 + 11}{2} \quad \text{or} \quad 9.$$

The median divides the variates in the population into two equal or approximately equal halves. Half of the variates are less than or equal to the median, while the other half are greater than or equal to the median. Thus, if the median income in Los Angeles is $14,000, then 50% of all Los Angeles families have annual earnings of $14,000 or less, and 50% of all families have an annual income of $14,000 or more.

Unlike the arithmetic mean, the median is generally not affected by the existence of either extremely high or extremely low variates in the population. The mean and median of the set (4, 5, 6, 7, 8) are the same, namely 6. However, the mean of the set (4, 5, 6, 7, 198) is 44 and the median is 6. While the median continues to represent the majority of the variates in the set, the mean does not represent any of the variates. Thus, in a population having variates with extreme values, the median describes the typical or average variate more accurately than the arithmetic mean does. For example, such populations of variates as family income in Chicago, hourly earnings at Lockheed Aircraft Corporation, and store retail sales in Los Angeles are likely to contain a few extremely high variates. The arithmetic mean will then be more extreme than the median, which can therefore be considered the better "average."

The Mode

Like the median and mean, the mode is another parameter that describes the typical variate. The mode is the most frequently appearing variate or attribute in the population. In our population of the ages of five houses (2, 2, 4, 5, 2), the mode is 2 years. Similarly, the modal annual family income in Los Angeles is $14,800 in 1980 if a greater number of families earned $14,800 than any other amount in that year.

Example 6 illustrates the use of the mode as a basis for decisions.

Example 6

A clothing manufacturer wishes to market three standard sizes of a "Hollywood Flair" sport jacket: large, medium, and small. He must decide on a sleeve length for each of the standard models. His decision is based on the modal sleeve length (mode) of the potential buyers of each size.

Selecting the Appropriate Average

Let us begin our discussion by considering some characteristics of a group of 12 people. These characteristics are

Age (*in years*): 22, 24, 25, 26, 27, 28,
28, 29, 30, 31, 33, 34

Annual income: $8000, $8200, $9000, $10,000, $11,000, $12,000,
$12,300, $12,500, $13,000, $13,500, $86,000, $97,000

Shoe size: $8\frac{1}{2}$, 9, $9\frac{1}{2}$, $9\frac{1}{2}$, $9\frac{1}{2}$, $9\frac{1}{2}$,
10, 10, $10\frac{1}{2}$, $10\frac{1}{2}$ 11, $11\frac{1}{2}$

What is the average age, average annual income, and average shoe size for this group of 12 people?

The average age may well be described by the arithmetic mean. Although the median and mode can also describe the average age in this case, we have chosen the arithmetic mean for two reasons. First, more people are familiar with the arithmetic mean than with either the median or mode. Second, there is no advantage to choosing the median or mode when the values of the three measures are almost the same (mean = 28.1, median = 28, mode = 28).

Examining annual incomes, we see that the annual income of two members of the group is extremely high. We know that extreme values affect the arithmetic mean much more than the median: the mean annual income is $24,375, the median is $12,150. Thus the median represents the majority of the variates, while the arithmetic mean represents none of the variates. The median is therefore the better average for incomes.

As to shoe size, both the mean and median are somewhat meaningless (mean = 9.92, median = 9.75). In fact, neither of these two measures represents any actual shoe size. The mode, on the other hand, represents the actual shoe size used by four members in the group. Thus the mode is the better average for shoe sizes.

There is no doubt that the foregoing illustration is oversimplified in that the choice of measure is rather obvious. However, it does emphasize the fact that the choice among the three measures depends on the purpose for which the data are selected as well as on the nature of the data gathered.

In many statistical problems, the choice among mean, median, and mode may not be obvious. Thus some general comments concerning the advantages and disadvantages of the three measures may be helpful.

First, we must recognize that of the three measures the arithmetic mean is most easily understood by the general public. Second, but all-important, the arithmetic mean lends itself much more readily to further statistical analysis. It is for these two reasons that most statistical studies report the arithmetic mean as the average.

The arithmetic mean has the disadvantage, however, of being more affected by extreme value than either the median or mode. Thus, whenever there are either extremely high variates or extremely low variates, the median or the mode is preferred over the arithmetic mean.

MEASURES OF VARIATION

While such parameters as the mean, median, and mode describe the typical variate, another set of parameters, namely the range, the standard deviation, and the average deviation, measures the disparity among values of the various variates comprising the population. These parameters, called *measures of variation*, are designed to indicate the degree of uniformity among the variates.

A population of variates has no variation when all the variates have the same value. In a population of five variates (3, 3, 3, 3, 3), each variate, as well as the arithmetic mean, is equal, and the population therefore has no variation. The set (1, 3, 3, 3, 5), on the other hand, has some variation since some of the variates have different values.

The variation among variates can be measured by the range, the average deviation, or the standard deviation. Each of these three parameters of variation is discussed separately.

The Range

The *range*, denoted by R, is the difference between the highest and lowest variate:

$$R = \text{maximum variate} - \text{minimum variate}.$$

The range of the five variates 1, 3, 3, 3, 5 is 4:

$$R = 5 - 1, \qquad R = 4.$$

The range as a measure of variation is deficient, since it considers only the highest and lowest values and neglects the variation in the remaining values. The two sets of variates,

$$\text{Set 1: 1, 3, 3, 3, 5} \qquad \text{Set 2: 1, 1, 4, 4, 5}$$

have the same range, 4, although there is less variation in set 1 than in set 2.

The Average Deviation

The *average deviation* is the *average absolute deviation* (when the sign, plus or minus, is ignored, the deviation is called *absolute* and is designated by two vertical bars) of each variate from the arithmetic mean. Symbolically,

$$AD = \frac{\sum\limits_{i=1}^{N} |X_i - \mu|}{N}.$$

For a population of ages of five houses 2, 2, 4, 5, and 2 years old, the average deviation is 1.2 years, as shown below.

X_i	$|X_i - \mu|$ $|X_i - 3|$
2	1
2	1
4	1
5	2
2	1
	$\sum |X - \mu| = 6$

$$AD = \frac{\sum |X - \mu|}{N}$$

$$AD = \tfrac{6}{5} = 1.2 \text{ years}$$

Thus the age of each house is different from the mean of the population by an average of 1.2 years. It must be emphasized that the average deviation is measured in the same unit as the variate itself. In our example, both are measured in years.

The Standard Deviation

The *standard deviation*, denoted by σ, is the most important measure of variation. In a broad sense, it measures the average deviation of each variate from the arithmetic mean. Symbolically,

$$\sigma = \sqrt{\frac{\sum\limits_{i=1}^{N} (X_i - \mu)^2}{N}}.$$

In words, the standard deviation is the square root of the average square deviation of each variate from the arithmetic mean.

For our population of five houses with ages 2, 2, 4, 5, and 2 years, the standard deviation is 1.26 years, as shown below.

X_i	$(X_i - \mu)$	$(X_i - \mu)^2$
2	-1	1
2	-1	1
4	1	1
5	2	4
2	-1	1
		$\Sigma (X - \mu)^2 = 8$

$$\sigma = \sqrt{\frac{\Sigma (X - \mu)^2}{N}}$$

$$\sigma = \sqrt{\tfrac{8}{5}} = \sqrt{1.6} = 1.26 \text{ years.}$$

Thus, on the average, the age of each house differs from the arithmetic mean of the population by 1.26 years. (Incidentally, the square of the standard deviation is called *variance* and is denoted by σ^2.)

Both the average deviation and standard deviation are good measures of the average deviation of each variate from the arithmetic mean. However, the standard deviation is the more relevant measure, since only the standard deviation lends itself to further statistical analysis and treatment.

In addition to its basic role in the development of the theory of statistics, the standard deviation has some important practical applications. Two examples are cited below.

Example 7

A food processor markets his instant coffee in 16-ounce jars. He considers that his filling operation performs satisfactorily if the average weight of all jars is 16.2 ounces and the standard deviation is .04 ounces. (The reader will understand later that if the average weight of a jar is 16.2 ounces and the standard deviation is .04 ounce, then virtually no jar will contain less than 16.0 ounces of coffee.)

Example 8

A machine shop accepts an order for 10,000 ball bearings of 2-inch diameter. The size specification of the product can be maintained only when the average diameter is 2 inches and the standard deviation is very small.

EXERCISES

1. A population consists of the weights, in pounds, of five babies born on August 28, 1965, in West Valley Hospital. These weights are 11, 5, 9, 11, 4.

   a) Compute the mean, mode, and median weights. Compare their values.
   b) Compute the range, the average deviation, and the standard deviation.

2. On December 31, 1980, 10 babies were born in West Valley Hospital. Their weights, in pounds, are 7, 8, 8, 6, 4, 9, 10, 11, 8, 9. Compute the mean, median, and mode. Compare their values.

3. Is it possible for the standard deviation to be larger than the arithmetic mean?

4. The blood pressures of 10 men were measured before and after smoking. The changes were $+10$, -5, $+7$, -4, $+2$, $+3$, -4, -5, -3, $+9$. Compute the mean change and the standard deviation, and compare their values.

5. The annual salaries of the 12 faculty members in the Department of Economics at Bradford College are $7200, $7200, $7200, $7500, $7500, $7800, $8100, $8100, $8400, $8400, $9000, $16,800. Compute the mean and the median annual salary. Which is a better "average" and why?

6. The Department of English at Bradford College, reporting on the annual salaries of its faculty members, states that $\mu = \$15,000$, and $\sigma = 0$. What is the median annual salary? What is the mode?

7. An agronomist reporting on the weight of a new variety of apples states that $\mu = 8$ ounces, and $\sigma = 2$ inches. What is wrong?

8. The 10 members of the "Play Toy Club" have the following characteristics:

   Annual income: $12,000, $11,000, $85,000, $14,000, $12,000,
   $13,000, $15,000, $13,000, $10,000, $12,500

   I.Q.: 102, 97, 103, 137, 105,
   108, 107, 102, 96, 111

   Weight (pounds): 150, 157, 162, 170, 155,
   163, 167, 172, 165, 173

   Hat size (inches): $6\frac{5}{8}$, $6\frac{7}{8}$, $7\frac{1}{4}$, $6\frac{7}{8}$, $6\frac{3}{4}$,
   $6\frac{7}{8}$, $6\frac{7}{8}$, $7\frac{1}{2}$, $6\frac{7}{8}$, $6\frac{7}{8}$

   What is the appropriate average for each of the above characteristics of the members of the "Play Toy Club"?

9. Each of the Sullivan and O'Connor families has 5 children. The ages of the Sullivan children show a mean of 11 years, a standard deviation of 3.16 years, and a median of 10 years. The O'Connor children, on the other hand, have a mean age of 9 years, a standard deviation of 3.16 years, and a median of 10 years. Compare the ages of the children in the two families and justify your conclusions.

10. Given the two following populations of variates.

$$\text{Population } A: 8, 9, 10, 11, 17$$
$$\text{Population } B: 3, 9, 10, 11, 12$$

Compute the mean, median, and the standard deviation for each population. Compare the results for the two populations and explain the differences if any.

11. Compare the means, the ranges, and the standard deviation of the two populations below. Explain the differences, if any.

$$\text{Population } A: 1, 4, 5, 6, 9$$
$$\text{Population } B: 1, 2, 5, 8, 9$$

12. A population of variates has a mean of 10 and a standard deviation of 3.

a) If each variate is increased by 2, what are the new mean and standard deviation of the population?

b) If each variate is multiplied by 2, what are the new mean and standard deviation of the population?

13. The final examination grades of 20 students in an elementary statistics class are

$$50, 55, 61, 60, 71, 73, 53, 54, 67, 67$$
$$54, 77, 72, 76, 81, 83, 87, 44, 48, 67$$

Determine the percentage of grades that fall in the interval

a) $\mu \pm 1\sigma$,   b) $\mu \pm 2\sigma$,   c) $\mu \pm 3\sigma$.

14. Upon receiving your accounting degree from the university, you are offered a junior accountant position by two public accounting firms: A and B. Your starting salary in both firms is $12,000 per year. Further investigation revealed that for those accountants who have been employed by firm A for a period of five years, the mean annual salary is $20,000 with a standard deviation of $5000. Similar accountants with firm B earn a mean salary of $20,000 but with a standard deviation of only $2000. Assuming that all other employment conditions in both firms are the same, which offer do you prefer? Explain.

15. The annual sales, in millions of dollars, for the 25 department stores in Santa Barbara, California are

38	35	43	27	24
40	38	21	34	27
35	36	24	33	25
33	44	45	23	33
33	24	33	27	25

a) Compute the mean, the median, and modal sales.

b) Compute the range, average deviation, and the standard deviation.

c) Determine the percentage of sales that fall in the interval (1) $\mu \pm 1\sigma$; (2) $\mu \pm 2\sigma$; (3) $\mu \pm 3\sigma$.

3
Probability

Probability

We have already learned that statistics consists of tools and methods that enable us to evaluate the reliability of conclusions derived from sample data. Of all the tools of statistics, the concept of probability is the most important one. Modern statistics may, indeed, be regarded as an application of the theory of probability. It is for this reason that two chapters of this text are devoted exclusively to the treatment of probability.

OBJECTIVE AND SUBJECTIVE PROBABILITIES

The concept of probability can be approached in two ways: an objective approach and a subjective approach. In the objective approach, the *probability* of an event is defined as the *relative frequency of its occurrence in the long run*. To clarify this notion, let us assume that a balanced coin is tossed 50 times and the event "head" appears 23 times. The relative frequency of the event "head," denoted by $RF(H)$, is

$$RF(H) = \frac{23}{50} = .46$$

Now, the probability of the event H, denoted by $P(H)$, is the limit of its relative frequency as the coin is tossed an infinite number of times. Since the relative frequency of the event "head" approaches .5 as the number of tosses approaches infinity, we may then conclude that the probability of getting a head on a single toss of a balanced coin is .5. Similarly, we can also conclude that the probability of getting "a four" in a single throw of a fair die is 1/6.

Initially, probability theory was developed in connection with the games of chance; and here the objective approach to probability is most appropriate indeed. In contrast, let us now consider some other situations, namely, business, where the objective interpretation of probability may no longer be possible. We may state, for example, that the probability of having a recession next year is .3 or that the probability that the price of a given stock will rise tomorrow is .9. Clearly, these probabilities can hardly be viewed as relative frequencies. More appropriate, however, these probabilities measure the degree of belief we attach to the occurrence of such events. Hence, in contrast to the objective approach, which defines probability as a relative frequency of an event, a subjective or a personalistic approach to probability defines the probability of an event as a measure of the degree of belief that an individual attaches to the occurence of that event.

Subjective probabilities are individualistic. An individual assesses the probability of an event based on whatever information (past and current) is available to him together with his personal biases, experience, judgment, and expectations. Subjective probabilities are also personal probabilities in the sense that two individuals may assign different probabilities to the same event due to differences in their attitudes, experiences, and outlooks.

Before concluding this discussion, however, the reader should clearly note that all probabilities, whether objective or subjective in nature, must satisfy the same theorems and obey the identical rules.

SETS

Our understanding of probability will be greatly enhanced if we first understand some elementary notions of sets.

A set is a collection of objects. Our 50 states comprise a set, the positive integers 1, 2, 3, 4, ... form a set, and the students in this college are a set. A set is a well-defined collection of objects, and these objects are called *elements* or *members* of the set. Thus the state of California is a member of the set of our 50 states and the integer 4 is an element of the set of positive integers of the number system.

A set may contain a limited or an infinite number of elements. The set of states in the United States, for example, contains only 50 elements. However, there are sets, such as the set of positive integers, that contain an infinite number of elements. In this context, a set of a particular interest is the *null* or *empty set*. Denoted by $\varnothing$, the null set contains no elements. The set of all men with 22 fingers, for example, is a null set. We will have more to say about the null set later on.

A set is usually denoted by a capital letter such as *A*, *B*, or *C*. A member of the set is denoted by a lower-case letter such as *a*, *b*, or *c*. When the object x is a member of set *A*, we write

$$x \in A,$$

which reads "x is an element of A." We also write

$$y \notin A$$

to mean that the element y is not a member of set A.

A set can be described by listing all its elements. The set A which consists of the first seven positive integers can, for example, be described as

$$A = \{1, 2, 3, 4, 5, 6, 7\}.$$

Note that the elements of a set are separated by commas and enclosed in braces. Similarly, the set B which consists of a student, a book, and a pen can be described as

$$B = \{\text{student, book, pen}\}.$$

In some cases, it is either impossible or inconvenient to describe a set by listing all its elements. It is impossible, for example, to list the elements of the set of all positive integers. A set may then be alternatively described by a statement which specifies the elements comprising the set. Thus the set of all positive integers, denoted by C, can be described as

$$C = \{x \mid x \text{ is a positive integer}\},$$

which reads "the set C is the set of all x such that x is a positive integer." Note that the vertical bar is read "such that." Similarly, the set D which consists of all families in New York City can be described as

$$D = \{x \mid x \text{ is a family in New York City}\}.$$

Subsets

Let us consider the set $A = \{x, y, z\}$. The set $B = \{x, y\}$ is a subset of A. This can be written as $B \subset A$, which reads "B is contained in A." Similarly, if $S = \{x \mid x \text{ is a family in Los Angeles County}\}$, and $S_1 = \{x \mid x \text{ is a family in Los Angeles County with annual income exceeding \$10,000}\}$, then the set S_1 is a subset of S. In general, the set S_1 is a subset of set S if every element in S_1 is also an element in S. According to this general definition of a subset we may also conclude that $S \subset S$, and $\emptyset \subset S$. In other words, every set is a subset of itself, and the null set $\emptyset$ is a subset of every set.

In a set of n elements, there are $(2)^n$ subsets. Thus, a set of three elements contains $(2)^3$ or 8 subsets. The set $S = \{x, y, z\}$, for example, contains the following eight subsets:

1. $S = \{x, y, z\}$	5. $D = \{x\}$
2. $A = \{x, y\}$	6. $E = \{y\}$
3. $B = \{x, z\}$	7. $F = \{z\}$
4. $C = \{y, z\}$	8. $\emptyset$

Note that the first subset is the set itself, and the last subset is the null set.

The Universal Set *U*

We have already learned that in many statistical problems the investigator is interested in studying some characteristic of a population. For example, he may be interested in determining the average age of the students in a given college. The set which consists of the ages of *all* students is our population. This set is called the *universal set* and is denoted by *U*. Similarly, if the problem at hand is to determine the average annual family income in Los Angeles, then our statistical population is the universal set *U*, where

$$U = \{x \mid x \text{ is the annual income of a family in Los Angeles}\}.$$

The set *A*,

$$A = \{x \mid x \text{ is the annual income of a Los Angeles family}$$
$$\text{earning } \$10,000 \text{ or more}\}$$

is a subset of the universal set *U*. In the same manner, the set *B*,

$$B = \{x \mid x \text{ is the annual income of a Los Angeles family}$$
$$\text{earning less than } \$5000\}$$

is also a subset of the universal set *U*.

It is, at times, convenient to represent sets and subsets by diagrams. In these diagrams, known as *Venn diagrams*, the universal set *U* is represented by a rectangle, and subsets are represented by circles inside the rectangle. In the diagram (Fig. 3–1), *A* and *B* are subsets of the universal set *U*. Furthermore, *A* and *B* have no elements in common.

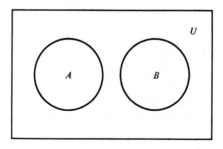

Figure 3–1

Equality of Two Sets

Sets *A* and *B* are equal only when both sets contain exactly the same elements. Sets *A* = {3, 4} and *B* = {4, 3}, for example, are equal. Note that these two sets are considered to be equal although the order of the elements in the two sets is not the same.

Operations with Sets

Let us consider the two sets A and B, where $A = \{1, 2, 3\}$ and $B = \{3, 4\}$. These two sets may be considered as subsets of a universal set $U = \{1, 2, 3, 4, 5\}$. Several other new sets can now be formed by certain manipulation of the two sets A and B. Some of these new sets are

1. The set $\{1, 2, 3, 4\}$, which is obtained by combining the elements of the two sets A and B, is called the union of A and B.
2. The set $\{3\}$, which is composed of the elements common to sets A and B, is called the intersection or product of A and B.
3. The set $\{4, 5\}$ is composed of all the elements of the universal set U which *are not* elements of set A. This set is called the *complement of A.*

There are many other sets that can be obtained by manipulating, or by performing certain operations on, sets A and B. The set $\{1, 2\}$, for example, consists of all the elements in set A which are not elements of set B. This set is called the *difference between A and B.* Of all these new sets, however, only the union, the intersection, and the complement will be discussed in some detail.

The Union of A and B

The union of A and B, denoted by $A \cup B$, is the set of elements that belongs to either A or B or both. Thus, if $A = \{2, 7, 8, 5\}$ and $B = \{7, 8, 11\}$, then the union of A and B is the set $A \cup B = \{2, 7, 8, 5, 11\}$. The union of A and B is shown by the shaded area in Fig. 3–2.

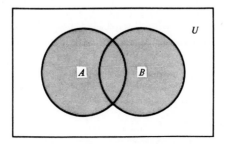

Fig. 3–2 Union of A and B, $A \cup B$.

The Intersection of A and B

The intersection of sets A and B, denoted by $A \cap B$, is the set of elements that belong to both A and B. Thus, if $A = \{2, 7, 8, 5\}$ and $B = \{7, 8, 11\}$, then the intersection of A and B is the set $A \cap B = \{7, 8\}$. The intersection of A and B is shown by the shaded area in Fig. 3–3. If the sets A and B have no elements in common, then the intersection of A and B is the empty set or the null set: $A \cap B = \varnothing$.

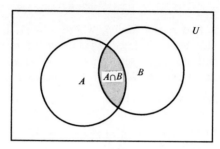

Fig. 3–3 *A* ∩ *B*, intersection of *A* and *B*.

Two sets that have no elements in common are called *mutually exclusive.* The two mutually exclusive sets *A* and *B* are shown in Fig. 3–4.

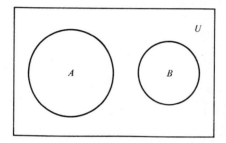

Fig. 3–4 Mutually exclusive sets.

Complement of Set *A*

The complement of set *A*, denoted by *A'*, is the set of elements of the universal set *U* which *do not* belong to set *A*.   Thus, if *A* = {1, 2, 3} and *U* = {1, 2, 3, 4, 5}, then the complement of set *A* is *A'* = {4, 5}. The complement of set *A* is shown by the shaded area in Fig. 3–5. We can even speak of a complement to the set *A* ∪ *B*. The complement to the set *A* ∪ *B*, denoted by (*A* ∪ *B*)', is shown by the shaded area in Fig. 3–6.

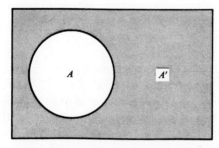

Fig. 3–5 *A'*, the complement of *A*.

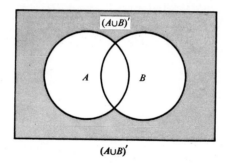

$(A \cup B)'$

Fig. 3–6 $(A \cup B)'$, the complement of $A \cup B$.

We shall now elaborate on the union, intersection, and complement by presenting a few illustrative examples.

Example 1

Given: $U = \{1, 2, 3, 4, 5, 6, 7\}$, $A = \{2, 3, 4\}$, $B = \{3, 4, 5, 6\}$.

Find: The sets a) $A \cup B$ b) $A \cap B$ c) A' d) B'

Solution:

a) $A \cup B = \{2, 3, 4, 5, 6\}$ b) $A \cap B = \{3, 4\}$
c) $A' = \{1, 5, 6, 7\}$ d) $B' = \{1, 2, 7\}$

Example 2

Given: $U = \{1, 2, 3, 4, 5\}$, $A = \{2, 3\}$, $B = \{3, 5, 4\}$.

Find: The sets a) $(A \cup B)'$ b) $(A \cap B)'$

Solution:

a) To determine the set $(A \cup B)'$, we must first determine the union of A and B:

$$A \cup B = \{2, 3, 5, 4\}.$$

The complement of $A \cup B$ is the elements of the universal set $\{1, 2, 3, 4, 5\}$ which do not belong to $A \cup B$. Accordingly, $(A \cup B)' = \{1\}$.

b) Similarly, to determine $(A \cap B)'$, we must first determine the set $A \cap B$. Since $A \cap B = \{3\}$, it follows that

$$(A \cap B)' = \{1, 2, 4, 5\}.$$

Example 3

Given: $U = \{a, b, c, d, e, f\}$ $A = \{a, b\}$, $B = \{c, d, e, f\}$.

Find: The set $(A \cup B)'$.

Solution:

$$A \cup B = \{a, b, c, d, e, f\}.$$

Therefore,

$$(A \cup B)' = \varnothing.$$

Example 4

Given:

$$A = \{x \mid x \text{ is an integer, and } 10 < x < 14\},$$
$$B = \{y \mid y \text{ is an integer, and } 12 < y < 16\}.$$

Find: $A \cap B$.

Solution: Since set A consists of all integers which are greater than 10 but less than 14, then $A = \{11, 12, 13\}$.

Similarly, set B consists of all integers which are greater than 12 but less than 16. Consequently, $B = \{13, 14, 15\}$, and $A \cap B = \{13\}$.

Example 5

Given:

$$A = \{x \mid x \text{ is an integer and } 2 \le x \le 5\},$$
$$B = \{y \mid y \text{ is an integer greater than } 3\}.$$

Find: $A \cap B$.

Solution:

$$A = \{2, 3, 4, 5\},$$
$$B = \{4, 5, 6, 7, \ldots, \infty\},$$
$$A \cap B = \{4, 5\}.$$

Example 6

Given: $A = \{x \mid 10 < x < 14\}$, $B = \{y \mid 12 < y < 16\}$.

Find: $A \cup B$.

Solution: Set A consists of an *infinite* number of elements which are greater than 10 but less than 14. Some of these elements are: 10.02, 10.717, 13.9897, ... Set B also consists of an infinite number of elements. The elements of B, however, are greater than 12 but less than 16. Some of these elements are: 12.001, 12.97, 15.76, ... The union $A \cup B$ is, therefore, an infinite set and $A \cup B = \{z \mid 10 < z < 16\}$.

Example 7

Given: $A = \{x \mid x^2 + x - 12 = 0\}$, $B = \{x \mid 2x - 6 = 0\}$.

Find: $A \cap B$.

Solution: To determine the elements of set A, we must solve the quadratic equation

$$x^2 + x - 12 = 0,$$
$$(x + 4)(x - 3) = 0,$$
$$(x + 4) = 0,$$
$$x = -4,$$

or

$$(x - 3) = 0,$$
$$x = 3.$$

Therefore,

$$A = \{-4, 3\}.$$

Similarly, to determine the elements of set B, we must solve the equation

$$2x - 6 = 0,$$
$$2x = 6,$$
$$x = 3.$$

Therefore

$$B = \{3\} \quad \text{and} \quad A \cap B = \{3\}.$$

EXERCISES, SET 1

1. List the elements of each of the following sets.
   a) Set of integers between 9 and 13
   b) Set of positive integers less than 9
   c) Set of integers between 1 and 20 divisible by 3

2. List the elements of each of the following sets.
   a) Set $A = \{x \mid x$ is an integer and $9 \leq x < 13\}$
   b) Set $B = \{x \mid x$ is an integer and $9 < x \leq 13\}$
   c) Set $C = \{x \mid x + 2 = 0\}$
   d) Set $D = \{y \mid y^2 + 3y = 28\}$

3. Let $U = \{x, y, 3\}$. List all subsets of U.

4. Let $U = \{$man, woman, baby, home$\}$. List all subsets of U.

5. *Given:* $U = \{10, 11, 12, 13, 14, 15\}$, $A = \{10, 11, 12, 13\}$, $B = \{12, 13, 14\}$.
   Find: a) $A \cup B$ b) $A \cap B$ c) A'

6. *Given:* $U = \{20, 22, 24, 26\}$, $A = \{20, 22\}$, $B = \{22, 24\}$.

    Find: a) $A \cup B$ b) $(A \cup B)'$ c) $(A \cap B)$ d) $(A \cap B)'$

7. *Given:* $U = \{v, w, x, y, z\}$, $A = \{v, w, x\}$, $B = \{y, z\}$.

    Find: a) $A \cup B$ b) $(A \cup B)'$ c) $A \cap B$ d) $(A \cap B)'$

8. *Given:*

$$A = \{x \mid x \text{ is an integer and } 12 < x < 20\},$$
$$B = \{x \mid x \text{ is an integer and } x > 17\}.$$

    Find: $A \cap B$.

9. *Given:* $A = \{x \mid x^2 - 7x = 8\}$, $B = \{x \mid 2x - 16 = 0\}$.

    Find: $A \cap B$.

10. *Given:* $X = \{x \mid 2x^2 + 4x - 16 = 0\}$, $Y = \{y \mid y^2 = 4\}$.

    Find: a) $X \cup Y$ b) $X \cap Y$

11. Let

$$U = \{x \mid x \text{ is a } UCLA \text{ student}\},$$
$$A = \{x \mid x \text{ is a } UCLA \text{ student 20 years of age or older}\},$$
$$B = \{y \mid y \text{ is a } UCLA \text{ student under 20 years}\}.$$

    Find: a) $A \cup B$ b) $(A \cup B)'$ c) $A \cap B$

EXPERIMENT, SAMPLE SPACE, AND EVENT

Experiment

The term experiment is used in statistics in a much broader sense than in chemistry or physics. The tossing of a coin, for example, is considered a statistical experiment. Other examples of statistical experiments are: rolling a die, selecting one fuse from a lot and observing whether the fuse is defective or not, and sending a manned vehicle to Mars.

Although the above-mentioned experiments seem dissimilar, they have two properties in common. One is that each experiment has several *possible outcomes* that can be specified in advance.

Experiment	Possible outcomes
1. Tossing a coin	Head, tail
2. Rolling a die	1, 2, 3, 4, 5, 6
3. Selecting a fuse	Defective, not defective
4. Sending a manned vehicle to Mars	Success, failure

The second is that we are uncertain about the outcome of each experiment. In tossing a coin, for example, we are not certain whether the outcome will be head or tail. Similarly, we are uncertain whether the fuse selected from the lot will be defective or not defective. Finally, in sending a manned vehicle to Mars, we are uncertain whether the venture will be a failure or a success.

Sample Space

The set S which consists of all possible outcomes of an experiment is called a *sample space*. Each element of a sample space is called a *sample point*. When we toss a coin, for example, our sample space is $S = \{H, T\}$, and each of the elements H and T is a sample point. Similarly, when we roll a die, our sample space is $S = \{1, 2, 3, 4, 5, 6\}$, and the element 1 or 4 is a sample point. We shall illustrate the concept of a sample space with some additional examples.

Example 8

Two coins are to be tossed once. The four possible outcomes of this experiment are shown below.

Coin 1 \ Coin 2	H	T
H	HH	HT
T	TH	TT

The sample space of this experiment is $S = \{HH, HT, TH, TT\}$.

Example 9

A pair of dice is to be cast once. The 36 possible outcomes of this experiment are shown below.

Outcome of one die	Outcome of other die					
	1	2	3	4	5	6
1	(1, 1)	(1, 2)	(1, 3)	(1, 4)	(1, 5)	(1, 6)
2	(2, 1)	(2, 2)	(2, 3)	(2, 4)	(2, 5)	(2, 6)
3	(3, 1)	(3, 2)	(3, 3)	(3, 4)	(3, 5)	(3, 6)
4	(4, 1)	(4, 2)	(4, 3)	(4, 4)	(4, 5)	(4, 6)
5	(5, 1)	(5, 2)	(5, 3)	(5, 4)	(5, 5)	(5, 6)
6	(6, 1)	(6, 2)	(6, 3)	(6, 4)	(6, 5)	(6, 6)

The sample space of this experiment is

$$S = \{(1, 1), (1, 2), (1, 3), (1, 4), (1, 5), (1, 6),$$
$$(2, 1), (2, 2), (2, 3), (2, 4), (2, 5), (2, 6),$$
$$(3, 1), (3, 2), (3, 3), (3, 4), (3, 5), (3, 6),$$
$$(4, 1), (4, 2), (4, 3), (4, 4), (4, 5), (4, 6),$$
$$(5, 1), (5, 2), (5, 3), (5, 4), (5, 5), (5, 6),$$
$$(6, 1), (6, 2), (6, 3), (6, 4), (6, 5), (6, 6)\}.$$

Example 10

Three coins are tossed once. The eight possible outcomes of this experiment can best be displayed by a tree diagram. A tree diagram describing all the possible outcomes of this experiment is shown in Fig. 3–7.

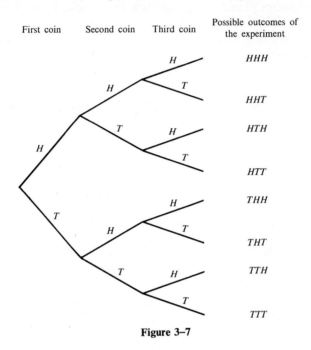

Figure 3–7

Note that each of the two possible outcomes of tossing the first coin is represented by a branch. These two branches are labeled *H* and *T*. Each branch of the first coin sprouts two branches that represent the two possible outcomes of the second coin. Thus we associate four branches with the second coin. Similarly, each branch of the second coin sprouts two branches representing the two possible outcomes of tossing the third coin. The eight possible outcomes of the experiment are then represented by the eight possible paths of the diagram. The first of these eight paths (or outcomes) is *HHH*, the second path is *HHT*, and the last path is *TTT*. The sample space or the set of all possible outcomes of this experiment is

$$S = \{HHH, HHT, HTH, HTT, THH, THT, TTH, TTT\}.$$

Example 11

Four registered voters chosen at random are asked whether or not they prefer gubernatorial candidate *X*. List the elements of the sample space *S*.

Solution: Let Y stand for "yes" and N stand for "no." Figure 3–8 shows the tree diagram representing the various possible outcomes of this experiment. The sample space of this experiment is

$$S = \{YYYY, \ YYYN, \ YYNY, \ YYNN, \ YNYY, \ YNYN, \ YNNY, \ YNNN,$$
$$NYYY, \ NYYN, \ NYNY, \ NYNN, \ NNYY, \ NNYN, \ NNNY, \ NNNN\}.$$

First voter	Second voter	Third voter	Fourth voter	Possible outcomes

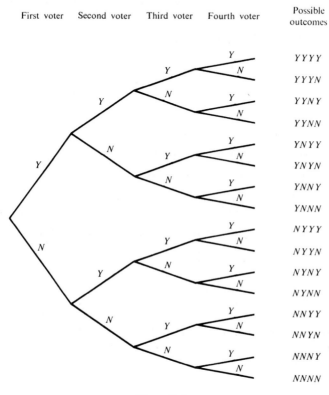

Figure 3–8

Event

We know that when three coins are tossed once, the sample space of this experiment is

$$S = \{HHH, \ HHT, \ HTH, \ HTT, \ THH, \ THT, \ TTH, \ TTT\}.$$

Let us now assume that we are interested in the outcomes in which the number of heads is exactly 2. These outcomes comprise the subset $A = \{HHT, HTH, THH\}$. Next we are interested in the outcomes in which the number of heads is exactly 3. There is only one outcome in which the number of heads is exactly 3, and the subset $B = \{HHH\}$ represents this outcome.

A subset of a sample space is called an event. The subset $A = \{HHT,$ $HTH, THH\}$, for example, is the event that exactly two heads appear when three coins are tossed once. The subset $B = \{HHH\}$ is the event that exactly three heads appear when three coins are tossed once.

Example 12

A pair of dice is to be rolled once. The sample space of this experiment is

$$S = \{(1, 1), (1, 2), (1, 3), (1, 4), (1, 5), (1, 6),$$
$$(2, 1), (2, 2), (2, 3), (2, 4), (2, 5), (2, 6),$$
$$(3, 1), (3, 2), (3, 3), (3, 4), (3, 5), (3, 6),$$
$$(4, 1), (4, 2), (4, 3), (4, 4), (4, 5), (4, 6),$$
$$(5, 1), (5, 2), (5, 3), (5, 4), (5, 5), (5, 6),$$
$$(6, 1), (6, 2), (6, 3), (6, 4), (6, 5), (6, 6)\}.$$

We may now consider the following events:

1. The event that the sum of the two faces is 7, i.e., the subset

$$A = \{(1, 6), (2, 5), (3, 4), (4, 3), (5, 2), (6, 1)\}.$$

2. The event that the sum of the two faces is 11, i.e., the subset

$$B = \{(5, 6), (6, 5)\}.$$

3. The event that the sum of the two faces is either 7 or 11, i.e., the union of sets A and B or the subset

$$A \cup B = \{(1, 6), (2, 5), (3, 4), (4, 3), (5, 2), (6, 1), (5, 6), 6, 5)\}.$$

EXERCISES, SET 2

1. A die is to be tossed once.
   a) List all the elements of the sample space S.
   b) List the elements of S contained in the event that the outcome is even.
   c) List the elements of S contained in the event that the outcome is greater than 4.

2. An experiment consists of tossing two coins once.
   a) List the elements of the sample space S.
   b) List the elements of S contained in the event that there is exactly one head.
   c) List the elements of S contained in the event that there is at least one head.

3. A pair of dice is to be cast once.

   a) List the elements of the sample space S.

   b) List the elements of S contained in the event that the sum is 9.

   c) List the elements of S contained in the event that the sum is either 4 or 5.

4. An experiment consists of selecting three parts from a manufacturing process and observing whether or not the part is defective.

   a) Denoting a defective part by D and a nondefective part by D', list all elements of the sample space S of this experiment.

   b) List the elements of S contained in the event that the number of defective parts is exactly zero.

   c) How do you define event $A = \{DDD', DD'D, D'DD\}$?

5. An experiment involves tossing a coin and casting a die at the same time. List the 12 elements of the sample space S.

PROBABILITY

Each sample point (each possible outcome of an experiment) is assigned a weight which measures the likelihood of its occurrence. This weight is called the probability of the sample point.

In assigning weights or probabilities to the various sample points, statisticians have agreed on two rules:

1. The probability assigned to any sample point ranges from 0 to 1.

2. The sum of probabilities assigned to *all* sample points in the sample space must be equal to 1.

A weight or probability close to 0 is assigned to an outcome (sample point) which is not likely to occur. A probability closer to 1 is assigned to an outcome which is likely to occur.

In many games of chance, the various possible outcomes or sample points are equally likely to occur. Consequently, the probability assigned to each sample point is the same. In tossing a balanced coin, for example, we assign a probability of $\frac{1}{2}$ to each sample point in the sample space $S = \{H, T\}$. Similarly, when a fair die is rolled, we assign a probability of $\frac{1}{6}$ to each sample point in the sample space $S = \{1, 2, 3, 4, 5, 6\}$.

The above examples are in realm of objective probabilities. How can we assign probabilities that are subjective in nature? For example, suppose a firm is planning to market a new product. The two possible outcomes of this experiment are success and failure. That is, $S = \{\text{success, failure}\}$. Now, how can we assign a probability to each of these sample points? One way is to utilize the firm's past experience in marketing similar products. For example, if 90% of all similar products that have been marketed in the past were successful, it would seem reasonable to assign a probability of .90 to the point "success" and a probability of .10 to the point "failure."

Probability of an Event

The probability of event A, denoted by $P(A)$, is the sum of the probabilities of all sample points in A. This definition is illustrated by the following two examples.

Example 13

What is the probability of getting *exactly* two heads when three balanced coins are tossed once?

Solution: The sample space of this experiment is

$$S = \{HHH, HHT, HTH, HTT, THH, THT, TTH, TTT\},$$

and the probability of each sample point is $\frac{1}{8}$. Since the event "getting exactly two heads" is the subset $A = \{HHT, HTH, THH\}$, the probability of getting exactly two heads is therefore

$$P(A) = \tfrac{1}{8} + \tfrac{1}{8} + \tfrac{1}{8} = \tfrac{3}{8}.$$

Example 14

What is the probability of getting a 7 when a fair pair of dice is rolled once?

Solution: The sample space of this experiment consists of 36 sample points, and the probability of each sample point is $\frac{1}{36}$. Since the event getting a 7 is the subset

$$A = \{(1, 6), (2, 5), (3, 4), (4, 3), (5, 2), (6, 1)\},$$

the probability of getting a 7 is therefore

$$P(A) = \tfrac{1}{36} + \tfrac{1}{36} + \tfrac{1}{36} + \tfrac{1}{36} + \tfrac{1}{36} + \tfrac{1}{36} = \tfrac{1}{6}.$$

Adhering to the two rules that the probability of a sample point ranges from 0 to 1, and that the sum of probabilities assigned to all sample points in a sample space must be 1, then, according to our definition of the probability of an event, we now arrive at several important conclusions:

1. $P(S) = 1$. This is because a sample space is an event which contains *all* sample points.
2. $P(\varnothing) = 0$. This is because the null set is an event that contains no sample point.
3. $P(A') = 1 - P(A)$. The probability that event A will not occur, denoted by $P(A')$, is equal to 1 minus the probability that A will occur. Thus, if the probability of getting a 5 on one roll of a die is $\frac{1}{6}$, then the probability of not getting a five is $(1 - \frac{1}{6})$ or $\frac{5}{6}$. Similarly, if the probability that it will rain on a certain day is $\frac{1}{4}$, then the probability that it *will not* rain on that day is $(1 - \frac{1}{4})$ or $\frac{3}{4}$.

Rules of Probability

The solution to many problems involving probabilities requires a thorough understanding of some basic rules which govern the manipulation of probabilities. In general, these rules enable us to determine the probability of an event once the probabilities of some other related events are known. Most important of these basic rules are the Rule of Addition, Rule of Multiplication, and Bayes' Rule. The three rules are discussed in great detail next.

Rule of Addition

Let us consider a game in which a card is to be selected from an ordinary deck of playing cards. We will win $10 if the card selected is either black or a king. What is the probability of winning in this game?

First, we must define the sample space of this experiment. Since there are 52 cards in an ordinary deck, the sample space of this experiment consists of 52 sample points with a probability of $\frac{1}{52}$ assigned to each point. Next, we must determine the number of sample points contained in the event "black, or king, or both."

An ordinary deck of cards contains 26 black cards and 4 cards which are kings. However, the two black kings are counted both as black cards and as kings. Hence, to determine the number of sample points in the event "black, or king, or both," we cannot simply add the number of black cards and kings in the deck $(26 + 4 = 30)$. If we did, the two black kings would be counted twice. To prevent this double counting, we must subtract 2 from the sum $(26 + 4)$. Thus in the event "black, or kings, or both,"

$$\text{Number of sample points} = \begin{array}{c} \text{number of} \\ \text{black cards} \end{array} + \begin{array}{c} \text{number of} \\ \text{kings} \end{array} - \begin{array}{c} \text{number of} \\ \text{black kings} \end{array}$$
$$= \quad 26 \quad + \quad 4 \quad - \quad 2$$
$$= 28.$$

Since the event "black, or kings, or both" contains 28 sample points with a probability of $\frac{1}{52}$ assigned to each point, the probability of winning $10 in our game is $\frac{28}{52}$.

Let us summarize and restate our solution in a more rigorous fashion:

$$P(B \cup K) = P(B) + P(K) - P(B \cap K)$$
$$= \tfrac{26}{52} + \tfrac{4}{52} - \tfrac{2}{52} = \tfrac{28}{52}.$$

The solution to our game can now be generalized into the *rule of addition*. This rule states that if A and B are events belonging to the sample space S, then

$$P(A \cup B) = P(A) + P(B) - P(A \cap B).$$

In words, the rule of addition states that the probability of event A, or B, or both occurring is equal to the probability that A occurs plus the probability that B occurs, minus the probability that both event A and B occur together.

Example 15

A fair die is to be rolled once. You win \$5 if the outcome is either even or divisible by 3. What is the probability of winning the game?

Solution: Let A = the event that the outcome is even, and B = the event that the outcome is divisible by 3. To determine the probability of winning, we describe the following events:

$$S = \{1, 2, 3, 4, 5, 6\}, \quad A = \{2, 4, 6\}, \quad B = \{3, 6\};$$
$$A \cap B = \{6\}.$$

Then, according to the rule of addition, the probability of the event $A \cup B$ is

$$P(A \cup B) = P(A) + P(B) - P(A \cap B)$$
$$= \tfrac{3}{6} + \tfrac{2}{6} - \tfrac{1}{6}$$
$$= \tfrac{2}{3} \text{ probability of winning the game.}$$

Example 16

A customer enters a supermarket. The probability that the customer buys (a) bread is .60, (b) milk is .50, and (c) both bread and milk is .30. What is the probability that the customer would buy either bread or milk or both?

Solution: Let B = the event that the customer buys bread, M = the event that the customer buys milk. Then, according to the rule of addition, we have

$$P(B \cup M) = P(B) + P(M) - P(B \cap M)$$
$$= .60 + .50 - .30 = .80.$$

Rule of Addition for Mutually Exclusive Events

We have described events A and B as mutually exclusive when $A \cap B = \varnothing$. Consequently,

$$P(A \cap B) = P(\varnothing) = 0.$$

Thus, when events A and B are mutually exclusive (in the sense that they cannot occur at the same time), then the rule of addition becomes

$$P(A \cup B) = P(A) + P(B).$$

A few illustrations are now in order.

1. When one card is drawn from an ordinary deck of cards, the card cannot be *both* an ace and a king. Thus the event ace, denoted by A, and the event king, denoted by K, are mutually exclusive. Furthermore,

$$P(A) = \tfrac{4}{52} \quad \text{and} \quad P(K) = \tfrac{4}{52},$$

since there are 4 aces and 4 kings in the deck. Now, if one card is drawn from the deck, what is the probability that the card drawn will be either an ace or a king? According to the rule of addition for mutually exclusive events,

$$P(A \cup K) = P(A) + P(K) \quad = \frac{4}{52} + \frac{4}{52} = \frac{8}{52}.$$

Thus, the probability that the card drawn is either an ace or a king is $\frac{8}{52}$.

2. Let us determine the probability of getting either a 4 or a 5 when a die is rolled once. Getting a 4 and getting a 5 in one roll of a die are mutually exclusive events because a die cannot show a 4 and a 5 in one roll. According to the rule of addition, therefore,

$$P(4 \cup 5) = P(4) + P(5)$$
$$= \frac{1}{6} + \frac{1}{6} = \frac{1}{3}.$$

Thus the probability of getting either a 4 or a 5 on one roll of a die is $\frac{1}{3}$.

The rule of addition also applies to more than two events.* Thus, if A, B, and C are mutually exclusive, the probability that A, or B, or C will take place is equal to the sum of their respective probabilities. Symbolically,

$$P(A \cup B \cup C) = P(A) + P(B) + P(C).$$

EXERCISES, SET 3

1. A fair die is to be cast once. What is the probability of getting
   a) an odd number?
   b) a number greater than 3?

2. Two fair coins are to be tossed once. What is the probability of getting
   a) exactly one head?
   b) at least one head?

3. A fair pair of dice is to be cast once. What is the probability of getting
   a) 7?
   b) 11?
   c) 7 or 11?
   d) a sum divisible by 3?

4. One card is selected from an ordinary deck of playing cards. What is the probability of getting
   a) a queen?
   b) a jack?
   c) either a queen or a jack?
   d) a queen or a red card?
   e) a face card?

---

* The rule of addition can also be generalized for events that are not mutually exclusive. The general rule, however, is beyond the scope of this text.

5. The probability that on July 4th it will rain is .10, it will thunder is .05, and it will rain and thunder is .03. What is the probability that it will either rain or thunder on that day?

6. In a certain community, the probability that a family has a television set is .80, a washing machine is .50, both a television set and a washing machine is 0.45. What is the probability that a family has either a television set or a washing machine or both?

7. The probability that a car salesman will sell at least 3 cars in a day is .20. What is the probability that he will sell 0, or 1, or 2 cars in a day? [*Hint:* $P(S) = 1$.]

8. The probability that Mrs. Chatter will receive at most 5 telephone calls a day is .20, at least 9 telephone calls a day is .50. What is the probability that Mrs. Chatter will receive 6, or 7, or 8 calls a day?

9. A box contains 100 television tubes. The probability that there is at least one defective tube in the box is .05. The probability that there are at least 2 defective tubes in the box is .01. What is the probability

   a) that the box will contain no defective tubes? [*Hint:* $P(S) = 1$.]
   b) that the box will contain exactly one defective tube?
   c) that the box will contain at most one defective tube?

Conditional Probability

A box contains black and white balls. Each ball is labeled either A or Z. The composition of the box is shown below:

	B(black)	W(white)	Total
A	5	3	8
Z	1	2	3
TOTAL	6	5	11

Let us now assume that a ball is to be selected at random from this box. We may then speak of the following probabilities:

1. $P(B)$ = probability of getting a black ball = $\frac{6}{11}$.

   There are 6 black balls among a total of 11 balls.

2. $P(B|A)$ = probability of getting a black *assuming* that the ball selected is labeled $A = \frac{5}{8}$.

   $P(B|A)$ (read "probability of B given A") is called *conditional probability*, because it refers to the probability of getting a black ball subject to the condition that the ball selected is labeled A. This conditional probability is $\frac{5}{8}$ since there are only 8 balls labeled A and 5 of these balls are black.

3. $P(A)$ = probability of getting a ball labeled A = $\frac{8}{11}$.

   There are 8 balls labeled A among a total of 11 balls.

4. $P(A \cap B)$ = probability of getting a ball that is *both*—labeled A and *black* = $\frac{5}{11}$.

   There are 5 balls that are both labeled A and black among a total of 11 balls.

Now let us summarize some of these findings:

$$P(B|A) = \tfrac{5}{8}, \qquad P(A) = \tfrac{8}{11}, \qquad P(A \cap B) = \tfrac{5}{11}.$$

Using these results, we can show that

$$P(B|A) = \frac{P(A \cap B)}{P(A)}$$

$$= \frac{\frac{5}{11}}{\frac{8}{11}} = \frac{5}{8},$$

which is exactly the same result as that obtained earlier. We now introduce a general definition of conditional probability: If events A and B belong to the sample space S, and if $P(A) \neq 0$, then the conditional probability of B relative to A, denoted by $P(B|A)$, is defined as

$$P(B|A) = \frac{P(A \cap B)}{P(A)}.$$

Example 17

A fair die is to be rolled once. Given that the outcome is even, what is the probability of getting a number greater than 3?

Solution: Let us first define the following sets:

$$
\begin{array}{ll}
S = \{1, 2, 3, 4, 5, 6\} & \text{(Sample space)} \\
A = \{2, 4, 6\} & \text{(Set of even outcomes)} \\
B = \{4, 5, 6\} & \text{(Set of numbers greater than 3)} \\
A \cap B = \{4, 6\} & \text{(Set of even numbers which are greater than three)}
\end{array}
$$

Since the die is fair, we assign a probability of $\frac{1}{6}$ to each sample point. Consequently,

$$P(A) = \tfrac{3}{6}, \qquad P(A \cap B) = \tfrac{2}{6}.$$

Using the definition of conditional probability, we can now determine the probability of getting a number greater than 3 given that the outcome is even:

$$P(B|A) = \frac{P(A \cap B)}{P(A)}$$

$$= \frac{\frac{2}{6}}{\frac{3}{6}} = \frac{2}{3}.$$

This result can be verified by the fact that out of three even outcomes {2, 4, 6} only two are greater than 3.

Example 18

In Honolulu, the probability that it will rain on the first day of December is .50. The probability that it will rain on both the first and second days of December is .40. Given that December 1 is a rainy day in Honolulu, what is the probability of rain on the next day?

Solution: Given: $P(R_1) = .50$, $P(R_1 \cap R_2) = .40$.

Using our definition of conditional probability, we have

$$P(R_2|R_1) = \frac{P(R_1 \cap R_2)}{P(R_1)}$$
$$= \frac{.40}{.50} = .80.$$

Rule of Multiplication

We begin with the definition of conditional probability

$$P(B|A) = \frac{P(A \cap B)}{P(A)}.$$

Now let us multiply both sides of the above equation by $P(A)$. The resulting new equation,

$$P(A \cap B) = P(A)P(B|A),$$

is called the *Rule of Multiplication*. In words, the rule of multiplication states that the probability that both events A and B occur is equal to the probability that A occurs multiplied by the probability that B occurs, given that event A has taken place.

Example 19

Two cards are to be drawn without replacement from an ordinary deck of playing cards. What is the probability that both of the cards drawn are aces?

Solution: Let A_1 be the event that the first card drawn is an ace and let A_2 be the event that the second card is an ace. Then, according to the rule of multiplication, the probability of the event $A_1 \cap A_2$ is equal to the probability of obtaining an ace on the first draw multiplied by the probability of obtaining an ace on the second draw, given that an ace was obtained on the first draw. Symbolically,

$$P(A_1 \cap A_2) = P(A_1)P(A_2|A_1).$$

The probability of obtaining an ace on the first draw, or $P(A_1)$, is $\frac{4}{52}$ since there are 4 aces in a deck of 52 cards. The probability of getting an ace on the second draw, given that an ace was obtained on the first draw, or $P(A_2/A_1)$, is $\frac{3}{51}$ since of the remaining 51 cards in the deck only 3 are aces. Thus,

$$P(A_1 \cap A_2) = P(A_1)P(A_2|A_1)$$
$$= (\tfrac{4}{52})(\tfrac{3}{51}) = \tfrac{12}{2652}.$$

Once again, the general rule of multiplication states

$$P(A \cap B) = P(A) \cdot P(B/A);$$

now, $P(A \cap B)$ is called the joint probability of A and B; $P(A)$ is called the marginal probability of event A, and $P(B|A)$ is the conditional probability of B with respect to A.

Rule of Multiplication for Independent Events

Events A and B are considered to be independent when the occurrence of one does not influence the probability of the occurrence of the other. This means that regardless of whether event A has or has not happened, the probability assigned to B is going to be the same. Stated differently, when A and B are independent, the conditional probability of B relative to A is the same as the unconditional probability of B, that is, $P(B/A) = P(B)$. If this is the case, the rule of multiplication for independent events becomes

$$P(A \cap B) = P(A) \cdot P(B).$$

To illustrate this rule, let us determine the probability of obtaining two heads in two successive tosses of a balanced coin:

$$P(H_1 \cap H_2) = P(H_1) \cdot P(H_2)$$
$$= (\tfrac{1}{2}) \cdot (\tfrac{1}{2}) = \tfrac{1}{4}.$$

As an alternative illustration, let us determine the probability of getting a 6 and a 5 in succession when a die is rolled twice. Since the two events are independent, it follows that

$$P(6 \cap 5) = P(6) \cdot P(5)$$
$$= (\tfrac{1}{6}) \cdot (\tfrac{1}{6}) = \tfrac{1}{36}.$$

The multiplication rule, whether for dependent or for independent events, can be extended to cover the case of more than two events. Although we shall not state the extended rules rigorously, we shall treat them informally in some of the illustrative examples below. These problems are meant to illustrate the various characteristics and rules of probabilities studied so far.

Example 20

An urn contains 6 white and 4 black marbles. Two marbles are to be selected in succession and without replacement from this urn.

    a) What is the probability that the two selected marbles are white?

    b) What is the probability that the first marble is white and the second marble is black?

    c) What is the probability that the first marble is black and the second marble is white?

    d) What is the probability that the two selected are black?

Solution:

    a) $P(W_1 \cap W_2) = P(W_1) \cdot P(W_2 | W_1)$

$$= \tfrac{6}{10} \cdot \tfrac{5}{9} = \tfrac{30}{90}$$

    b) $P(W_1 \cap B_2) = P(W_1) \cdot P(B_2 | W_1)$

$$= \tfrac{6}{10} \cdot \tfrac{4}{9} = \tfrac{24}{90}$$

    c) $P(B_1 \cap W_2) = P(B_1) \cdot P(W_2 | B_1)$

$$= \tfrac{4}{10} \cdot \tfrac{6}{9} = \tfrac{24}{90}$$

    d) $P(B_1 \cap B_2) = P(B_1) \cdot P(B_2 | B_1)$

$$= \tfrac{4}{10} \cdot \tfrac{3}{9} = \tfrac{12}{90}$$

    In this example, we have actually considered all possible outcomes of our experiment. Furthermore, we have used the general rule of multiplication to determine the probability of each of these possible outcomes. Let us summarize our results below.

Possible outcome	Marginal probability $P(A)$	Conditional probability $P(B\|A)$	Joint probability = Marg. prob. × Cond. prob. $P(A \cap B) = P(A) \cdot P(B\|A)$
$W_1 \cap W_2$	$P(W_1) = \tfrac{6}{10}$	$P(W_2\|W_1) = \tfrac{5}{9}$	$\tfrac{6}{10} \cdot \tfrac{5}{9} = \tfrac{30}{90}$
$W_1 \cap B_2$	$P(W_1) = \tfrac{6}{10}$	$P(B_2\|W_1) = \tfrac{4}{9}$	$\tfrac{6}{10} \cdot \tfrac{4}{9} = \tfrac{24}{90}$
$B_1 \cap W_2$	$P(B_1) = \tfrac{4}{10}$	$P(W_2\|B_1) = \tfrac{6}{9}$	$\tfrac{4}{10} \cdot \tfrac{6}{9} = \tfrac{24}{90}$
$B_1 \cap B_2$	$P(B_1) = \tfrac{4}{10}$	$P(B_2\|B_1) = \tfrac{3}{9}$	$\tfrac{4}{10} \cdot \tfrac{3}{9} = \tfrac{12}{90}$

The same results are also displayed in a tree diagram in Fig. 3–9.

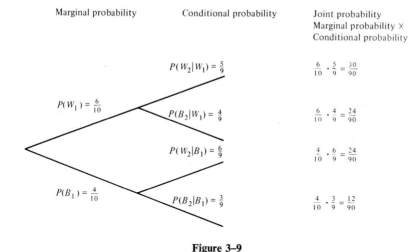

Figure 3–9

Example 21

Urn *A* contains 6 white and 4 red marbles. Urn *B* contains 3 white and 7 red marbles. Two marbles are to be selected, one from *A* and one from *B*. What is the probability that the two marbles selected are of the same color?

Solution:

$$P(\text{same color}) = P(\text{both white}) + P(\text{both red})$$
$$= P(W_1 \cap W_2) + P(R_1 \cap R_2)$$
$$= P(W_1) \cdot P(W_2) + P(R_1) \cdot P(R_2)$$
$$= \left(\tfrac{6}{10}\right) \cdot \left(\tfrac{3}{10}\right) + \left(\tfrac{4}{10}\right) \cdot \left(\tfrac{7}{10}\right) = \tfrac{18}{100} + \tfrac{28}{100} = \tfrac{46}{100}.$$

Example 22

An assembly consists of three independent components *A*, *B*, and *C*. The assembly is considered as defective if one or more of its components are defective. The probability that component *A* is defective is .01, that component *B* is defective is .02, and that component *C* is defective is .10.

a) What is the probability that the assembly is defective?

b) What is the probability of a defective assembly due to the failure of component *C* only?

Solution:

a) Let

$$P(A) = \text{probability that component } A \text{ is defective,}$$
$$P(A') = \text{probability that component } A \text{ is not defective,}$$
$$P(B) = \text{probability that component } B \text{ is defective,}$$

$P(B')$ = probability that component B is not defective,
$P(C)$ = probability that component C is defective,
$P(C')$ = probability that component C is not defective.

Now let us calculate $P(A')$, $P(B')$, and $P(C')$.

$$P(A') = 1 - P(A) \qquad P(B') = 1 - P(B) \qquad P(C') = 1 - P(C)$$
$$= 1 - .01 \qquad\qquad = 1 - .02 \qquad\qquad = 1 - .10$$
$$= .99; \qquad\qquad\quad = .98; \qquad\qquad\quad = .90.$$

The assembly is in good order only if all the components are in good order. Thus

$$P(\text{good assembly}) = P(A' \cap B' \cap C')$$
$$= P(A') \cdot P(B') \cdot P(C')$$
$$= (.99) \cdot (.98) \cdot (.90) = .87318.$$

But

$$P(\text{defective assembly}) = 1 - P(\text{good assembly})$$
$$= 1 - .87318 = .12682.$$

b) The probability of a defective assembly due to failure of component C alone is

$$P(A' \cap B' \cap C) = P(A') \cdot P(B') \cdot P(C)$$
$$= (.99)(.98)(.10) = .09702.$$

Example 23

An aircraft is equipped with three engines that operate independently. The probability of an engine failure is .01. What is the probability of a successful flight if only one engine is needed for the successful operation of the aircraft?

Solution: Let

$$P(S) = \text{the probability of a successful flight,}$$
$$P(S') = \text{the probability of an unsuccessful flight,}$$
$$P(F) = \text{the probability of an engine failure.}$$

Since the flight is unsuccessful only when all three engines fail, then the probability of unsuccessful flight is

$$P(S') = P(F \cap F \cap F)$$
$$= (.01)(.01)(.01) = (.01)^3.$$

But

$$P(S) = 1 - P(S')$$
$$= 1 - (.01)^3 = 1 - .000001 = .999999.$$

Example 24

Five cards are to be drawn in succession and without replacement from an ordinary deck of playing cards.

a) What is the probability that there will be no ace among the five cards drawn?

b) What is the probability that the first three cards are aces and the last two cards are kings?

c) What is the probability that only the first three cards are aces?

d) What is the probability that an ace will appear only on the fifth draw?

Solution:

a) Denoting the probability of getting an ace by $P(A)$ and the probability of not getting an ace by $P(A')$, we find that the probability that there will be no ace among the five cards drawn is

$$P(A_1' \cap A_2' \cap A_3' \cap A_4' \cap A_5') = \left(\frac{48}{52}\right)\left(\frac{47}{51}\right)\left(\frac{46}{50}\right)\left(\frac{45}{49}\right)\left(\frac{44}{48}\right) = \frac{205,476,480}{311,875,200}.$$

b) Denoting the probability of getting a king by $P(K)$, we find that the probability that the first three cards are aces and the last two are kings is

$$P(A_1 \cap A_2 \cap A_3 \cap K_1 \cap K_2) = \left(\frac{4}{52}\right)\left(\frac{3}{51}\right)\left(\frac{2}{50}\right)\left(\frac{4}{49}\right)\left(\frac{3}{48}\right) = \frac{288}{311,875,200}.$$

c) The probability that only the first three cards are aces is

$$P(A_1 \cap A_2 \cap A_3 \cap A_1' \cap A_2') = \left(\frac{4}{52}\right)\left(\frac{3}{51}\right)\left(\frac{2}{50}\right)\left(\frac{48}{49}\right)\left(\frac{47}{48}\right) = \frac{54,144}{311,875,200}.$$

d) The probability that an ace will appear only on the fifth draw is

$$P(A_1' \cap A_2' \cap A_3' \cap A_4' \cap A) = \left(\frac{48}{52}\right)\left(\frac{47}{51}\right)\left(\frac{46}{50}\right)\left(\frac{45}{49}\right)\left(\frac{4}{48}\right) = \frac{18,679,680}{311,875,200}.$$

EXERCISES, SET 4

1. Given that $P(K) = .50$ and $P(K \cap L) = .30$, find $P(L/K)$.

2. Of all students attending this college, 40% are males, and 4% are males majoring in Art. A student is to be selected at random. Given that the selected student is a male, what is the probability that he is an Art major?

3. An urn contains 4 white and 3 red marbles.
   a) If two marbles are selected *without* replacement from this urn, what is the probability that the two marbles selected are white?

    b) If two marbles are selected *with* replacement from this urn, what is the probability that the two marbles selected are white?

4. Urn *A* contains 4 white and 3 red marbles and urn *B* contains 2 white and 5 red marbles. A marble is to be selected from urn *A* and another marble is to be selected from urn *B*. What is the probability that the two marbles selected are white?

5. Urn *A* contains 4 white and 3 red marbles, urn *B* contains 2 white and 5 red marbles, and urn *C* contains 3 white and 6 red marbles. A marble is to be selected from each one of the three urns. What is the probability that the three selected are of the same color?

6. Two cards are to be drawn, without replacement, from an ordinary deck of playing cards. What is the probability
    a) that the first card to be drawn is a queen and the second card a king?
    b) of drawing a combination of queen and king?
    c) that neither of the two cards will be a queen?
    d) that neither of the two cards will be a queen nor a king?

7. Two cards are to be drawn, without replacement, from a deck of playing cards from which the face cards have already been removed. What is the probability that the sum of the two cards drawn will be 19?

8. Five cards are to be drawn without replacement from an ordinary deck of playing cards. What is the probability
    a) that the first three cards are queens and the last two cards are kings?
    b) that only the first three cards are queens?
    c) that the first three cards are queens?

9. Cards are to be drawn, in succession and without replacement, from an ordinary deck of playing cards. What is the probability
    a) that the first queen would appear on the third draw?
    b) that a queen would appear on the third draw?

10. Three cards are to be drawn without replacement from an ordinary deck of playing cards. What is the probability of getting at least one king among the three cards to be drawn?

11. A die is cast three times. What is the probability
    a) of not getting three 6's in succession?
    b) that the same face appears three times?

12. One face of a die is colored red, two faces are colored green, and the remaining three faces are colored black. The die is thrown four times. What is the probability
    a) that the first three throws are red and the last throw is green?
    b) that only the first three throws are red?
    c) that the first three throws are red?

13. A die is thrown three times. What is the probability
    a) that the sum of the three faces shown is either 3 or 4?
    b) that the sum of the three faces shown is greater than 4?

14. A hunter fires 7 consecutive bullets at the angry tiger.   If the probability that one bullet will kill is 0.6, what is the probability that the *hunter* is still alive?

15. Mr. Ho boards a 6-engine aircraft to attend a summit meeting in Paris. The probability of an engine failure is .10, and each engine operates independently of the others. If at least one operating engine is needed on each side of the aircraft, what is the probability that Mr. Ho will be absent from the summit meeting due to a crash of his aircraft?

16. A lot of 100 fuses is known to contain 2 defective fuses. If the fuses are tested one at a time, what is the probability that the last defective fuse is found on the third test?

17. Marbles are drawn, one at a time and without replacement, from an urn that contains 6 black marbles and 8 white marbles. The drawing is terminated as soon as marbles from both colors are obtained.

   a) What is the probability that the drawing is terminated by the fourth draw?

   b) What is the probability that exactly three drawings are needed to terminate the selection?

18. A rifleman fires one shot at a time at a given target, until the target is hit twice. The probability that any single shot would hit the target is .2.

   a) What is the probability that *exactly* three shots are needed to terminate the firing?

   b) What is the probability that the firing is terminated by the third shot?

19. A box contains four fuses, two of which are known to be defective. Two fuses are selected at random and without replacement from the box.

   a) What is the probability that the two defective fuses are selected?

   b) What is the probability that there is at least one defective fuse among the two selected fuses?

20. A voucher must be approved by two members of the purchasing department before payment is made. The probability that an erroneous voucher is approved by the first member is .01. Given that an erroneous voucher is approved by the first member, the probability that it is approved by the second member is .20. What is the probability that a payment is made on an erroneous voucher?

21. Twenty students in a business statistics class are categorized according to age and sex as follows:

	Male	Female
Under 21 years	4	2
21 years or older	11	3

Assume that two students are selected at random from this class.

   a) What is the probability that both students are male?

   b) What is the probability that both students are under 21 years old?

c) What is the probability that the two selected students are of the same sex?

22. Phil E. Buster, the budget analyst in Flat-brook, California, estimates his city's revenues and expenditures for the next fiscal year as follows:

Revenues, billions of dollars	Probability
50	1/4
60	1/2
70	1/4

Expenditures, billions of dollars	Probability
60	1/6
70	3/6
80	2/6

a) What is the probability of a balanced budget?

b) What is the probability of a 10-billion-dollar budget surplus?

Bayes' Theorem

Let us consider the following game. We have two urns. Urn A_1 contains 8 black and 2 white marbles, and urn A_2 contains 3 black and 7 white marbles. One of the urns is to be selected at random and a marble is to be drawn from the chosen urn. We win a sum of $10 if the marble drawn is black. What is the probability of winning in this game?

There are two ways of winning in this game (each is represented by a path in Fig. 3–10). First, we will win if urn A_1 is chosen and a black marble is drawn from this urn. The probability of winning along this path is

$$P(A_1 \cap B) = P(A_1)P(B|A_1)$$
$$= \tfrac{1}{2} \cdot \tfrac{8}{10} = \tfrac{8}{20}.$$

We also win if urn A_2 is chosen and a black marble is drawn from this urn. The probability of winning along this second path is

$$P(A_2 \cap B) = P(A_2)P(B|A_2)$$
$$= \tfrac{1}{2} \cdot \tfrac{3}{10} = \tfrac{3}{20}.$$

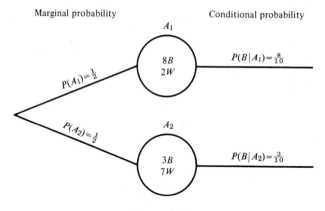

Figure 3–10

Thus we win when either of two events occur: (1) when A_1 is selected and a black marble is drawn from it; (2) when A_2 is selected and a black marble is drawn from it. These two events are mutually exclusive. Thus, according to the rule of addition, the probability of winning, or the probability of drawing a black marble regardless of its origin, is equal to the sum of the probabilities of the two events:

$$P(B) = P(A_1 \cap B) + P(A_2 \cap B)$$
$$= P(A_1)P(B|A_1) + P(A_2)P(B|A_2)$$
$$= \tfrac{1}{2} \cdot \tfrac{8}{10} + \tfrac{1}{2} \cdot \tfrac{3}{10} = \tfrac{8}{20} + \tfrac{3}{20} = \tfrac{11}{20}.$$

(See Fig. 3–11.)

Now let us assume that the game was played once and that we have won the game. What is the probability that urn A_1 was selected?

First, since the probability contribution of the upper path in the diagram above is $\tfrac{8}{20}$ while the probability contribution of the second path is only $\tfrac{3}{20}$, it is only reasonable to assume that the selection of A_1 is more likely than that of A_2. Second, it also seems reasonable to conclude that the probability that A_1 is chosen is

$$\tfrac{8}{20} \div \tfrac{11}{20} = \tfrac{8}{11},$$

while the probability that A_2 is selected is

$$\tfrac{3}{20} \div \tfrac{11}{20} = \tfrac{3}{11}.$$

We can now formally approach this intuitive solution to the problem of determining the probability of selecting each urn after the game has been

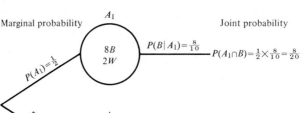

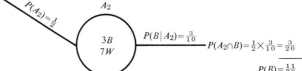

Figure 3–11

won. According to the definition of conditional probability, the probability that A_1 is chosen given that the marble selected is black is

$$P(A_1|B) = \frac{P(A_1 \cap B)}{P(B)}.$$

But since we have already shown that

$$P(B) = P(A_1 \cap B) + P(A_2 \cap B),$$

it follows that

$$P(A_1|B) = \frac{P(A_1 \cap B)}{P(A_1 \cap B) + P(A_2 \cap B)}$$

$$= \frac{\frac{8}{20}}{\frac{8}{20} + \frac{3}{20}} = \frac{8}{11}.$$

Similarly, the probability that A_2 is chosen given that the marble selected is black is

$$P(A_2|B) = \frac{P(A_2 \cap B)}{P(B)}$$

$$= \frac{P(A_2 \cap B)}{P(A_1 \cap B) + P(A_2 \cap B)}$$

$$= \frac{\frac{3}{20}}{\frac{8}{20} + \frac{3}{20}} = \frac{3}{11}.$$

Example 25

Urn A_1 contains 8 black and 2 white marbles. Urn A_2 contains 3 black and 7 white marbles, and urn A_3 contains 5 white and 5 black marbles. A fair die is to be cast. If the die turns up 1, 2, or 3, then a marble will be selected from A_1. If the die turns up 4 or 5, a marble will be selected from A_2. Finally, a marble will be selected from A_3 if the die turns up 6. Given that the marble selected is black, what is the probability that the marble was chosen from A_2?

Solution: We refer to Fig. 3–12. Let us assume that the marble selected is black. Then we find the probability that the marble was chosen from urn A_2 is

$$P(A_2|B) = \frac{P(A_2 \cap B)}{P(B)}$$

$$= \frac{P(A_2 \cap B)}{P(A_1 \cap B) + P(A_2 \cap B) + P(A_3 \cap B)}$$

$$= \frac{P(A_2)P(B|A_2)}{P(A_1)P(B|A_1) + P(A_2)P(B|A_2) + P(A_3)P(B|A_3)}$$

$$= \frac{\left(\frac{2}{6} \times \frac{3}{10}\right)}{\left(\frac{3}{6} \times \frac{8}{10}\right) + \left(\frac{2}{6} \times \frac{3}{10}\right) + \left(\frac{1}{6} \times \frac{5}{10}\right)}$$

$$= \frac{\frac{6}{60}}{\frac{24}{60} + \frac{6}{60} + \frac{5}{60}} = \frac{\frac{6}{60}}{\frac{35}{60}} = \frac{6}{35}.$$

We will generalize the foregoing solution, obtaining *Bayes' Theorem.*

Let $A_1, A_2, \ldots, A_k$ be mutually exclusive events which occupy the entire sample space S. If each of these events has nonzero probability and one of them must occur, then for any event B in the sample space S,

$$P(A_1|B) = \frac{P(A_1)P(B|A_1)}{P(A_1)P(B|A_1) + P(A_2)P(B|A_2) + \cdots + P(A_k)P(B|A_k)},$$

$$P(A_2|B) = \frac{P(A_2)P(B|A_2)}{P(A_1)P(B|A_1) + P(A_2)P(B|A_2) + \cdots + P(A_k)P(B|A_k)},$$

and finally

$$P(A_k|B) = \frac{P(A_k)P(B|A_k)}{P(A_1)P(B|A_1) + P(A_2)P(B|A_2) + \cdots + P(A_k)P(B|A_k)}.$$

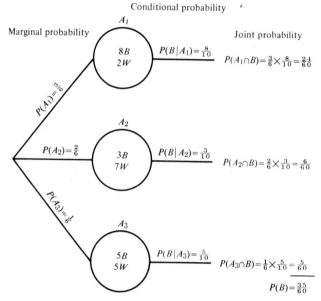

Figure 3–12

Example 26

Urn *A* contains 6 green and 4 red marbles, and urn *B* contains 2 green and 7 red marbles. A marble is to be selected at random from *A* and placed in *B*. One marble is then selected from *B*. Given that the marble selected from *B* is green, what is the probability that the marble selected from *A* will also be green?

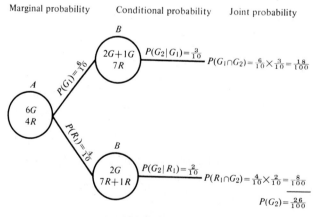

Figure 3–13

Solution: Refer to Fig. 3–13.   Let

$P(G_1)$ = probability that the marble moved from A is green,
$P(R_1)$ = probability that the marble moved from A is red,
$P(G_2)$ = probability that the marble selected from B is green.

Then, according to Bayes' Theorem,

$$P(G_1|G_2) = \frac{P(G_1)P(G_2|G_1)}{P(G_1)P(G_2|G_1) + P(R_1)P(G_2|R_1)}$$

$$= \frac{\left(\frac{6}{10}\cdot\frac{3}{10}\right)}{\left(\frac{6}{10}\cdot\frac{3}{10}\right) + \left(\frac{4}{10}\cdot\frac{2}{10}\right)} = \frac{\frac{18}{10}}{\frac{18}{100}+\frac{8}{100}} = \frac{\frac{18}{100}}{\frac{26}{100}} = \frac{18}{26}.$$

Example 27

The ABC Company is considering the marketing of an electronic printing calculator.   According to the market research director, the probability that the product would succeed is 0.80 if a foreign competing firm did not introduce the same product into the market.   However, the probability of success is only 0.30 if the foreign firm marketed the same product.   The company further believes that the probability is 0.40 that the competing firm will market the electronic printing calculator.   Given that the ABC Company product is successful, what is the probability that the foreign firm has marketed their electronic printing calculator?

Solution: Refer to Fig. 3–14.   Let

$P(M)$ = Probability that the foreign firm markets the same product.
$P(M')$ = Probability that the foreign firm does not market the same product.
$P(S)$ = Probability that the ABC Company succeeds in marketing its electronic calculator.

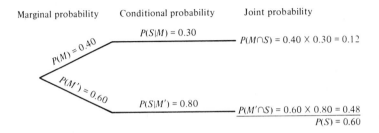

Figure 3–14

Then, according to Bayes' Theorem,

$$P(M|S) = \frac{P(M)P(S|M)}{P(M)P(S|M) + P(M')P(S|M')}$$

$$= \frac{(0.40 \times 0.30)}{(0.40 \times 0.30) + (0.60 \times 0.80)} = \frac{0.12}{0.12 + 0.48} = \frac{12}{60} = 0.20.$$

Example 28

A manufacturer is considering the purchase of a lot of 10,000 parts from a supplier. The manufacturer estimates the proportion of defective parts in the lot as follows:

Proportion of defective parts (π)	Probability $P(\pi)$
$\pi_1 = 0.10$	$P(\pi_1) = 0.20$
$\pi_2 = 0.15$	$P(\pi_2) = 0.30$
$\pi_3 = 0.25$	$P(\pi_3) = 0.50$

This means that the manufacturer is not certain about the proportion of defective parts in the lot. However, based on his past experience, he personally believes that there is 0.20 chance that the lot contains 10 percent defective parts; there is 0.30 chance that the lot contains 15 percent defective parts; and finally, there is 0.50 chance that the lot contains 25 percent defective parts.

Now, assuming that one part is selected at random from the lot,

A) What is the probability of selecting a defective part?
B) Assuming that the selected part is defective, what is the probability that the lot contains 25 percent defective parts?

Solution to A: There are three possible ways of obtaining a defective part from the lot (each is represented by a path in Fig. 3-15).   The three ways are

a) $P(\pi_1 \cap D) = P(\pi_1) \times P(D|\pi_1) = 0.20 \times 0.10 = 0.0200.$
b) $P(\pi_2 \cap D) = P(\pi_2) \times P(D|\pi_2) = 0.30 \times 0.15 = 0.0450.$
c) $P(\pi_3 \cap D) = P(\pi_3) \times P(D|\pi_3) = 0.50 \times 0.25 = 0.125.$

Hence, the probability of getting a defective part, no matter whether the lot contains 10, 15, or 25 percent defective parts, is

$$P(D) = P(\pi_1 \cap D) + P(\pi_2 \cap D) + P(\pi_3 \cap D)$$
$$= \quad 0.0200 \quad + \quad 0.0450 \quad + \quad 0.1250 \quad = 0.19.$$

Solution to B: According to Bayes' Theorem, the probability that the lot contains 25 percent defective parts, given that the selected part was defective, is

$$P(\pi_3 \mid D) = \frac{P(\pi_3 \cap D)}{P(D)} = \frac{0.1250}{0.1900} = 0.6579.$$

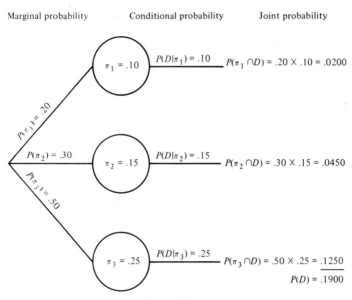

Marginal probability Conditional probability Joint probability

$\pi_1 = .10$ $P(D\mid\pi_1) = .10$ $P(\pi_1 \cap D) = .20 \times .10 = .0200$

$P(\pi_1) = .20$

$P(\pi_2) = .30$ $\pi_2 = .15$ $P(D\mid\pi_2) = .15$ $P(\pi_2 \cap D) = .30 \times .15 = .0450$

$P(\pi_3) = .50$

$\pi_3 = .25$ $P(D\mid\pi_3) = .25$ $P(\pi_3 \cap D) = .50 \times .25 = .1250$

$P(D) = .1900$

Figure 3–15

EXERCISES, SET 5

1. Urn *A* contains 4 white and 3 red marbles, and urn *B* contains 2 white and 5 red marbles. One of the urns is to be chosen at random and a marble is to be selected from the chosen urn.

   a) What is the probability of drawing a white marble?

   b) Given that the marble drawn was white, what is the probability that urn *A* was selected?

2. Urn *A* contains 4 white and 3 red marbles, and urn *B* contains 2 white and 5 red marbles. A fair die is to be cast once. If the die turns up 1 or 2, then two marbles will be selected, without replacement, from urn *A*. If the die turns up 3, 4, 5, or 6, then two marbles will be selected, without replacement, from urn *B*.

   a) What is the probability that the two marbles selected are wnite?

   b) Given that the two marbles selected were white, what is the probability that urn *A* was selected?

3. Urn *A* contains 4 white and 3 red marbles, and urn *B* contains 2 white and 3 red marbles. A marble is to be selected from *A* and placed in *B*. Two marbles are then to be selected from *B*.

a) What is the probability that the two marbles selected from urn *B* are white?

b) Given that the two marbles selected from *B* were white, what is the probability that the marble selected from urn *A* was also white?

4. Twenty percent of all students at UCLA are graduates and 80% are under-graduates. The probability that a graduate student is married is .50 and the probability that an undergraduate student is married is .10. One student is selected at random from UCLA.

a) What is the probability that the student selected is married?

b) Given that the student selected is married, what is the probability that he is a graduate student?

5. The probability that Miss Jones will study for her final examination in Economics is .20. If she studies, the probability is .80 that she will pass her finals. If she does not study, the probability of passing the finals is only .50.

a) What is the probability that Miss Jones will pass her Economics finals?

b) Given that Miss Jones passed her final examination in Economics, what is the probability that she studied for the finals?

6. Every night Mr. Smith arrives home late. Being a good wife, Mrs. Smith leaves the porch light on. The probability that Mr. Smith arrives home drunk is .60. If he arrives home drunk, the probability is .90 that he forgets to turn the porch light off. If Mr. Smith arrives home sober, the probability is only .05 that he forgets to turn the light off.

a) What is the probability that Mr. Smith will turn the porch light off on a given night?

b) Given that Mr. Smith turned the porch light off on a certain night, what is the probability that he arrived home drunk?

7. A real-estate development company is considering the possibility of building a shopping center in a Los Angeles suburb. A proposed freeway route linking the suburb to the Los Angeles central freeway system is a vital element in the company's consideration of the project. The probability that the real-estate company would build the shopping center is 0.90, if the State Department of Public Works approves the proposed freeway route. If the new route is not approved, the probability is only 0.20 that the company would build the shopping center. Based on all available information, the president of the company believes that there is 0.60 chance that the proposed freeway route would be approved.

a) What is the probability that the real estate company would build the shopping center?

b) Given that the shopping center was built, what is the probability that the state Department of Public Works approved the new freeway route?

8. Irving Aircraft has submitted a bid on a contract to build one hundred F-13 planes for the United States Air Corps. The probability that Irving Aircraft

would get the contract is 0.90 if a competing firm, Helga Aircraft, did not submit a bid on the same contract. However, the probability is only 0.20 that Irving Aircraft would get the contract if Helga Aircraft submits a bid. Mr. Irving, president of Irving Aircraft, personally believes that there is 0.80 chance that Helga Aircraft would bid against his company.

a) What is the probability that Irving Aircraft would get the contract?
b) Given that Irving Aircraft got the contract to build the F-13 plane, what is the probability that Helga Aircraft submitted a bid on the same contract?

9. The general manager of a nationally franchised neighborhood food store chain estimates the proportion of his stores which will achieve a million dollar annual sales as follows:

Proportion of Stores (π)	Probability $P(\pi)$
0.60	0.2
0.70	0.5
0.80	0.3

Two stores are to be selected at random from all stores.

a) What is the probability that both stores have achieved a million dollar annual sales?
b) Given that both stores achieved a million dollar annual sales, what is the probability that 80 percent of all stores achieved a million dollar annual sales?

10. The ABC Company undertakes a market survey in order to assess the profitability of each of its newly developed products. Past surveys indicated that 90 percent of the newly developed products would be profitable products. Subsequent analysis of the reliability of such surveys has shown however, that only 70 percent of the products predicted to be profitable by a market survey yielded any profit. By contrast, of those products predicted to be unprofitable by a survey, 20 percent produced some profit when marketed. The company has marketed a newly developed product called X. Given that product X was actually profitable, what is the probability that the market survey predicted failure for product X?

11. Bob drives his automobile to work 40% of the time. He takes the city bus 50% of the time, and he hitchhikes his way to work 10% of the time. The probability he would arrive on time to work is .9 if he drives his automobile, .6 if he takes the bus, and .3 if he hitchhikes. Given that Bob arrived late to work on a given day, what is the probability that he took the bus?

12. Richard knows only two girls: Pat and Jan. He asks Pat for a date $\frac{3}{4}$ of the time and asks Jan $\frac{1}{4}$ of the time. The probability that Pat would accept his invitation to a show is .2, and the probability that Jan would is .3. Given that Richard and one of the two girls were at the show last night, what is the probability that Richard was accompanied by Pat? What is the probability that he was accompanied by Jan?

13. The Dixon Detective Agency specializes in solving burglary cases. The Agency employs two detectives: Holman and Ericman. Holman is assigned 60% of all cases, and he solves 10% of the cases assigned to him. Ericman is assigned 40% of all cases, and he solves 20% of the cases assigned to him. If a burglary case has just been solved by the Dixon Agency, what is the probability that the case was solved by Ericman?

14. Johnny Carson engages any of three comedy writers, Tom, Dick, or Jerry, to write his nightly TV monologue. Tom is assigned this task 50% of the time, Dick 5% of the time, and Jerry 45% of the time. The probability of writing a successful monologue is .20, .10, and .15 for Tom, Dick, and Jerry, respectively. Last night's monologue bombed. What is the probability that the monologue was written by Tom?

15. Millionaire oilman Ty Coon has purchased the petroleum rights on a piece of property. Ty's initial estimate is that there is only a .05 chance of oil being present. Ty is thinking about having a seismic survey made. From past experience Ty knows that if oil is on the land, the seismic survey will show a reservoir 90% of the time. Also, if no oil is present, the seismic survey will indicate a reservoir 20% of the time. Suppose that the survey was made and a reservoir was found; find the probability that oil is present.

Random Variable

Let us consider the possible number of heads that can appear when three balanced coins are tossed once. The eight possible outcomes of this experiment, together with the number of heads associated with each outcome, are listed below.

Outcome	Number of heads
TTT	0
HTT	1
THT	1
TTH	1
THH	2
HTH	2
HHT	2
HHH	3

We have assigned a value (0, 1, 2, or 3) to each of the eight possible outcomes or sample points. The value assigned to any sample point refers to the number of heads associated with that sample point. Thus, we have assigned the value 0 to the sample point *TTT* (no head), and the value 2 to each of the sample points *THH*, *HTH*, *HHT* (each of these points has exactly 2 heads).

Now, let the variable *X* represent the number of heads in this experiment. Depending on the outcome of our experiment, the variable *X* can assume any of the values 0, 1, 2, or 3. When the numerical value of a variable is determined by the outcome of an experiment, the variable is called a *random variable*. Thus the variable *X* which represents the number of heads when three balanced coins are tossed is a random variable.

It is useful to distinguish between the random variable itself and the numerical values which can be assumed by the random variable. The random variable is designated by the capital letter *X*, and any value assumed by the random variable is designated by the lower-case letter *x*.

Mathematical Expectation

Once again, let us consider the possible number of heads that can appear when three coins are tossed once. The eight possible outcomes of this experiment are listed at the left below.

Outcome	Number of Heads	Probability
TTT	0	$\frac{1}{8}$
HTT	1	$\frac{1}{8}$
THT	1	$\frac{1}{8}$
TTH	1	$\frac{1}{8}$
THH	2	$\frac{1}{8}$
HTH	2	$\frac{1}{8}$
HHT	2	$\frac{1}{8}$
HHH	3	$\frac{1}{8}$

x (number of heads)	$P(x)$
0	$\frac{1}{8}$
1	$\frac{3}{8}$
2	$\frac{3}{8}$
3	$\frac{1}{8}$

Denoting each of the possible number of heads by *x*, and its probability of occurrence by $P(x)$, we list, in the table at the right above, all possible numbers of heads, together with their respective probabilities of occurrence.

Thus, when three coins are tossed once, we obtain 0, 1, 2, or 3 heads. Furthermore, it is more likely to obtain 1 or 2 heads than it is to obtain 0 or 3 heads.

Now let us assume that the three coins are to be tossed an infinite number of times. Although the number of heads that can appear in any of the trials is 0, 1, 2, or 3, in this infinite number of trials we expect to obtain an average of 1.5 heads per toss. This long-run average of 1.5 heads per toss is called the *mathematical expectation.*

Although we derived the mathematical expectation of the number of heads in a toss of three balanced coins in an intuitive fashion, there is a systematic procedure for determining it. To determine the mathematical expectation of the random variable X, we must first multiply each possible value of X by its probability of occurrence. The mathematical expectation of X, denoted by $E(X)$, is the sum of all these cross products. This procedure is illustrated below.

(1) x (number of heads)	(2) $P(x)$	(3) $xP(x)$
0	$\frac{1}{8}$	0
1	$\frac{3}{8}$	$\frac{3}{8}$
2	$\frac{3}{8}$	$\frac{6}{8}$
3	$\frac{1}{8}$	$\frac{3}{8}$

$$\sum xP(x) = \tfrac{12}{8}$$
$$E(X) = \sum xP(x)$$
$$= \tfrac{12}{8}$$
$$= 1.5 \text{ heads.}$$

Column (1) above lists all possible values of random variable X, and column (2) lists the probability of occurrence of each of these values. Each possible value of X in column (1) is multiplied by its corresponding probability from column (2), and the resulting cross product is placed in column (3). The sum of all the cross products in column (3) is the mathematical expectation of X, or $E(X)$.

We shall now illustrate the application of the mathematical expectation concept.

Example 29

A box contains 3 black and 7 white marbles. One marble is to be selected from this box. You get $2 if the marble selected is black and you lose $1 if the marble selected is white. What is the mathematical expectation of this game?

Denoting any possible gain or loss by x, and denoting its probability of occurrence by $P(x)$, we calculate the mathematical expectation as follows:

x (amount of gain or loss)	$P(x)$	$xP(x)$
+ $2	$\frac{3}{10}$	$\frac{6}{10}$
− $1	$\frac{7}{10}$	$-\frac{7}{10}$
		$E(X) = -\frac{1}{10} = -\$.10$

Thus the mathematical expectation of this game is a loss of $.10.

Let us assume that you are to play this game several thousand times. At any one play you will either win $2 or lose $1. However, in all these thousand plays, you can expect an average loss of $.10 per play.

Example 30

A box contains 4 red, 6 black, and 8 green marbles. One marble is to be selected from this box. If the marble selected is red, you get $3, and if it is black, you get $2. How much should you pay if the marble selected is green in order to ensure an equitable game?

A game is considered to be equitable whenever its mathematical expectation is zero. To make our game equitable, we must therefore determine the loss that is to be associated with drawing a green marble if the mathematical expectation of the game is to be zero. Denoting the loss associated with drawing a green marble by L, we calculate the mathematical expectation of the game.

x (amount of gain or loss)	$P(x)$	$xP(x)$
$3	$\dfrac{4}{18}$	$\dfrac{12}{18}$
$2	$\dfrac{6}{18}$	$\dfrac{12}{18}$
$L	$\dfrac{8}{18}$	$\dfrac{8L}{18}$
		$E(X) = \sum xP(x),$ $= \dfrac{8L}{18} + \dfrac{24}{18}.$

We may solve for L now by setting the mathematical expectation equal to zero.

$$\frac{8L}{18} + \frac{24}{18} = 0,$$

$$8L = -24,$$

$$L = -\tfrac{24}{8} = -\$3.$$

Thus, if you pay \$3 when the selected ball is green, the mathematical expectation of the game becomes zero, and the game is equitable.

Example 31

The hourly sales of a cigarette vending machine at the Rendezvous Room are 20, 21, or 22 packs, with probabilities of .3, .5, and .2 respectively. What are the expected hourly sales of this vending machine?

x (number of packs sold)	$P(x)$	$xP(x)$
20	.3	6.0
21	.5	10.5
22	.2	4.4
		$E(X) = 20.9$ packs

Example 32

A florist estimates his daily sales of carnations as follows:

Estimated daily sales, dozens	Probability of estimated sales
10	.60
11	.30
12	.10

The florist must order the carnations one day in advance. Any carnations left unsold at the end of the day are worthless. If the cost of carnations is \$1 per dozen and the selling price is \$3 per dozen, how many dozens should the florist order in order to maximize his expected daily profit?

Solution: The florist could order 10, 11, or 12 dozens a day. The profit from each order depends on the number of dozens sold. The table below shows all possible levels of profit that the florist can make.

All possible levels of profit

Possible sales per day	Number of dozens ordered		
	10	11	12
10 dozens	$20 (1)	$19 (4)	$18 (7)
11 dozens	$20 (2)	$22 (5)	$21 (8)
12 dozens	$20 (3)	$22 (6)	$24 (9)

The nine entries in the table are interpreted as follows: The first entry, $20, is the profit obtained when the florist orders 10 dozens and sells 10 dozens the same day. The second entry is also $20 because the florist ordered only 10 dozens, although he could have sold 11 dozens; and similarly for entry (3).

The fourth entry in the table, $19, is the profit obtained by stocking 11 dozens and selling only 10 dozens. This profit is computed as follows:

$$\text{Profit} = \text{total revenue} - \text{total cost}$$
$$= (10 \times \$3) - (11 \times \$1) = \$19.$$

All other entries in the table are determined in the same manner. The seventh entry, $18, is the profit obtained by stocking 12 units and selling only 10 units.

$$\text{Profit} = \text{total revenue} - \text{total cost}$$
$$= (10 \times \$3) - (12 \times \$1) = \$18.$$

Before proceeding with the next step in the solution, we summarize the results of our previous computations.

All possible levels of profit

Probability	Order 10	Order 11	Order 12
.60	$20	$19	$18
.30	$20	$22	$21
.10	$20	$22	$24

The expected profit of *each* of the three levels of "order" is the sum of the cross products obtained by multiplying each possible profit by its corresponding probability of occurrence. The computations are carried out below.

Order 10		Order 11		Order 12	
$20 \times .60 =$	12.00	$19 \times .60 =$	11.40	$18 \times .60 =$	10.80
$20 \times .30 =$	6.00	$22 \times .30 =$	6.60	$21 \times .30 =$	6.30
$20 \times .10 =$	2.00	$22 \times .10 =$	2.20	$24 \times .10 =$	2.40
Expected profit	$20.00		$20.20		$19.50

Thus the expected profit is maximized by ordering 11 dozens per day.

Example 33

A radio manufacturer receives shipments of tubes in lots of 20. The receiving department uses the following rule of inspection. Two tubes are tested from each lot. If none of the tubes examined is defective, no further inspection is made on the remaining tubes in the lot. If any of the tubes examined is defective, the entire lot will be inspected. What is the expected number of tubes examined per lot if each lot contains exactly 25% defectives (i.e., 5 defective tubes per lot).

Solution: A lot is partially inspected if both tubes selected are nondefectives. Denoting a defective tube by D and nondefective tube by D', we find that the probability of a partial inspection is

$$P(D'_1 \cap D'_2) = P(D'_1) \cdot P(D'_2/D'_1)$$

$$= \tfrac{15}{20} \times \tfrac{14}{19} = \tfrac{210}{380} = \tfrac{21}{38}.$$

In all other circumstances, the entire lot is inspected. The probability of complete inspection therefore is $1 - \tfrac{21}{38} = \tfrac{17}{38}$. Thus there is $\tfrac{21}{38}$ chance that only two tubes will be inspected and a $\tfrac{17}{38}$ chance that all 20 tubes will be examined. Denoting the number of inspected tubes from a lot by X, we compute the expected number of tubes examined per lot, or $E(X)$, as follows:

x	$P(x)$	$xP(x)$
2	$\tfrac{21}{38}$	$\tfrac{42}{38}$
20	$\tfrac{17}{38}$	$\tfrac{340}{38}$
		$E(X) = \tfrac{382}{38}$
		$= 10.05$ tubes per lot

EXERCISES, SET 6

1. One face of a die is red, two faces are green, and the remaining three faces are black. The die is to be rolled once. You get $2 if red is thrown and $.50 if green is thrown. How much should you pay when black is thrown in order to make this game equitable?

2. One face of a die is red, two faces are green, and the remaining three faces are black. The die is to be rolled twice. If the same color appears on both rolls, you get $11; otherwise you lose $7. What is the expected value of this game?

3. An urn contains 4 red and 6 blue marbles. Three marbles are to be drawn in succession and *with* replacement. If you get $2 for each red marble and $1 for each blue marble, how much should you pay for the right to play in order to make the game equitable?

4. An urn contains 4 red and 6 blue marbles. Three marbles are to be drawn in succession and *without* replacement. If you get $2 for each red marble and $1 for each blue marble, how much should you pay for the right to play in order to make the game equitable?

5. A radio manufacturer receives tubes in boxes, each box containing 100 tubes. The receiving department uses the following rule of inspection. Four tubes are tested from each box. If none of the tubes examined is defective, no further inspection is made on the remaining tubes in that box. If any of the tubes examined is defective, the entire content of the box is inspected. Determine the expected number of tubes examined per box if each box contains exactly 10% defectives.

6. The Geneva Weaving Company wants to purchase 10 specialized machines to be used in its weaving process. The company has previously purchased such machines from two different manufacturers. Based on experience, the useful lives of the two brands of machines can be estimated as follows:

Machine *A*

Estimated useful life, hours	Probability of useful life
2000	.60
3000	.30
4000	.10

Machine *B*

Estimated useful life, hours	Probability of useful life
2000	.50
3000	.45
4000	.05

Which brand should the Geneva Weaving Company purchase if the cost of the two brands is the same?

7. A television manufacturer uses a special type of electronic component in assembling color television sets. Each television set requires 6 such components. A defective component cannot be detected until a television set has been completely assembled. The cost of detecting, repairing, and replacing a defective component is $15.

The television manufacturer has been purchasing these components in lots of 100 components from two different suppliers. The cost of purchasing a lot from supplier A is $100, while the cost of a lot from supplier B is $120. Based on past experience, the comparative qualities of the lots purchased from the two suppliers are as follows:

Supplier A

Estimated number of defective components per lot	Probability of number of defective components
1	.30
2	.25
3	.20
4	.15
5	.10

Supplier B

Estimated number of defective components per lot	Probability of number of defective components
1	.60
2	.30
3	.10

From which supplier should the television manufacturer purchase his electronic components in order to minimize the cost of these components?

8. The Land Investor Corporation owns a 10-acre lot in a Los Angeles suburb. The corporation has petitioned the City Planning Commission for rezoning the lot from "Agricultural" to "Multiple Dwellings." If the rezoning permit is granted, the lot is worth one million dollars. If the rezoning permit is denied, the lot is worth only $200,000.

Prior to the decision on rezoning by the Planning Commission, a large real estate developer offered to buy the lot for $500,000 outright.

a) Should the Land Investor Corporation sell the lot to the real estate developer if the probability of approving the request for rezoning is $\frac{1}{2}$?

b) For the Land Investor Company to have no preference for one of the two alternatives (selling the lot outright or waiting for the decision on rezoning), what probability must be assigned to the chance of rezoning approval?

9. A grocer estimates his daily sales of "Slim & Trim" bread as follows:

Estimated daily sales, loaves	Probability of estimated sales
4	.50
5	.40
6	.10

The cost per loaf is 25c and the selling price is 50c per loaf. The bread must be ordered one day in advance, and any bread left unsold at the end of the day is turned over to the Thrift Bakery for a salvage price of 10c per loaf. How many loaves of this bread should the grocer order every day in order to maximize his expected daily profit?

10. The Associated Students in the college are planning a concert. Based on similar concerts given on this campus, attendance is estimated as follows:

Number of persons attending (X)	Probability $P(X)$
2000	0.2
3000	0.5
4000	0.3

If total cost of the concert is $2,000 and the price of admission is set at $1 per person, what is the expected profit (or loss) from the concert?

11. A manufacturer is considering the production of a seasonal novelty. The manufacturer estimates the demand for the novelty as follows:

Number of units demanded (X)	Probability $P(X)$
1000	$\frac{1}{4}$
2000	$\frac{1}{2}$
3000	$\frac{1}{4}$

The cost of producing and marketing the novelty consists of a fixed cost of $5,000 and a variable cost of $1 per unit. If the selling price of the novelty is $5 per unit, what is the manufacturer's expected profit?

12. Mr. Smith is a used car salesman. For each car he sells, Smith receives a commission of either $200 or $100, with a probability of .2 and .8, respec-

tively. The following is a probability distribution of the number of auto-mobiles sold by Mr. Smith on any given day:

Number of cars sold	Probability
0	.20
1	.60
2	.15
3	.05

a) What is the expected number of cars sold by Mr. Smith each day?

b) What is his expected commission *per car*?

c) What is his expected *daily* sales commission?

d) What is the probability that Mr. Smith sells two cars each day in two con-secutive days?

13. A florist estimates his daily demand for carnations as follows:

Number of dozens demanded	Probability
10	.3
11	.6
12	.1

a) What is the probability that the florist would sell 12 dozen carnations in three consecutive days?

(b) Find the expected number of dozens of carnations sold per day.

4
Probability
(continued)

Probability
(continued)

Rules of Counting

The solution to many problems involving probabilities may require some formal rules of counting. Such rules enable us, for example, to determine the number of sample points in a sample space or in a given event without having to list these sample points. Only rules which are needed in our subsequent discussion are treated here.

Rule 1. If operation A can be performed in n number of ways, and operation B in m number of ways, then the two operations can be performed together in n·m ways.

In order to illustrate the above rule, let us determine the number of arrangements in which a man can wear a suit and a tie if he has three suits and five ties.

The man can choose one of three suits and, with each chosen suit, he may wear any of the five ties. According to Rule 1, he can therefore come up with suit and tie combinations in

$$3 \times 5 = 15 \text{ ways.}$$

Example 1

A coin is thrown and a die is cast. Determine the number of sample points in the sample space of this experiment.

Solution: The coin can land in 2 different ways (H, T), and the die can land in 6 different ways (1, 2, 3, 4, 5, 6). The two operations can therefore be performed in $2 \times 6 = 12$ ways. Thus, our sample space consists of 12 points.

This rule can be extended to cover the case where there are more than two operations, as illustrated by the following two examples:

85

Example 2

A die is tossed three times. Determine the number of sample points in the sample space of this experiment.

Solution: $6 \times 6 \times 6 = 216$ sample points.

Example 3

How many three-digit numbers can be formed from the digits 2, 4, 6, 7, 9?

Solution: There are three positions to be filled in a three-digit number. Each position can be filled in 5 different ways. Thus, the three positions can be filled in

$$5 \times 5 \times 5 = 125 \text{ ways.}$$

Now, how many *even* three-digit numbers can be formed from the digits 2, 4, 6, 7, 9? The unit position can now be filled only in three different ways (2, 4, or 6), but each of the remaining two positions can still be filled in five different ways. Thus we can form

$$3 \times 5 \times 5 = 75 \text{ even three-digit numbers.}$$

Rule 2. The number of permutations of n distinct objects is n!.

In order to illustrate the above rule, let us consider three distinct objects: *A*, *B*, and *C*. These objects can be arranged in a variety of ways. Some of these arrangements are: *ABC, BCA, ACB, ...* Each of these arrangements of our three letters is called a permutation.

How many possible arrangements or permutations can be formed? According to Rule 2, there are 3! possible permutations; that is,

$$3! = 3 \times 2 \times 1 = 6 \text{ permutations.}$$

In arranging three distinct objects, we must fill three positions. The first position can be filled by any one of the three objects (Fig. 4–1).

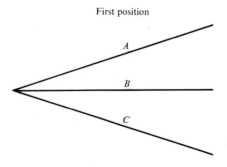

Figure 4–1

Having filled the first positions by a certain object, say *A*, we can fill the second position by any one of the remaining two objects (Fig. 4–2).

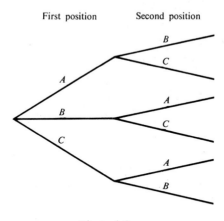

Figure 4–2

Having filled the first and second positions by two of the objects, we see that the third position can now be occupied only by the remaining third object.

The six possible paths in the tree diagram of Fig. 4–3 represent the six possible permutations or arrangements of three letters *A*, *B*, and *C*. Thus, the number of possible permutations of our three distinct objects is $3 \times 2 \times 1 = 3!$.

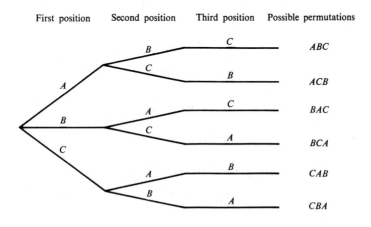

Figure 4–3

Example 4

The number of possible permutations of the four letters A, B, C, and D is

$$4! = 4 \times 3 \times 2 \times 1$$
$$= 24.$$

Example 5

In how many ways can six different books be arranged on a shelf?

Solution:

$$6! = 6 \times 5 \times 4 \times 3 \times 2 \times 1$$
$$= 720.$$

Rule 3. The number of permutations of r objects taken from n distinct objects, denoted by $_nP_r$, is $n!/(n - r)!$.

To illustrate the above rule, let us determine the number of possible permutations of three objects taken from the five objects A, B, C, D, E ($r = 3$, $n = 5$).

According to Rule 3, the number of possible permutations is

$$_nP_r = \frac{n!}{(n - r)!},$$

$$_5P_3 = \frac{5!}{(5 - 3)!} = \frac{5!}{2!}$$

$$= \frac{5 \times 4 \times 3 \times 2 \times 1}{2 \times 1}$$

$$= 5 \times 4 \times 3 = 60.$$

This result can be justified as follows: There are three positions to be filled. The first position can be filled by any of the five objects. Once the first position is occupied by one object, the second position can be filled by any of the remaining four objects. Finally, with the first and second positions being occupied by two of our objects, the third position can now be occupied by any of the remaining three objects. Consequently, the three positions can be filled in $5 \times 4 \times 3 = 60$ ways.

Example 6

How many four-digit numbers can be formed from the digits 1, 2, 3, 4, 5, 6 if each digit *cannot* appear more than once in each number?

Solution:

$$_nP_r = \frac{n!}{(n-r)!},$$

$$_6P_4 = \frac{6!}{(6-4)!} = \frac{6!}{2!}$$

$$= \frac{6 \times 5 \times 4 \times 3 \times 2 \times 1}{2 \times 1}$$

$$= 360 \text{ four-digit numbers.}$$

Rule 4. The number of combinations in which r objects can be selected from n distinct objects, is

$$\binom{n}{r} = \frac{n!}{(n-r)!\,r!}.$$

To illustrate this rule, let us determine the number of three-man committees which can be selected from a total of four people *A, B, C,* and *D* ($r = 3$, $n = 4$). According to Rule 4, the number of possible committees or combinations is

$$\binom{n}{r} = \frac{n!}{(n-r)!\,r!},$$

$$\binom{4}{3} = \frac{4!}{(4-3)!\,3!} = \frac{4!}{1!\,3!}$$

$$= 4 \text{ committees.}$$

These four possible committees or combinations are

$$A\ B\ C$$
$$A\ B\ D$$
$$A\ C\ D$$
$$B\ C\ D$$

Let us see how this rule is derived. According to Rule 3, the number of permutations of three objects taken from the four objects A, B, C, and D is

$$_nP_r = \frac{n!}{(n-r)!},$$

$$_4P_3 = \frac{4!}{(4-3)!} = \frac{4!}{1!}$$

$$= 24 \text{ permutations.}$$

These 24 permutations are listed in four columns below.

(1)	(2)	(3)	(4)
ABC	ABD	ACD	BCD
ACB	ADB	ADC	BDC
BAC	BAD	CAD	CBD
BCA	BDA	CDA	CDB
CAB	DAB	DAC	DBC
CBA	DBA	DCA	DCB

The reader may have already observed that the six permutations listed in column (1) represent the same committee (ABC). Stated differently, there are $3! = 6$ possible arrangements or permutations of the objects A, B, and C. However, all these six permutations represent one and the same committee (ABC). Similarly, the six permutations of column (2) represent combination or committee (ABD), the six permutations of column (3) represent the committee (ACD), and finally the six permutations of column (4) represent combination or committee (BCD).

Since we have reduced each group of $3! = 6$ permutations into one committee or combination, we may now generalize and say that each group of $r!$ permutations is reduced to one combination. In other words, the number of combinations of r objects that can be selected from n distinct objects is equal to the number of permutations of the r objects divided by r' Symbolically, we can then write

$$\binom{r}{n} = _nP_r \div r!,$$

$$= \frac{n!}{(n-r)!} \div r!$$

$$= \frac{n!}{(n-r)!\, r!}.$$

Example 7

In how many ways can a five-man basketball team be selected from seven men?

Solution:

$$\binom{n}{r} = \frac{n!}{(n-r)!\,r!},$$

$$\binom{7}{5} = \frac{7!}{(7-5)!\,5!} = \frac{7!}{2!\,5!}$$

$$= 21 \text{ possible ways.}$$

Rule 5. The number of permutations of n objects of which r are of one kind and n − r are of a second kind is

$$\frac{n!}{r!\,(n-r)!}.$$

To illustrate this rule let us determine the number of permutations of the five objects:

$$\underbrace{A,\ A,\ A,}_{r}\ \underbrace{B,\ B.}_{n-r}$$

Given that

$$n = 5: \quad A, A, A, B, B,$$
$$r = 3: \quad A, A, A,$$
$$n - r = 2: \quad B, B,$$

then, according to Rule 5, the number of permutations of the objects *A, A, A, B, B* is

$$\frac{n!}{r!\,(n-r)!} = \frac{5!}{3!\,(5-3)!}$$

$$= \frac{5!}{3!\,2!} = 10.$$

These 10 possible permutations are shown below.

$$
\begin{array}{ll}
A\,A\,A\,B\,B & A\,B\,A\,A\,B \\
A\,A\,B\,B\,A & A\,A\,B\,A\,B \\
A\,B\,B\,A\,A & A\,B\,A\,B\,A \\
B\,B\,A\,A\,A & B\,A\,A\,B\,A \\
B\,A\,A\,A\,B & B\,A\,B\,A\,A
\end{array}
$$

In order to see how this rule is derived, let us assume that we have n objects, say A, A, A, B, B, where r of these objects are of one kind,

$$\underbrace{A\ A, A, B, B,}_{r}$$

and $n - r$ of these objects are of a second kind,

$$A,\quad A,\quad A,\quad \underbrace{B,\quad B.}_{n-r}$$

Let us furthermore assume that these n objects can be arranged in X number of ways (10 possible arrangements in our special case).

Next, let us observe what happens to the number of possible arrangements X when instead of r objects of the same kind, we have r objects that are different from one another. To be specific, let us see what happens to the 10 possible arrangements of the objects A, A, A, B, B when we replace the three like objects A, A, A by K, L, M. The first of these 10 permutations, $A\,A\,A\,B\,B$, will become $K\,L\,M\,B\,B$, and from this permutation we can now generate $3! = 6$ distinct permutations by simply changing the positions of K, L, *and* M. These six permutations are listed below.

$$K\,L\,M\,B\,B \qquad L\,K\,M\,B\,B \qquad M\,K\,L\,B\,B$$
$$K\,M\,L\,B\,B \qquad L\,M\,K\,B\,B \qquad M\,L\,K\,B\,B$$

Similarly, when the three like objects A, A, A are replaced by K, L, M, the second of our 10 permutations, $A\,A\,B\,B\,A$, becomes $K\,L\,B\,B\,M$, and from this permutation we can again generate $3! = 6$ distinct permutations by simply changing the positions of objects K, L, and M. The resulting six permutations are listed below.

$$K\,L\,B\,B\,M \qquad L\,K\,B\,B\,M \qquad M\,K\,B\,B\,L$$
$$K\,M\,B\,B\,L \qquad L\,M\,B\,B\,K \qquad M\,L\,B\,B\,K$$

Thus from each of the 10 original permutations of the objects A, A, A, B, B we can generate $3! = 6$ distinct permutations by replacing the three like objects A, A, A by the three distinct letters K, L, M. Stated differently, when the three like objects A, A, A are made different, the number of possible permutations will be $3! = 6$ times as large as before.

Returning to our general case of n objects of which r are alike and $n - r$ are alike, we can thus say that when the r like objects are replaced by r distinct objects, the number of possible arrangements or permutations will be $r!$ times as large as the number of permutations before replacement. In other words, when instead of r like objects, we have r objects that differ from one another, the number of previous arrangements, X, will increase to $Xr!$

Similarly, when the $n - r$ objects of the second kind are replaced by $n - r$ distinct objects, then from *each* arrangement or permutation before the change we can now generate $(n - r)!$ permutations, and the total number of arrangements will be $(n - r)!$ times as large as the previous number of arrangements.

Thus we may conclude that once the r like objects (A, A, A), as well as the $n - r$ like objects (B, B), are made to differ from one another, then the total number of possible permutations will be $r! \, (n - r)!$ times as large as the number of permutations before the change. In other words, when all like objects are made distinct, then the number of original arrangements, X, will increase to $Xr! \, (n - r)!$.

But if all n objects become distinct, then, according to Rule 2, the number of permutations of these n distinct objects is $n!$. Equating our two results, we obtain

$$Xr! \, (n - r)! = n! \quad \text{or} \quad X = \frac{n!}{r! \, (n - r)!}$$

as the number of permutations of n objects of which r are alike and $n - r$ are also alike, but different from r.

Example 7

In how many possible orders can two boys and three girls be born to a family having five children?

Solution:

$$n = 5, \quad r = 3, \quad n - r = 2;$$

$$\frac{n!}{r! \, (n - r)!} = \frac{5!}{3! \, 2!} = 10 \text{ ways.}$$

This rule can now be extended to cover the case where the n objects consist of r_1 objects of one kind, r_2 objects of a second kind, r_3 objects of a third kind, etc. The number of possible arrangements or permutations of these n objects is

$$\frac{n!}{r_1! \, r_2! \, r_3! \cdots}.$$

Example 8

How many possible permutations can be formed from the word SUCCESS?

Solution:

$$n = 7 \quad \text{SUCCESS}$$
$$r_1 = 3 \quad \text{S S S}$$
$$r_2 = 1 \quad \text{U}$$
$$r_3 = 2 \quad \text{C C}$$
$$r_4 = 1 \quad \text{E}$$

Thus the number of distinct permutations that can be formed from the word SUCCESS is

$$\frac{n!}{r_1! \, r_2! \, r_3! \, r_4!} = \frac{7!}{3! \, 1! \, 2! \, 1!}$$

$$= 420 \text{ distinct permutations.}$$

EXERCISES, SET 1

1. In how many ways can a customer order a sandwich and a drink if there are five sandwiches and four drinks on a menu?

2. A coin is tossed four times. Determine the number of sample points in the sample space of this experiment.

3. How many possible permutations can be formed from the word HISTORY?

4. How many possible permutations can be formed from the word SUCCES-SIVE?

5. How many 4-member committees can be formed from a group of 7 people?

6. How many three-letter license plates can be formed

    a) if it is permitted that a letter appear more than once?
    b) if it is not permitted that a letter appear more than once?

7. How many two-digit numbers can be formed from the digits 4, 6, 8, 2? How many odd two-digit numbers can be formed from the same digits?

8. The Department of Economics at Pacific Hope University has the following members: Adams, Baker, Campbell, Daniel, and Edwards.

    a) How many possible three-member committees can be formed from the faculty of this department?
    b) How many possible three-member committees can be formed if Adams and Baker cannot serve on the same committee?

9. How many five-member committees can be formed from a group of 5 Republicans and 4 Democrats

    a) if no restriction is imposed on the composition of the committee?
    b) if the committee must contain two Democrats?

10. In how many different ways can five people be seated at a round table?

The Binomial Formula

The basic rules of addition and multiplication used in determining probabilities are supplemented by many specialized formulas. Most important of these formulas is the binomial formula. In this section, we will develop the binomial formula first; then, we will illustrate its use by several examples, and, finally, we will conclude our discussion by stating the necessary conditions for the use of the binomial formula.

To develop the binomial formula, let us now consider the following problem: Assuming that the probability of a "success" on any single trial is $\frac{1}{4}$, what is the probability of getting *exactly* two successes in five consecutive trials?

There are ten possible ways of getting exactly two successes in a series of five trials. Denoting the outcome "success" by S, and the outcome "failure" by F, we find the ten possible ways or permutations are

Permutation	Probability
$(S_1 \cap S_2 \cap F_3 \cap F_4 \cap F_5)$	$\frac{1}{4} \cdot \frac{1}{4} \cdot \frac{3}{4} \cdot \frac{3}{4} \cdot \frac{3}{4} = (\frac{1}{4})^2(\frac{3}{4})^3$
$(S_1 \cap F_2 \cap S_3 \cap F_4 \cap F_5)$	$\frac{1}{4} \cdot \frac{3}{4} \cdot \frac{1}{4} \cdot \frac{3}{4} \cdot \frac{3}{4} = (\frac{1}{4})^2(\frac{3}{4})^3$
$(S_1 \cap F_2 \cap F_3 \cap S_4 \cap F_5)$	$\frac{1}{4} \cdot \frac{3}{4} \cdot \frac{3}{4} \cdot \frac{1}{4} \cdot \frac{3}{4} = (\frac{1}{4})^2(\frac{3}{4})^3$
$(S_1 \cap F_2 \cap F_3 \cap F_4 \cap S_5)$	$\frac{1}{4} \cdot \frac{3}{4} \cdot \frac{3}{4} \cdot \frac{3}{4} \cdot \frac{1}{4} = (\frac{1}{4})^2(\frac{3}{4})^3$
$(F_1 \cap S_2 \cap S_3 \cap F_4 \cap F_5)$	$\frac{3}{4} \cdot \frac{1}{4} \cdot \frac{1}{4} \cdot \frac{3}{4} \cdot \frac{3}{4} = (\frac{1}{4})^2(\frac{3}{4})^3$
$(F_1 \cap S_2 \cap F_3 \cap S_4 \cap F_5)$	$\frac{3}{4} \cdot \frac{1}{4} \cdot \frac{3}{4} \cdot \frac{1}{4} \cdot \frac{3}{4} = (\frac{1}{4})^2(\frac{3}{4})^3$
$(F_1 \cap S_2 \cap F_3 \cap F_4 \cap S_5)$	$\frac{3}{4} \cdot \frac{1}{4} \cdot \frac{3}{4} \cdot \frac{3}{4} \cdot \frac{1}{4} = (\frac{1}{4})^2(\frac{3}{4})^3$
$(F_1 \cap F_2 \cap S_3 \cap S_4 \cap F_5)$	$\frac{3}{4} \cdot \frac{3}{4} \cdot \frac{1}{4} \cdot \frac{1}{4} \cdot \frac{3}{4} = (\frac{1}{4})^2(\frac{3}{4})^3$
$(F_1 \cap F_2 \cap S_3 \cap F_4 \cap S_5)$	$\frac{3}{4} \cdot \frac{3}{4} \cdot \frac{1}{4} \cdot \frac{3}{4} \cdot \frac{1}{4} = (\frac{1}{4})^2(\frac{3}{4})^3$
$(F_1 \cap F_2 \cap F_3 \cap S_4 \cap S_5)$	$\frac{3}{4} \cdot \frac{3}{4} \cdot \frac{3}{4} \cdot \frac{1}{4} \cdot \frac{1}{4} = (\frac{1}{4})^2(\frac{3}{4})^3$

Each of the ten permutations above represents a way of getting exactly two successes in five trials. The probability of each permutation is $(\frac{1}{4})^2(\frac{3}{4})^3$. The ten permutations, furthermore, are mutually exclusive. Hence, the probability of getting exactly two successes in five trials is the sum of the probabilities of all ten permutations, or ten times the probability of any one permutation:

$$P(\text{exactly 2 successes in 5 trials}) = 10(\tfrac{1}{4})^2(\tfrac{3}{4})^3$$
$$= \tfrac{270}{1024}.$$

Now, let us develop a general formula to solve our problem. We will begin by defining the following symbols:

π = probability of a "success" on any *single* trial,
$(1 - \pi)$ = probability of a "failure" on any single trial,
n = number of trials,
x = number of successes,
$(n - x)$ = number of failures.

The probability of any permutation is

$$\pi^x(1 - \pi)^{(n-x)} = (\tfrac{1}{4})^2(\tfrac{3}{4})^3.$$

The number of possible permutations is*

$$\frac{n!}{x!(n-x)!} = \frac{5!}{2!(5-2)!} = \frac{5!}{2!3!}$$

$$\frac{5 \times 4 \times 3 \times 2 \times 1}{(2 \times 1) \cdot (3 \times 2 \times 1)} = 10$$

The probability of getting *exactly* x successes in n trials, denoted by $P(X = x)$ is equal to the product obtained by multiplying the probability of any permutation by the number of possible permutations:

$$P(X = x) = \frac{n!}{x!(n-x)!}\, \pi^x(1 - \pi)^{n-x},$$

$$P(X = 2) = \frac{5!}{2!3!}\left(\frac{1}{4}\right)^2\left(\frac{3}{4}\right)^3$$

$$= 10(\tfrac{1}{4})^2(\tfrac{3}{4})^3 = \tfrac{270}{1024}.$$

Similarly, assuming that the probability of a "success" on any given trial is $\tfrac{1}{4}$, the probability of getting 3 successes in seven consecutive trials is

$$P(X = x) = \frac{n!}{x!(n-x)!}\, \pi^x(1 - \pi)^{n-x},$$

$$P(X = 3) = \frac{7!}{3!4!}\left(\frac{1}{4}\right)^3\left(\frac{3}{4}\right)^4$$

$$= \frac{7 \times 6 \times 5 \times 4 \times 3 \times 2 \times 1}{(3 \times 2 \times 1) \cdot (4 \times 3 \times 2 \times 1)}\left(\frac{1}{4}\right)^3\left(\frac{3}{4}\right)^4$$

$$= 35(\tfrac{1}{4})^3(\tfrac{3}{4})^4 = \tfrac{2835}{16384}.$$

The formula

$$P(X = x) = \frac{n!}{x!(n-x)!}\, \pi^x(1 - \pi)^{n-x},$$

called the *binomial formula*, is an important tool used in determining the probabilities of certain events. The following examples illustrate some of its applications.

---

*   According to the rule of permutations, this is the number of distinct permutations of n objects of which x are alike (successes) and $n - x$ are alike (failures).

Example 9

What is the probability of getting exactly 3 heads in 5 flips of a balanced coin?

Solution:

$\pi = \frac{1}{2}$ (probability of obtaining a head),
$n = 5$ (number of flips),
$x = 3$ (number of required heads).

According to the binomial formula,

$$P(X = x) = \frac{n!}{x!(n-x)!}\pi^x(1-\pi)^{n-x},$$

$$P(X = 3) = \frac{5!}{3!2!}\left(\frac{1}{2}\right)^3\left(\frac{1}{2}\right)^2$$

$$= \frac{5 \times 4 \times 3 \times 2 \times 1}{(3 \times 2 \times 1)\cdot(2 \times 1)}\left(\frac{1}{2}\right)^5 = 10\left(\frac{1}{2}\right)^5 = \frac{10}{32}.$$

Example 10

What is the probability of getting exactly 4 "sixes" when a die is rolled 7 times?

Solution:

$\pi = \frac{1}{6}$ (probability of getting a six),
$n = 7$ (number of times a die is rolled),
$x = 4$ (number of required sixes).

According to the binomial formula,

$$P(X = x) = \frac{n!}{x!(n-x)!}\pi^x(1-\pi)^{n-x},$$

$$P(X = 4) = \frac{7!}{4!3!}\left(\frac{1}{6}\right)^4\left(\frac{5}{6}\right)^3 = 35\left(\frac{1}{6}\right)^4\left(\frac{5}{6}\right)^3 = \frac{4375}{279936}.$$

Example 11

In a family of 4 children, what is the probability that there will be exactly two boys?

Solution:

$\pi = \frac{1}{2}$ (probability that a child is a boy),
$n = 4$ (number of children),
$x = 2$ (number of boys).

According to the binomial formula,

$$P(X = x) = \frac{n!}{x!(n-x)!} \, \pi^x (1 - \pi)^{n-x},$$

$$P(X = 2) = \frac{4!}{2!\,2!} \left(\frac{1}{2}\right)^2 \left(\frac{1}{2}\right)^2 = 6\left(\frac{1}{2}\right)^4 = \frac{6}{16}.$$

Having illustrated the use of the binomial formula by several examples, we must now state the necessary conditions for the use of the formula. Upon examining the foregoing illustrative problems, we observe the following common characteristics:

1. The experiment in each illustrative problem consists of a series of repeated trials.

2. Each trial has only two possible outcomes. The two outcomes are generally referred to as a "success" and a "failure."

3. The probability of a success, denoted by π is the same on each trial. The probability of a failure, $(1 - \pi)$ is also constant.

4. The consecutive trials are independent, i.e., the outcome of any trial in the series is independent of the outcome of the preceding trials.

An experiment satisfying each of these four conditions is called a *binomial experiment*.

In a toss of a balanced coin, for example, the two possible outcomes of a toss are a "head" or a "tail." In a series of repeated tosses, furthermore, the outcome of each toss is independent of the outcomes of the preceding trial. In the same manner, when a die is rolled several times, the outcome of each roll can be either "a six" or "not a six," and the probability of getting a six in any roll is independent of the outcomes of the preceding rolls (always $\frac{1}{6}$). Finally, a child can be either a boy or a girl, and the probability that a baby would be a boy is independent of the sex of the previous children.

As a final remark, let us examine the applicability of the binomial formula to the general problem of sample selection. The processes of selecting a sample of n elements from a given population may be viewed as an experiment consisting of n repeated trials (a series of n selections). The binomial formula is applicable to this sampling process only when the outcome of each selection is independent of the outcome of preceding selections. The series of n independent selections are considered independent, and hence the binomial formula is applicable under two situations.

1. The sample elements are selected with or without replacement from an infinite population.

2. The sample elements are selected *with replacement* from a finite population.

In the real world, a sample is often selected without replacement from a finite population, a situation in which the assumption of independent selections is clearly violated, and the binomial formula appears to be not applicable. However, when the sample size is small relative to the population size (the sample size is 10% or less of the population size), the binomial formula can still be used and the results obtained are reasonably accurate.* The following example is an illustration of this case.

Example 12

A sample of 4 fuses is selected without replacement from a lot consisting of 5000 fuses. Assuming that 20% of the fuses in the lot are known to be defective, what is the probability that the sample would contain exactly 2 defective fuses?

Solution:

Since the sample constitutes a small fraction of the population, the probability of getting a defective fuse, π, is approximately the same on each selection, i.e., $\pi = 0.20$. Using the binomial formula we obtain

$$P(X = x) = \frac{n!}{x!(n-x)!} \pi^x (1-\pi)^{n-x},$$

$$P(X = 2) = \frac{4!}{2!2!} (0.20)^2 (0.80)^2. = 0.1536.$$

EXERCISES, SET 2

1. A machine is producing parts that contain, on the average, 5% defectives. In a random sample of five such parts, what is the probability

   a) of getting exactly one defective part?

   b) of obtaining *at least* one defective part?

2. Two-thirds of the registered voters in a certain congressional district are Democrats. What is the probability that in a random sample of 6 registered voters from this district there will be exactly 5 Democrats?

3 It is known that 60% of all students attending XYU college are smokers. In a random sample of four students from this college, what is the probability

   a) that there are exactly two smokers?

   b) that *only* the first two students interviewed are smokers?

4. What is the probability that the sum 9 would appear exactly twice when a pair of dice is rolled four times?

---

* Assuming that the value of π is not very close to zero or one.

5. Five marbles are to be drawn, with replacement, from an urn that contains 4 white, 4 green, and 12 black marbles. What is the probability of drawing exactly 3 marbles that are not white?

6. Machine A and Machine B produce, on the average, 5% and 10% defective parts, respectively. A random sample of four parts is to be selected from the output of each machine. What is the probability that the sample selected from A's output would contain exactly one defective part, and the sample selected from B's output would contain exactly two defective parts?

7. A bookkeeping firm employs two part-time file clerks: Jan and Jane. While Jan works on Monday, Wednesday, and Friday, Jane works on Tuesday and Thursday. Jan misfiles one out of every four documents and Jane misfiles one out of every five documents. A day of the week is to be selected at random, and a random sample of four documents from the documents filed on that day is to be examined.
   a) What is the probability that the sample will show exactly two misfiled documents?
   b) Given that the sample shows two misfiled documents, what is the probability that the two documents were filed by Jan?

8. Cards are drawn with replacement from a well-shuffled deck of 52 playing cards.
   a) What is the probability of getting at least one black card when two cards are drawn?
   b) What is the probability of drawing at least two black cards when four cards are drawn?

The Binomial Distribution

Let us assume that a balanced coin is tossed four times. On these four tosses, heads could come up 0, 1, 2, 3, or 4 times. The binomial formula can be used to determine the probability of each of these possible outcomes (see Table 4–1).

The data in Table 4–1 which list all possible numbers of heads, together with their probabilities, are called a *probability distribution*, in this case a probability distribution of the number of heads in four tosses of a balanced coin.

Next let us consider the probability distribution of the number of boys in a family of four children. The family can have 0, 1, 2, 3, or 4 boys. The binomial formula can be used to determine the probability of each of these possible outcomes. Table 4–2 shows this distribution.

As you may already have noticed, the probability distribution of the number of heads in four tosses of a balanced coin and the probability distribution of the number of boys in a family of four children are identically the same. This is by no means a coincidence. Both are binomial distributions with $\pi = \frac{1}{2}$ and $n = 4$.

Table 4–1

x (number of heads)	$P(X = x)$
0	$\dfrac{4!}{0!\,4!}\,(\tfrac{1}{2})^0(\tfrac{1}{2})^4 = \tfrac{1}{16}$
1	$\dfrac{4!}{1!\,3!}\,(\tfrac{1}{2})^1(\tfrac{1}{2})^3 = \tfrac{4}{16}$
2	$\dfrac{4!}{2!\,2!}\,(\tfrac{1}{2})^2(\tfrac{1}{2})^2 = \tfrac{6}{16}$
3	$\dfrac{4!}{3!\,1!}\,(\tfrac{1}{2})^3(\tfrac{1}{2})^1 = \tfrac{4}{16}$
4	$\dfrac{4!}{4!\,0!}\,(\tfrac{1}{2})^4(\tfrac{1}{2})^0 = \tfrac{1}{16}$

Let us consider another binomial distribution with $\pi = \tfrac{1}{4}$ and $n = 3$. The possible values of X in this distribution are 0, 1, 2, and 3. This distribution is shown in Table 4–3.

The binomial distribution in Table 4–3 can be the probability distribution of the number of spade cards when three cards are drawn, with replacement, from an ordinary deck of playing cards. It can also be the probability distribution of the number of defective parts in a sample of three parts

Table 4–2

Probability distribution of the number of boys in a family of four children

x (number of boys)	$P(x)$
0	$\dfrac{4!}{0!\,4!}\,(\tfrac{1}{2})^0(\tfrac{1}{2})^4 = \tfrac{1}{16}$
1	$\dfrac{4!}{1!\,3!}\,(\tfrac{1}{2})^1(\tfrac{1}{2})^3 = \tfrac{4}{16}$
2	$\dfrac{4!}{2!\,2!}\,(\tfrac{1}{2})^2(\tfrac{1}{2})^2 = \tfrac{6}{16}$
3	$\dfrac{4!}{3!\,1!}\,(\tfrac{1}{2})^3(\tfrac{1}{2})^1 = \tfrac{4}{16}$
4	$\dfrac{4!}{4!\,0!}\,(\tfrac{1}{2})^4(\tfrac{1}{2})^0 = \tfrac{1}{16}$

Table 4–3

Binomial distribution ($\pi = \tfrac{1}{4}$, $n = 3$)

x	$P(X = x)$
0	$\dfrac{3!}{0!\,3!}\,(\tfrac{1}{4})^0(\tfrac{3}{4})^3 = \tfrac{27}{64}$
1	$\dfrac{3!}{1!\,2!}\,(\tfrac{1}{4})^1(\tfrac{3}{4})^2 = \tfrac{27}{64}$
2	$\dfrac{3!}{2!\,1!}\,(\tfrac{1}{4})^2(\tfrac{3}{4})^1 = \tfrac{9}{64}$
3	$\dfrac{3!}{3!\,0!}\,(\tfrac{1}{4})^3(\tfrac{3}{4})^0 = \tfrac{1}{64}$

produced by a machine whose output contains, on the average, 25% defectives. Each of these two experiments is a binomial experiment with $\pi = \frac{1}{4}$ and $n = 3$. Finally, let us consider the binomial distribution with $\pi = \frac{1}{6}$ and $n = 5$.

Table 4–4

Binomial distribution

$(\pi = \frac{1}{6}, n = 5)$

x	$P(X = x)$
0	$\dfrac{5!}{0!\,5!}\,(\frac{1}{6})^0(\frac{5}{6})^5 = \frac{3125}{7776}$
1	$\dfrac{5!}{1!\,4!}\,(\frac{1}{6})^1(\frac{5}{6})^4 = \frac{3125}{7776}$
2	$\dfrac{5!}{2!\,3!}\,(\frac{1}{6})^2(\frac{5}{6})^3 = \frac{1250}{7776}$
3	$\dfrac{5!}{3!\,2!}\,(\frac{1}{6})^3(\frac{5}{6})^2 = \frac{250}{7776}$
4	$\dfrac{5!}{4!\,1!}\,(\frac{1}{6})^4(\frac{5}{6})^1 = \frac{25}{7776}$
5	$\dfrac{5!}{5!\,0!}\,(\frac{1}{6})^5(\frac{5}{6})^0 = \frac{1}{7776}$

The binomial distribution in Table 4–4 can be interpreted as the probability distribution of the number of "sixes" showing when a die is rolled five times.

The Mean and the Standard Deviation of the Binomial Distribution

Let us consider again the binomial distribution with $\pi = \frac{1}{2}$ and $n = 4$. This can very well be the probability distribution of the number of heads showing in a toss of four balanced coins.

x	0	1	2	3	4
$P(X = x)$	$\frac{1}{16}$	$\frac{4}{16}$	$\frac{6}{16}$	$\frac{4}{16}$	$\frac{1}{16}$

To determine the mean of the above distribution, we must first multiply each possible value of x by its respective probability. The mean of the distribution, denoted by the Greek letter μ, is the sum of all these cross products.   Symbolically, $\mu = \sum xP(x)$.*

Table 4–5

x	$P(x)$	$xP(x)$
0	$\frac{1}{16}$	0
1	$\frac{4}{16}$	$\frac{4}{16}$
2	$\frac{6}{16}$	$\frac{12}{16}$
3	$\frac{4}{16}$	$\frac{12}{16}$
4	$\frac{1}{16}$	$\frac{4}{16}$

$$\sum xP(x) = \frac{32}{16}$$
$$\mu = \sum xP(x) = \frac{32}{16} = 2 \text{ heads.}$$

Thus the mean of our binomial distribution (Table 4–5) is two heads, that is, although we may get 0, 1, 2, 3, or 4 heads when four balanced coins are tossed, we do expect an average of two heads per trial if this binomial experiment is repeated an infinite number of times.

The binomial distribution ($\pi = \frac{1}{2}$, $n = 4$) has a mean of 2. The standard deviation of this distribution, denoted by σ, is defined as

$$\sigma = \sqrt{\sum (x - \mu)^2 P(x)}.$$

The computations are carried out in Table 4–6.

Table 4–6

x	$P(x)$	$(x - 2)$ $(x - \mu)$	$(x - \mu)^2$	$(x - \mu)^2 P(x)$
0	$\frac{1}{16}$	-2	4	$\frac{4}{16}$
1	$\frac{4}{16}$	-1	1	$\frac{4}{16}$
2	$\frac{6}{16}$	0	0	0
3	$\frac{4}{16}$	$+1$	1	$\frac{4}{16}$
4	$\frac{1}{16}$	$+2$	4	$\frac{4}{16}$

$$\sum (x - u)^2 P(x) = \frac{16}{16}$$
$$\sigma = \sqrt{\sum (x - \mu)^2 P(x)} = \sqrt{\frac{16}{16}} = 1 \text{ head.}$$

Thus the binomial distribution ($\pi = \frac{1}{2}$, $n = 4$) has a mean of 2 and a standard deviation of 1. This means that although we may get either 0, 1, 2,

---

* To simplify the formulas in this section, the symbol $P(X = x)$ is simply written as $P(x)$.

3, or 4 heads when four balanced coins are tossed, we expect an average of two heads if this experiment is repeated an infinite number of times. Furthermore, since the standard deviation is equal to 1, we expect the number of heads in each of these infinite experiments to differ from the mean, on the average, by 1 head.

Statisticians have found that they can determine the mean and the standard deviation of a binomial distribution without having to calculate the entire distribution, i.e. without having to list all possible values of X and their probabilities. The mean and the standard deviation of a binomial distribution can indeed be determined from the values of π and n, using the following two formulas:

$$\mu = n\pi, \qquad \sigma = \sqrt{n\pi(1 - \pi)}.$$

Thus in our experiment of tossing four balanced coins where $\pi = \frac{1}{2}$ and $n = 4$, we have

$$\mu = n\pi$$
$$= 4(\tfrac{1}{2}) = 2 \text{ heads}$$

and

$$\sigma = \sqrt{n\pi(1 - \pi)}$$
$$= \sqrt{4(\tfrac{1}{2})(\tfrac{1}{2})} = 1 \text{ head}.$$

Now, let us demonstrate the use of these formulas in another example.

Example 13

A die is to be rolled 180 times. Determine the mean and the standard deviation of the number of "sixes" in this experiment. This is a binomial distribution with $\pi = \frac{1}{6}$ and $n = 180$. Therefore,

$$\mu = n\pi$$
$$= 180(\tfrac{1}{6}) = 30 \text{ sixes}$$

and

$$\sigma = \sqrt{n\pi(1 - \pi)}$$
$$= \sqrt{180(\tfrac{1}{6})(\tfrac{5}{6})} = 5 \text{ sixes}.$$

Thus, when a die is rolled 180 times, we may get 0, 1, 2, ..., or 180 sixes. However, if this experiment is repeated an infinite number of times, we expect an average of 30 sixes per trial. Moreover, the number of sixes in each of these infinite trials is expected to differ from 30 sixes, on the average, by 5 sixes.

EXERCISES, SET 3

1. Given the binomial distribution with $\pi = \frac{1}{3}$ and $n = 2$,
   a) compute the entire binomial distribution;
   b) determine the mean and the standard deviation of the distribution, using the data obtained in part (a);
   c) verify the results obtained in part (b) by using the appropriate short-cut formula for determining the mean and the standard deviation of a binomial distribution.

2. Compute the mean and the standard deviation of each of the following binomial distributions:
   a) $\pi = \frac{1}{2}$ and $n = 1600$
   b) $\pi = .10$ and $n = 900$
   c) $\pi = .64$ and $n = 400$.

3. A pair of dice is rolled 180 times. Determine the expected number of times in which the sum 5 appears. Determine the standard deviation.

4. If the probability of a defective rivet is .01, what is the expected number of defective rivets in an airplane that has 10,000 such rivets?

Cumulative Binomial Distribution

Once again, let us consider the probability distribution of the number of heads appearing in four tosses of a balanced coin. This distribution is a binomial distribution with $n = 4$ and $\pi = 0.50$ (See Table 4–7).

Table 4–7

Binomial Distribution
$(n = 4, \pi = 0.50)$

x	$P(X = x)$
0	0.0625
1	0.2500
2	0.3750
3	0.2500
4	0.0625

Table 4–7 shows the probability of getting exactly 0 heads, exactly 1 head, exactly 2 heads, etc. For example, the probability of getting *exactly* two heads is 0.3750.

Now, let us assume that we are interested in finding the probability of getting *2 or more* heads. In contrast to the probability of getting exactly 2 heads, which is $P(X = 2)$, the probability of getting 2 heads *or more* is

$P(X \geq 2)$, where

$$P(X \geq 2) = P(X = 2) + P(X = 3) + P(X = 4)$$
$$= 0.3750 \quad + 0.2500 \quad + 0.0625$$
$$= 0.6875.$$

Similarly, the probability of getting 3 or more heads is

$$P(X \geq 3) = P(X = 3) + P(X = 4)$$
$$= 0.2500 \quad + 0.0625$$
$$= 0.3125.$$

Both $P(X \geq 2)$ and $P(X \geq 3)$ are called *cumulative probabilities.* Similar cumulative probabilities for other values of X are computed in the same fashion and are shown in column (3) of Table 4–8.

Table 4–8

Binomial Distribution with Cumulative Probabilities
($n = 4$, $\pi = 0.50$)

x	$P(X = x)$	$P(X \geq x)$
0	0.0625	1.0000
1	0.2500	0.9375
2	0.3750	0.6875
3	0.2500	0.3125
4	0.0625	0.0625

Let us further assume that we are interested in finding the probability of getting *2 or less* heads.   The probability of getting two heads or less, denoted by $P(X \leq 2)$ is determined as follows:

$$P(X \leq 2) = P(X = 2) + P(X = 1) + P(X = 0)$$
$$= 0.3750 \quad + 0.2500 \quad + 0.0625$$
$$= 0.6875.$$

Similarly, the probability of obtaining 3 heads or less is

$$P(X \leq 3) = P(X = 3) + P(X = 2) + P(X = 1) + P(X = 0)$$
$$= 0.2500 \quad + 0.3750 \quad + 0.2500 \quad + 0.0625$$
$$= 0.9375.$$

$P(X \geq 2)$ and $P(X \leq 2)$ are both examples of cumulative probability. $P(X \geq 2)$ is a cumulative probability of the "equal or greater than" type, and $P(X \leq 2)$ is a cumulative probability of the "equal or less than" type. Table 4–9 shows three different probabilities for each value of X.

Table 4–9

Binomial Distribution with Cumulative Probabilities
$(n = 4, \pi = 0.50)$

x	$P(X = x)$	$P(X \geq x)$	$P(X \leq x)$
0	0.0625	1.0000	0.0625
1	0.2500	0.9375	0.3125
2	0.3750	0.6875	0.6875
3	0.2500	0.3125	0.9375
4	0.0625	0.0625	1.0000

Now the entries in column (2) of Table 4–9 refer to the probability of getting exactly 0 heads, exactly 1 head, exactly 2 heads, etc.   The entries in column (3), on the other hand, refer to the probability of getting 0 or more heads, 1 or more heads, 2 or more heads, etc., and finally, the entries in column (4) refer to the probability of getting 0 or *less* heads, 1 or less heads, 2 or less heads, etc.

It is important to note that the probabilities recorded in column (2) of Table 4–9 (that is, $P(X = x)$) can be indirectly obtained from the cumulative probabilities in column (3) or column (4).   For example, the probability of getting *exactly* two heads, 0.3750, is obtained from the cumulative probabilities in column (3) as follows:

$$P(X \geq 2) = P(X = 2) + P(X = 3) + P(X = 4) \qquad \text{(A)}$$

$$P(X \geq 3) = \qquad\qquad P(X = 3) + P(X = 4) \qquad \text{(B)}$$

Subtracting equation (B) from equation (A) we obtain

$$P(X \geq 2) - P(X \geq 3) = P(X = 2).$$

Hence,

$$P(X = 2) = P(X \geq 2) - P(X \geq 3)$$
$$= 0.6875 \quad - 0.3125$$
$$= 0.3750.$$

The probability of getting exactly two heads can also be obtained from the cumulative probabilities of column (4) as follows:

$$P(X \leq 2) = P(X = 2) + P(X = 1) + P(X = 0) \qquad \text{(C)}$$

$$P(X \leq 1) = \qquad\qquad P(X = 1) + P(X = 0) \qquad \text{(D)}$$

Subtracting equation (D) from equation (C), we obtain

$$P(X = 2) = P(X \leq 2) - P(X \leq 1)$$
$$= 0.6875 - 0.3125 = 0.3750.$$

Some additional examples are in order.

Example 14   .

Given the following cumulative binomial distribution, find (a) $P(X = 1)$; (b) $P(X = 4)$; (c) $P(X = 5)$.

$(n = 5, \pi = 0.31)$

x	$P(X \geq x)$
0	1.0000
1	0.8436
2	0.4923
3	0.1766
4	0.0347
5	0.0029

Solution:

a) $P(X = 1) = P(X \geq 1) - P(X \geq 2)$ b) $P(X = 4) = P(X \geq 4) - P(X \geq 5)$
$\qquad = 0.8436 \quad - 0.4923 \qquad\qquad\qquad = 0.0347 \quad - 0.0029$
$\qquad = 0.3513. \qquad\qquad\qquad\qquad\qquad = 0.0318.$

c) $P(X = 5) = P(X \geq 5) = 0.0029.$

Example 15

Given the following binomial distribution, find (a) $P(X = 4)$; (b) $P(X = 1)$; (c) $P(X = 0)$.

$(n = 5, \pi = 0.69)$

x	$P(X \leq x)$
0	0.0029
1	0.0347
2	0.1766
3	0.4923
4	0.8436
5	1.0000

Solution:

a) $P(X = 4) = P(X \leq 4) - P(X \leq 3)$ b) $P(X = 1) = P(X \leq 1) - P(X \leq 0)$
          $= 0.8436 - 0.4923$ $= 0.0347 - 0.0029$
          $= 0.3513$ $= 0.0318$

c) $P(X = 0) = P(X \leq 0) = 0.0029$.

The reader may perhaps have noticed that examples 14 and 15 have the same solution. The similarity between the two answers can be easily explained if the binomial formula is used to solve both examples. In example 14, where $n = 5$ and $\pi = 0.31$,

$$P(X = 1) = \frac{5!}{1!4!} (0.31)^1(0.69)^4 = 0.3513.$$

In example 15, where $n = 5$ and $\pi = 0.69$,

$$P(X = 4) = \frac{5!}{4!1!} (0.69)^4(0.31)^1 = 0.3513.$$

Hence, the answer to part (a) of both examples is the same. This is also true for part (b) and part (c).

Cumulative Binomial Tables

Let us assume that we wish to determine the probability of getting 47 heads or more in 100 tosses of a balanced coin. Since this is a binomial distribution with $n = 100$ and $\pi = 0.50$, we may use the binomial formula to determine the probability of getting 47, 48, 49, ..., 100 heads. The probability of getting 47 heads or more is the sum of all these individual probabilities.

$$P(X \geq 47) = P(X = 47) + P(X = 48) + \cdots + P(X = 100)$$

$$= \frac{100!}{47!53!} (0.5)^{47}(0.5)^{53} + \cdots + \frac{100!}{100!0!} (0.5)^{100}(0.5)^0$$

But as the reader may already have noticed, the use of the binomial formula in this case involves many computations. And it is for this reason that the electronic computer has been called upon to help.

Using the computer, cumulative probabilities have been computed and tabulated for selected values of n, and values of π ranging from 0.01 to 0.99 (see Table J at the end of the text). To facilitate our present discussion, we have reproduced in Table 4–10 the cumulative binomial probabilities for $n = 4$ and values of π ranging from 0.01 to 0.99. However, the table is used. in two different ways. The difference depends on whether $\pi \leq 0.50$ or $\pi \geq 0.50$. The two methods are discussed in turn.

Table 4–10

Left				Cumulative binomial probabilities						Right

$P(X \geq x)$ $n = 4$ $P(X \leq x)$

x	$\pi =$	01	02	03	04	05	06	07	08	09	10	
1		0394	0776	1147	1507	1855	2193	2519	2836	3143	3439	3
2		0006	0023	0052	0091	0140	0199	0267	0344	0430	0523	2
3				0001	0002	0005	0008	0013	0019	0027	0037	1
4										0001	0001	0
x		99	98	97	96	95	94	93	92	91	90	$= \pi$	x

n = 4

x	$\pi =$	11	12	13	14	15	16	17	18	19	20	
1		3726	4003	4271	4530	4780	5021	5254	5479	5695	5904	3
2		0624	0732	0847	0968	1095	1228	1366	1509	1656	1808	2
3		0049	0063	0079	0098	0120	0144	0171	0202	0235	0272	1
4		0001	0002	0003	0004	0005	0007	0008	0010	0013	0016	0
x		89	88	87	86	85	84	83	82	81	80	$= \pi$	x

n = 4

x	$\pi =$	21	22	23	24	25	26	27	28	29	30	
1		6105	6298	6485	6664	6836	7001	7160	7313	7459	7599	3
2		1963	2122	2285	2450	2617	2787	2959	3132	3307	3483	2
3		0312	0356	0403	0453	0508	0566	0628	0694	0763	0837	1
4		0019	0023	0028	0033	0039	0046	0053	0061	0071	0081	0
x		79	78	77	76	75	74	73	72	71	70	$= \pi$	x

n = 4

x	$\pi =$	31	32	33	34	35	36	37	38	39	40	
1		7733	7862	7985	8103	8215	8322	8425	8522	8615	8704	3
2		3660	3837	4015	4193	4370	4547	4724	4900	5075	5248	2
3		0915	0996	1082	1171	1265	1362	1464	1596	1679	1792	1
4		0092	0105	0119	0134	0150	0168	0187	0209	0231	0256	0
x		69	68	67	66	65	64	63	62	61	60	$= \pi$	x

n = 4

x	$\pi =$	41	42	43	44	45	46	47	48	49	50	
1		8788	8868	8944	9017	9085	9150	9211	9269	9323	9375	3
2		5420	5590	5759	5926	6090	6252	6412	6569	6724	6875	2
3		1909	2030	2155	2283	2415	2550	2689	2831	2977	3125	1
4		0283	0311	0342	0375	0410	0448	0488	0531	0576	0625	0
x		59	58	57	56	55	54	53	52	51	50	$= \pi$	x

Case I $(\pi \leq 0.50)$

When $\pi \leq 0.50$, the value of π is found at the head of the column, and the values of X are located in the *leftmost* column of the table. Furthermore, the probability in any cell in the table is the probability of getting x *or more*, that is, $P(X \geq x)$. (See Table 4–10)

Example 16

For $n = 4$ and $\pi = 0.23$, find (a) $P(X \geq 2)$ and (b) $P(X = 2)$.

Solution: Using Table 4–10,

a) $P(X \geq 2)$ $= 0.2285$.

b) $P(X = 2) = P(X \geq 2) - P(X \geq 3)$
$\qquad = 0.2285 - 0.0403$
$\qquad = 0.1882$.

Case II $(\pi \geq 0.50)$

When $\pi \geq 0.50$, the value of π is found at the *bottom* of the column, and the values of X are listed (in a decreasing order) in the far *right* column of the table. The probability in each cell is the probability of getting x *or less*, that is, $P(X \leq x)$.

Example 17

For $n = 4$ and $\pi = 0.73$, find (a) $P(X \leq 2)$ and (b) $P(X = 2)$

Solution: Using Table 4–10

a) $P(X \leq 2) = 0.2959$
b) $P(X = 2) = P(X \leq 2) - P(X \leq 1)$
$\qquad = 0.2959 - 0.0628$
$\qquad = 0.2331$.

EXERCISES, SET 4

1. Given the binomial distribution with $\pi = 0.26$ and $n = 7$; use the cumulative binomial table to determine

   a) $P(X \geq 2)$ c) $P(X \geq 4)$

   b) $P(X = 2)$ d) $P(X = 4)$

2. Given the binomial distribution with $\pi = 0.84$ and $n = 9$; use the cumulative binomial table to determine

   a) $P(X \leq 7)$ c) $P(X \leq 5)$

   b) $P(X = 7)$ d) $P(X = 5)$

3. Given the binomial distribution with $\pi = 0.37$ and $n = 8$; use the cumulative binomial table to determine

a) $P(X = 0)$ c) $P(X < 3)$

b) $P(X = 3)$ d) $P(X \leq 3)$

4. Given the binomial distribution with $\pi = 0.70$ and $n = 20$; use the binomial table to determine

a) $P(X = 0)$ c) $P(X > 12)$

b) $P(X = 12)$ d) $P(X \geq 12)$

The Poisson Distribution

The Poisson distribution is used as a model to describe the probability distribution of such events as the arrivals of cars at a service station, the arrivals of airplanes at an airport, and the arrivals of customers at a restaurant.

Regarding the arrivals of cars at a service station, let us now assume that the average number of cars, μ, that arrive at a particular service station is 20 cars per hour. If the Poisson distribution is used to describe car arrivals in this case, then we must further assume that

1. The probability of a single car arriving at the service station during a very short interval, say one second, is proportional to the length of the time interval; that is, $20/(60 \times 60) = 1/180$. That very short time interval is generally called a subinterval.

2. The probability that *two* or more cars arrive during a subinterval is virtually zero.

3. The number of cars that arrive during any subinterval is *independent* of the number of cars that arrive *outside* that subinterval.

The Poisson distribution is also used to describe such experiments as the number of typing errors per page, the number of defective rivets in an airplane, and the number of bacteria in a cubic centimeter of ocean water. In these experiments, it is assumed that the average number of occurrences (generally called "successes") in a specified region is known. The specified region could be a unit of length, an area, or a volume. It is also assumed that the probability of a single success occurring within a very small region (subregion) is proportional to the size of the subregion, and independent of the number of successes occurring outside this subregion. Finally, it is also assumed that the probability of two or more successes occurring in a subregion is approximately zero.

In any Poisson experiment, the probability of getting *exactly* x successes, denoted by $P(X = x)$, is

$$P(X = x) = \frac{\mu^x e^{-\mu}}{x!}$$

for $x = 0, 1, 2, \ldots$, where μ is the average number of successes during a given time interval (or in a specific region), and e is 2.71828 ... (the base of the natural logarithm system).

Example 18

The average number of claims filed against an insurance company is 2 claims per day. What is the probability that on any given day,

a) Exactly one claim is filed against the insurance company?

b) No claim is filed against the insurance company?

c) Exactly three claims are filed against the insurance company?

Solution (a):

$$\mu = 2 \quad \text{(average number of claims per day)},$$
$$x = 1 \quad \text{(exact number of claims filed)}.$$

According to the Poisson distribution,

$$P(X = x) = \frac{\mu^x e^{-\mu}}{x!},$$

$$P(X = 1) = \frac{(2)^1 (2.71828)^{-2}}{1!}$$

$$= \frac{2}{(2.71828)^2} = \frac{2}{7.38905} = 0.27067.$$

Solution (b):

$$\mu = 2 \quad \text{(average number of claims per day)},$$
$$x = 0 \quad \text{(exact number of claims filed)}.$$

According to the Poisson distribution,

$$P(X = x) = \frac{\mu^x e^{-\mu}}{x!},$$

$$P(X = 0) = \frac{(2)^0 (2.71828)^{-2}}{0!}$$

$$= \frac{1}{(2.71828)^2} = \frac{1}{7.38905} = 0.13534.$$

Solution (c):

Using a Poisson distribution with $\mu = 2$ and $x = 3$, we find

$$P(X = 3) = \frac{(2)^3(2.71828)^{-2}}{3!}$$

$$= \frac{8}{(6)(7.38905)} = 0.18045.$$

Example 19

A student makes, on the average, one typing error on each page. What is the probability that the student would make exactly 2 typing errors in a 3-page term paper?

Solution:

It is necessary here to define μ as the average number of typing errors per *three* pages. Hence,

$\mu = 3$ (average number of typing errors per 3 pages),
$x = 2$ (exact number of typing errors in a 3-page term paper).

According to the Poisson distribution

$$P(X = x) = \frac{\mu^x e^{-\mu}}{x!},$$

$$P(X = 2) = \frac{(3)^2(2.71828)^{-3}}{2!} = \frac{9}{2(2.71828)^3} = 0.22404.$$

Cumulative Poisson Tables

Cumulative Poisson probabilities have been computed and tabulated for values of μ ranging from 0.1 to 10 (see Table K). The value of μ is found at the head of the column, and the values of X ($x = 0, 1, 2, \ldots$) are located in the *left* column of the table. The probability in any cell in the table is the probability of getting x *or less* successes, that is, $P(X \le x)$. The use of Table K is illustrated with the following two examples:

Example 20

A car salesman sells, on the average, 2.5 cars per day. Use Table K to determine the probability that on a given day the salesman would sell (a) at most 4 cars, (b) exactly 4 cars.

Solution (a):

Since Table K shows the cumulative probabilities $P(X \le x)$, the probability of selling 4 cars or less is found *directly* from Table K. Thus with $\mu = 2.5$,

Table K shows $P(X \leq 4) = 0.8912$.

Solution (b):

The probability of selling *exactly* 4 cars, $P(X = 4)$, can be found as follows:

$$P(X = 4) = P(X \leq 4) - P(X \leq 3).$$

Hence, with $\mu = 2.5$, according to Table K, we find

$$P(X = 4) = 0.8912 - 0.7576 = 0.1336$$

Example 21

The average number of traffic accidents that take place on the Hollywood Freeway on a week day between 7:00 A.M. and 8:00 A.M. is 0.7 accident per hour.   Use Table K to determine the probability that more than 2 traffic accidents would occur on the Hollywood Freeway on Tuesday morning between 7:00 A.M. and 8:00 A.M.

Solution:

Using a Poisson distribution with $\mu = 0.7$, and according to Table K, we find

$$\begin{aligned} P(X > 2) &= 1 - P(X \leq 2) \\ &= 1 - 0.9659 = 0.0341 \end{aligned}$$

The Poisson Approximation to the Binomial Distribution

Let us consider a binomial distribution with $\pi = 0.02$ and $n = 100$.   Let us now assume that we are interested in finding the probability that $X = 3$. Using the binomial formula, we can find the *exact* probability as

$$P(X = 3) = \frac{100!}{3!97!}(0.02)^3(0.98)^{97} = 0.1823.$$

But as the reader may already have noticed, the use of the binomial formula in this case involves many tedious computations.   Fortunately, however, when the value of π is close to zero and n is very large (as it is in the case here), the binomial distribution can be approximated by a Poisson distribution with $\mu = n\pi$.   Hence, using a Poisson distribution with

$$\mu = n\pi = 100(0.02) = 2,$$

we find the probability that $X = 3$ is

$$P(X = 3) = \frac{\mu^x e^{-\mu}}{x!}$$

$$= \frac{(2)^3(2.71828)^{-2}}{3!} = \frac{8}{(6)(2.71828)^2} = 0.1805.$$

Thus, using the binomial formula, we find that the probability that $X = 3$ is exactly 0.1823.   On the other hand, when we use the Poisson distribution as an approximation to the binomial, the probability that $X = 3$ is found to be approximately 0.1805.

The approximation of binomial probabilities by Poisson probabilities improves as the value of π gets closer to 0 and as the value of n becomes large. As a rule of thumb, the approximation is considered appropriate when $\pi \leq 0.05$ and $n \geq 20$.

Example 22

A machine is producing parts that contain, on the average, 0.02 defectives. In a random sample of 75 such parts, what is the probability of obtaining exactly 3 defective parts?

Solution:

This is a binomial distribution with $\pi = 0.02$ and $n = 75$.   Since π is close to 0 and n is large, we will use a Poisson distribution as an approximation to the binomial.   Hence, using a Poisson distribution with

$$\mu = n\pi = 75(0.02) = 1.5,$$

we find, from Table K,

$$P(X = 3) = 0.9344 - 0.8088 = 0.1256.$$

EXERCISES, SET 5

1. The average number of homicides per day in Los Angeles County is 2.   Use the Poisson distribution to determine the probability that on a given day

   a) there will be three or less homicides in Los Angeles County,
   b) there will be exactly three homicides.

2. The average number of earthquakes per year in California is 0.5.   Use the Poisson distribution to determine the probability that there will be no earthquakes in California in the next three years.

3. The average number of major fires per month in a certain city is 1.5.   Use the Poisson distribution to determine the probability that there will be exactly one major fire in a period of two months.

4. Telephone calls come into the Watergate Hotel switchboard at a rate of 10 calls per minute.   Use the Poisson distribution to determine the probability of receiving exactly 4 calls in a 30-second interval.

5. The average number of defects in a roll of a certain type of wallpaper is 2.5. Use the Poisson distribution to determine the probability that a roll will have 4 or more defects.

6. A machine is producing parts that contain, on the average, 2% defectives. Use the Poisson distribution to approximate the probability of obtaining exactly two defective parts in a random sample of 40 such parts.

7. The probability that a sales contact will result in a sale is $\frac{1}{20}$. If 200 sales contacts are made on any given day, what is the probability of making at least 12 sales? Use the Poisson approximation to the binomial.

The Normal Distribution

Let us consider first the following variables: number of workers in a given establishment, the number of heads in 14 tosses of a balanced coin, and the number of students in a statistics class. These variables can assume only the limited values 0, 1, 2, ... Consequently, these variables are *discrete* variables.

By contrast, let us consider the following variables: the weight of a jar of instant coffee, the height of a telephone pole, the temperature at a certain time of the day. Each of these variables can assume an infinite number of values on a measuring scale. The weight of a jar of instant coffee, for example, can be 4.5 ounces, 4.512 ounces, 4.512718 ounces, ... Similarly, the height of a telephone pole can be 103 feet, 103.178 feet, 103.178912 feet, ... Consequently, these variables are *continuous* variables.

As the reader may have already suspected, the binomial distribution and the Poisson distribution can represent the probability distribution of only a discrete variable. What about continuous variables? The probability distribution of some continuous variables can be described by a very common continuous distribution called the *normal distribution*.

Let us consider some continuous variables such as the height of a soldier in any army regiment, the life of a 100-watt electric bulb in a large lot of a certain brand, and the weight of a 10-ounce jar of instant coffee in a very large shipment. The probability distribution of each of these continuous variables may very well be described by the normal distribution (Fig. 4–4).

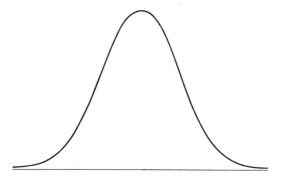

Fig. 4–4 The normal distribution

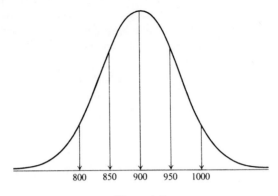

Figure 4–5

The normal distribution can best be explained by a concrete example. Therefore, let us consider the case of a manufacturer who receives a very large shipment of 100-watt electric bulbs.   Let us assume, furthermore, that the mean life μ of this bulb is 900 hours with a standard deviation σ of 50 hours, that is, μ = 900 hours and σ = 50 hours.   Now, if the life of the bulb is normally distributed, then we expect the majority of the bulbs to have a life of about 900 hours. Furthermore, there will be as many bulbs that last longer than 950 hours as there are bulbs lasting less than 850 hours. Similarly, there will be as many bulbs that last more than 1000 hours as there are that last less than 800 hours (Fig. 4–5).   These remarks about the normal distribution are of an informal nature.   They are made here to give us a general idea of what the normal distribution is all about.   Next, let us be more specific and rigorous.

The normal distribution is completely specified by its mean μ and standard deviation σ.   In other words, you will know all you want to know about life of these electric bulbs once the mean and the standard deviation of the distribution are known; for example, you will be able to determine the

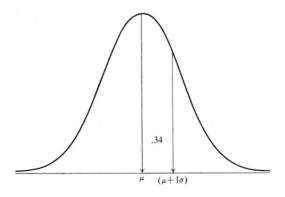

Figure 4–6

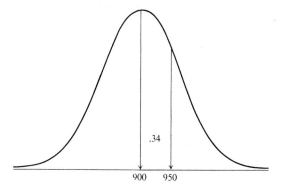

Figure 4–7

percentage of bulbs that should last beyond 917 hours, from 927 to 1071 hours, or less than 892 hours, and so on.

It is a fact, for example, that 34% of the area under the normal curve is between μ and $\mu + 1.0\sigma$ (Fig. 4–6). This means that the lives of 34% of all bulbs are expected to last between

$$\mu \text{ and } \mu + 1\sigma, \quad \text{or} \quad 900 \text{ and } 900 + 50,$$

that is, between 900 and 950 hours (Fig. 4–7). We can also conclude that the probability is .34 that any one bulb will last from 900 to 950 hours.

These facts concerning the area under the normal curve are obtained from Table A, which shows the area enclosed between the mean and any other value x (x can be either above or below the mean). The letter z labeling the first column in the table indicates how many standard deviations x is away from the mean. Table A shows, for example,

1. .4332 of the area under the normal curve is between μ and $(\mu + 1.5\sigma)$, Fig. 4–8.

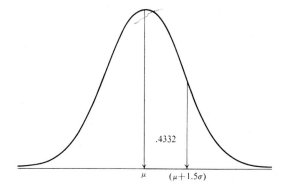

Figure 4–8

2. .3869 of the area under the normal curve is enclosed between μ and $(\mu - 1.21\sigma)$, Fig. 4–9.

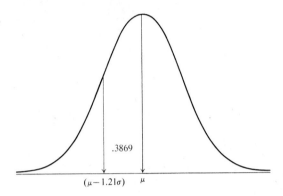

.3869

$(\mu - 1.21\sigma)$ μ

Figure 4–9

3. .4830 of the area under the normal curve is enclosed between either μ and $(\mu + 2.12\sigma)$ or between μ and $(\mu - 2.12\sigma)$, Fig. 4–10.

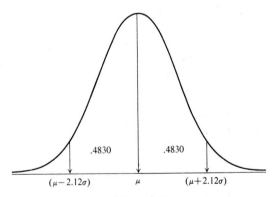

.4830   .4830

$(\mu - 2.12\sigma)$ μ $(\mu + 2.12\sigma)$

Figure 4–10

As another example, let us consider a plant with a very large number of workers. Let the random variable X indicate the age of a worker in the plant. Furthermore, let us assume that X has a normal distribution with mean $\mu = 50$ years and standard deviation $\sigma = 5$ years. The following questions can be answered using Table A.

1. What percentage of workers are between 50 and 52.5 years?

This problem can be solved in two steps:

a) $52.5 - 50 = 2.5$ years (the difference between x and μ in years)

b) $2.5 \div 5 = .5$ (x exceeds μ by $.5\sigma$).

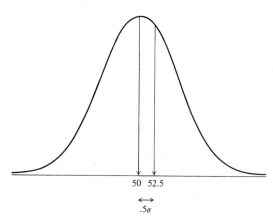

Figure 4–11

From Table A, we find that .1915 of the area under the normal curve is enclosed between μ and $(\mu + .5\sigma)$. Thus 19.15% of the workers are between 50 and 52.5 years old (Fig. 4–11). The two steps above can be combined in the following formula,

$$z = \frac{x - \mu}{\sigma} = \frac{52.5 - 50}{5} = .5,$$

where z is the deviation of x from the mean, measured in standard deviations. Thus 52.5 deviates from the mean by $.5\sigma$. Once again, 19.15% of the workers are between 50 and 52.5 years old.

2. What is the probability that any one worker is 45 years or younger?

Determining the probability that any one worker is 45 years or younger is the same as determining the percentage of workers who are younger than 45 years:

$$z = \frac{x - \mu}{\sigma} = \frac{45 - 50}{5} = -1.0.$$

Since .3413 of the area under the normal curve is between μ and $(\mu - 1\sigma)$, the area that lies to the left of $(\mu - 1\sigma)$ is

$$.5000 - .3413 = .1587.$$

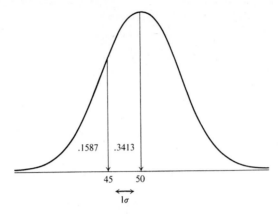

.1587   .3413

45   50

$\longleftrightarrow$
1σ

Figure 4–12

Thus 15.87% of the workers are younger than 45 years (Fig. 4–12). Stated differently, the probability is .1587 that any one worker is younger than 45. Symbolically, we may write $P(X < 45) = .1587$. Note that $P(X \leq 45)$ is also .1587. Because X is a continuous random variable, it can assume an infinite number of possible values, at least theoretically. Hence, the probability that X is equal to one given value, for instance 45 years, is zero—that is, $P(X = 45) = 0$. Consequently, $P(X \leq 45) = P(X < 45) = .1587$.

3. What is the probability that any one worker will be between 41 and 58 years?

Let us denote by A_1 the area between 58 and 50, and by A_2 the area between 41 and 50; then

$$P(41 \leq X \leq 58) = A_1 + A_2$$

(Fig. 4–13). We proceed as follows:

$$z_2 = \frac{x - \mu}{\sigma} \qquad\qquad z_1 = \frac{x - \mu}{\sigma}$$

$$= \frac{41 - 50}{5} \qquad\qquad = \frac{58 - 50}{5}$$

$$= -1.8; \qquad\qquad = 1.6;$$

$$\text{Area}_2 = .4641. \qquad \text{Area}_1 = .4452.$$

Therefore

$$P(41 \leq X \leq 58) = A_1 + A_2$$
$$= .4452 + .4641$$
$$= .9093.$$

Thus the probability that any one worker is between 41 and 58 years is .9093.

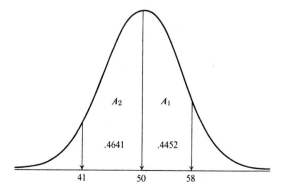

Figure 4–13

4. What is the probability that any one worker is between 55 and 60 years?

We denote by A_1 the area between 60 and 50, and by A_2 the area between 55 and 50; then

$$P(55 \leq X \leq 60) = A_1 - A_2$$

(Fig. 4–14). We calculate:

$$z_2 = \frac{x - \mu}{\sigma} \qquad\qquad z_1 = \frac{x - \mu}{\sigma}$$

$$= \frac{55 - 50}{5} \qquad\qquad = \frac{60 - 50}{5}$$

$$= 1.0; \qquad\qquad = 2.0;$$

$$\text{Area}_2 = .3413. \qquad \text{Area}_1 = .4772.$$

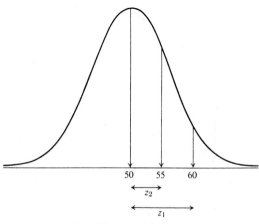

Figure 4–14

Therefore

$$P(55 \le X \le 60) = A_1 - A_2$$
$$= .4772 - .3413$$
$$= .1359.$$

Thus the probability that any one worker is between 55 and 60 years old is .1359.

5. Twenty percent of the workers are below a certain age. What is this age?

Table A shows that .2995 or .30 of the area under the normal curve is between μ and $(\mu - .84\sigma)$. Thus if x is located at a distance of $-.84\sigma$ from the mean, then .2005 of the workers will not have reached age x (Fig. 4–15). Therefore

$$x = \mu - .84\sigma$$
$$= 50 - .84(5) = 50 - 4.20$$
$$= 45.80 \text{ years.}$$

Thus about 20% of all workers are younger than 45.8 years.

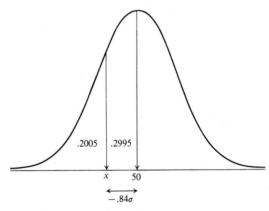

Figure 4–15

Example 23

The mean height of a soldier in an army regiment is 70 inches. Ten percent of the soldiers in the regiment are taller than 72 inches. Assuming that the heights of the soldiers in this regiment are normally distributed, what is the standard deviation?

Solution: Since 10% of the soldiers are taller than 72 inches, then 40% of the soldiers in the regiment are between 70 and 72 inches tall. But according to

Table A, approximately 40% (.3997) of the area under the normal curve is between μ and $\mu + 1.28\sigma$ (Fig. 4–16). Consequently,

$$\mu + 1.28\sigma = 72, \qquad 70 + 1.28\sigma = 72, \qquad 1.28\sigma = 2;$$

$$\sigma = \frac{2}{1.28}, \qquad \sigma = 1.56 \text{ inches.}$$

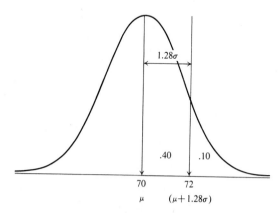

Figure 4–16

Example 24

A food processor packages instant coffee in small jars. The weights of the jars are normally distributed with a standard deviation of .3 ounce. If 5% of the jars weigh more than 12.492 ounces, what is the mean weight of the jars?

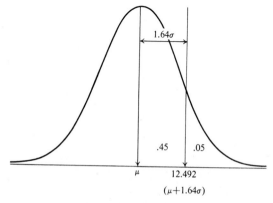

Figure 4–17

Solution: Since approximately 45% (.4495) of the jars weigh between μ and 12.492 ounces, then, according to Table A, the point 12.492 is at a distance of 1.64σ from μ (Fig. 4–17). In other words,

$$
\begin{aligned}
12.492 &= \mu + 1.64\sigma \\
&= \mu + 1.64(.3) \\
&= \mu + .492, \\
\mu &= 12.492 - .492 = 12 \text{ ounces.}
\end{aligned}
$$

Example 25

The heights of soldiers are normally distributed. If 13.57% of the soldiers are taller than 72.2 inches and 8.08% are shorter than 67.2 inches, what are the mean and the standard deviation of the heights of the soldiers?

Solution: Since .3643 of all the area under the normal curve is enclosed between μ and 72.2, then, according to Table A, the point 72.2 is at a distance of 1.1σ from μ (Fig. 4–18). In other words,

$$72.2 = \mu + 1.1\sigma.$$

Similarly, since .4192 of the area under the normal curve is between μ and 67.2, then, according to Table A, the point 67.2 is 1.4σ to the left of μ. In other words,

$$67.2 = \mu - 1.4\sigma.$$

Now we have two simultaneous equations containing the two unknowns σ and μ. Solving these two equations, we have

$$
\begin{array}{r}
72.2 = \mu + 1.1\sigma \\
67.2 = \mu - 1.4\sigma \\
\hline
5 = 0 + 2.5\sigma, \\
\sigma = \dfrac{5}{2.5} = 2 \text{ inches,}
\end{array}
$$

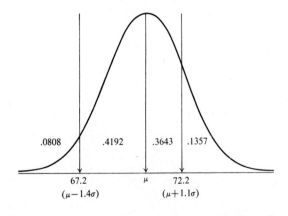

Figure 4–18

and the value of μ can be found from any of the two equations.

$$72.2 = \mu + 1.1\sigma$$
$$= \mu + 1.1(2)$$
$$= \mu + 2.2,$$
$$\mu = 72.2 - 2.2 = 70 \text{ inches.}$$

Example 26

The diameters of a large shipment of ball bearings are normally distributed with a mean of 2.0 inches and a standard deviation of .01 inch. If three ball bearings are selected at random from the shipment, what is the probability that exactly two of the selected ball bearings will have a diameter larger than 2.02 inches?

Solution: First, let us determine the probability that the diameter of any *one* ball bearing is greater than 2.02 inches (Fig. 4–19):

$$z = \frac{x - \mu}{\sigma}$$

$$= \frac{2.02 - 2.00}{.01} = 2.$$

Therefore

$$P(X > 2.02) = .5000 - .4772$$
$$= .0228.$$

Now, let

$\pi = .0228$ (the probability that the diameter of any one ball bearing is larger than 2.02 inches),

$n = 3$ (sample size),

$x = 2$ (number of ball bearings with a diameter larger than 2.02 inches).

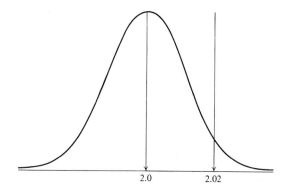

2.0   2.02   **Figure 4–19**

Using the binomial formula, the probability that exactly two of the selected ball bearings will have a diameter greater than 2.02 inches is

$$P(X=x) = \frac{n!}{x!\,(n-x)!}\,\pi^x (1-\pi)^{n-x}$$

$$= \frac{3!}{2!\,1!}\,(.0228)^2(.9772)^1$$

$$= 3(.00051984)(.9772) = .00152.$$

Example 27

The production manager believes that the life of machine K is normally distributed with a mean of 3000 hours. If the manager furthermore believes that there is a 0.50 chance that machine K could last less than 2632 or more than 3368 hours, what is the standard deviation?

Solution:

Since the manager believes that there is a 0.50 chance that machine K could last less than 2632 or longer than 3368 hours, then the probability is 0.50 that the life of machine K is between 2632 and 3368 hours. And because of the symmetry of the area under the normal curve, then the probability is 0.25 that the machine life is between 3000 and 3368 hours (Fig. 4–20).

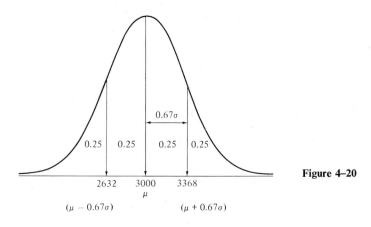

Figure 4–20

But according to Table A, approximately 0.25 (0.2486) of the area under the normal curve is between μ and $\mu + 0.67\sigma$ (Fig. 4–20). Consequently,

$$0.67\sigma = 3368 - 3000,$$

$$= 368 \text{ hours};$$

$$\sigma = \frac{368}{0.67} = 549.25 \text{ hours.}$$

EXERCISES, SET 6

1. The heights of soldiers in an Army regiment are normally distributed with a mean of 69 inches and a standard deviation of 2 inches.

   a) What is the probability that any one soldier is over 72 inches?
   b) What is the percentage of soldiers who are between 69 and 73 inches tall?
   c) If the tallest 20% of the soldiers in the regiment are eligible to perform a certain task, what is the minimum height required to perform such a task?

2. A ball bearing is considered defective and therefore is rejected if its diameter is greater than 2.02 inches or less than 1.98 inches. What is the expected number of rejected ball bearings if the diameters in a lot of 10,000 ball bearings are normally distributed with a mean of 2 inches and a standard deviation of .01 inches?

3. The final examination scores in a "Survey of Psychology" class are normally distributed. The mean score is 60 and the standard deviation is 10.

   a) If the lowest passing grade is 48, what percentage of the class is failing?
   b) If the highest 80% of the class are to pass, what is the lowest passing score?

4. A food processor markets flour in a "NET WT. 12 OZ." package. The filling process can be adjusted so that the mean fill can be set to any desired level. If the fill is normally distributed with a standard deviation of .2 ounce,

   a) at what level should the mean fill be set so that only .001 of the packages would have a net weight less than 12 ounces?
   b) at what level should the mean fill be set so that only .05 of the packages would have a net weight exceeding 12.4 ounces?

5. The mean weight of a pineapple in a very large shipment is 5 pounds. Ten percent of these pineapples weigh less than 4 pounds. Assuming that the weights are normally distributed, what is the standard deviation in the shipment?

6. The lives of a certain brand of nine-volt batteries are normally distributed. If 6.68% of the batteries last more than 56 hours and 30.85% last less than 52 hours, what are the mean and the standard deviation of the life of this brand of batteries?

7. Miss Mode, who is a statistical analyst for the Data Kontrol Corporation, drives to work every day. The time required for the trip from her apartment to the office can be approximated by a normal distribution with a mean of 20 minutes and a standard deviation of 5 minutes. At what time should Miss Mode leave her apartment to give herself a .95 chance that she will be at her office by 9 A.M.?

8. The lives of 40-watt electric bulbs of brand *A* are normally distributed with a mean of 850 hours and a standard deviation of 50 hours. In a random sample of 4 bulbs, what is the probability that exactly three bulbs will last 822 hours or more?

9. The lives of 40-watt electric bulbs of brand A have a mean of 850 hours with a standard deviation of 50 hours. The mean life of brand B 40-watt electric bulbs, on the other hand, is 820 hours with a standard deviation of 40 hours. The life of each brand of bulbs is normally distributed. If one bulb is to be selected from each brand at random, what is the probability that bulb A will last longer than 825 hours and bulb B less than 840 hours?

10. A magazine publisher is considering publishing a special Christmas edition. The publisher believes that sales of the special edition are normally distributed with a mean of 100,000 copies. If the publisher furthermore believes that the probability is 0.20 that sales could exceed 120,000 copies, what is the standard deviation?

11. The credit manager of a department store estimates bad debt losses in the current year as follows: the loss is normally distributed with a mean of $30,000; and the probability is 0.50 that the loss could be greater than $35,000 or less than $25,000. What is the standard deviation?

12. Arithmetic test scores of 1000 eleventh graders are normally distributed with a mean of 60 and a standard deviation of 8. Scores of the same test for 800 twelfth graders are also normally distributed with a mean of 68 and a standard deviation of 10.

   a) How many of the eleventh graders score above the average twelfth grader?

   b) How many of the twelfth graders score lower than the average eleventh grader?

13. Variable X is normally distributed with $\mu = 100$ and $\sigma = 10$. Find

   a) $P(X > 112)$ b) $P(95 < X < 110)$ c) $P(85 \leq X \leq 95)$

14. The life of tires used by a tucking company is normally distributed with a mean of 80,000 miles and a standard deviation of 2000 miles. To avoid a tire blow-out, the company replaces each tire after it has been used for a given number of miles.

   After how many miles of usage should a tire be replaced if the chance of a tire blowout is to be limited to .001?

15. An electronic device is used in an ocean-exploration apparatus. The operational life of the device is normally distributed with a mean of 500 hours and a standard deviation of 50 hours.

   a) What is the probability that the device would fail prior to 420 hours of operation time?

   b) After how many hours of operation time should the device be replaced if the probability of failure is to be approximately .05?

16. Weekly demand for product X is normally distributed with $\mu = 10$ units and $\sigma = 2$ units. Assume that weekly supply is fixed at 12 units per week. What is the probability that the quantity demanded in a given week would exceed supply?

The Normal Curve Approximation to the Binomial Distribution

Let us assume that we wish to determine the probability of getting 6 heads in 16 tosses of a balanced coin. Since this is a binomial experiment with $\pi = \frac{1}{2}$ and $n = 16$, we can use the binomial formula to determine the *exact* probability of obtaining 6 heads in this experiment:

$$P(X = 6) = \frac{16!}{6!\,10!} \left(\frac{1}{2}\right)^{6}\left(\frac{1}{2}\right)^{10}$$

$$= 8008\left(\frac{1}{2}\right)^{16} = \frac{8008}{65,536} = .122.$$

The probability of getting 6 heads in 16 tosses of a balanced coin can be approximated using the normal distribution as an approximation to the binomial distribution. The approximation is obtained as follows:

1. Determine the mean and the standard deviation of the binomial distribution.

The mean is

$$\mu = n\pi = 16(\tfrac{1}{2}) = 8 \text{ heads,}$$

and the standard deviation is

$$\sigma = \sqrt{n\pi(1 - \pi)} = \sqrt{16(\tfrac{1}{2})(\tfrac{1}{2})} = 2 \text{ heads.}$$

2. Approximate the binomial distribution by a normal distribution with $\mu = 8$ and $\sigma = 2$ (Fig. 4–21).

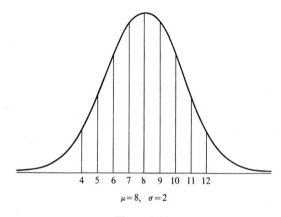

4 5 6 7 8 9 10 11 12

$\mu = 8, \quad \sigma = 2$

Figure 4–21

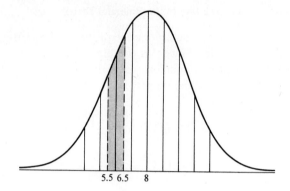

5.5 6.5   8

Figure 4–22

3. The probability of obtaining 6 heads is approximated by the area under the normal curve between $x = 5.5$ and $x = 6.5$ (see shaded area in Fig. 4–22). We have

$$z_1 = \frac{5.5 - 8}{2} \qquad z_2 = \frac{6.5 - 8}{2}$$

$$= \frac{-2.5}{2} \qquad = \frac{-1.5}{2}$$

$$= -1.25; \qquad = -.75;$$

$$A_1 = .3944. \qquad A_2 = .2734.$$

Therefore

$$P(5.5 \le X \le 6.5) = .3944 - .2734$$
$$= .3944 - .2734$$
$$= .121.$$

Thus, using the binomial distribution, the probability of getting 6 heads in 16 tosses of a balanced coin is exactly .122. On the other hand, when we use the normal curve as an approximation to the binomial distribution, the probability of getting 6 heads in 16 tosses of a balanced coin is found to be approximately .121.

The probability of getting any given number of heads in 16 tosses of a balanced coin can be approximated in a similar fashion. The probability of getting 8 heads, for example, is approximated by the area under the normal curve between $x = 7.5$ and $x = 8.5$. Similarly, the probability of getting 10 heads is approximated by the area under the normal curve between $x = 9.5$ and $x = 10.5$. In general, the probability of obtaining x number of heads is approximated by the area under the normal curve enclosed between $x - \frac{1}{2}$ and $x + \frac{1}{2}$.

The usefulness of the normal-curve approximation to the binomial distribution becomes evident when we consider the following example. Let us assume that we are interested in determining the probability of getting *6 heads or more* in 16 tosses of a balanced coin.

To determine the exact probability, we must use the binomial formula to determine the probability of getting 6 heads, the probability of getting 7 heads, the probability of getting 8 heads, etc., and finally the probability of getting 16 heads. The exact probability of getting 6 heads or more is the sum of all these individual probabilities:

$$P(X \geq 6) = P(X = 6) + P(X = 7) + \cdots + P(X = 16)$$

$$= \frac{16!}{6! \, 10!} \left(\frac{1}{2}\right)^6 \left(\frac{1}{2}\right)^{10} + \cdots + \frac{16!}{16! \, 10!} \left(\frac{1}{2}\right)^{16} \left(\frac{1}{2}\right)^0.$$

As the reader may already have observed, using the binomial formula in a case of this kind requires some lengthy computations. The probability of getting 6 heads or more in our binomial experiment can be easily approximated by the area under the normal curve lying to the right of 5.5 (see shaded area in Fig. 4–23):

$$z = \frac{x - \mu}{\sigma} = \frac{5.5 - 8}{2}$$

$$= \frac{-2.5}{2} = -1.25;$$

$$A = .3944.$$

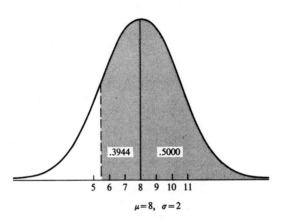

$\mu = 8, \quad \sigma = 2$

Figure 4–23

Therefore

$$P(X \geq 5.5) = .3944 + .5000$$
$$= .8944.$$

Thus the probability of getting 6 or more heads in 16 tosses of a balanced coin is approximately .8944.

Finally, let us approximate the probability of getting anywhere from 5 to 10 heads in 16 tosses of a balanced coin. This probability is approximated by the area under the normal curve between 4.5 and 10.5 (see shaded area in Fig. 4–24);

$$z_1 = \frac{4.5 - 8}{2} \qquad z_2 = \frac{10.5 - 8}{2}$$

$$= \frac{-3.5}{2} \qquad = \frac{2.5}{2}$$

$$= -1.75; \qquad = 1.25;$$

$$A_1 = .4599. \qquad A_2 = .3944.$$

Therefore

$$P(4.5 \leq X \leq 10.5) = A_1 + A_2$$
$$= .4599 + .3944$$
$$= .8543.$$

Thus the probability of getting anywhere from 5 to 10 heads in 16 tosses of a balanced coin is approximately .8543.

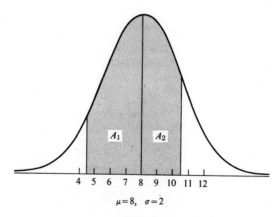

$$\mu = 8, \quad \sigma = 2$$

Figure 4–24

The reader may wonder about the accuracy of the normal-curve approximation to the binomial distribution. The degree of accuracy depends on the values of n and π. It improves as n gets larger and as the value of π becomes closer to $\frac{1}{2}$. As a general rule, the approximation is regarded as satisfactory if both $n\pi$ and $n(1 - \pi)$ are greater than 5.

Example 28

Congressional candidate X will base his decision on whether or not to endorse the Educational Opportunity Act on the results of a poll of a random sample of 200 registered voters from his district. He will endorse the Educational Opportunity Act only if 100 or more voters are in favor of it.

a) What is the probability that candidate X will endorse the Act if 45% of *all* registered voters in his district *are* in favor of it?

b) What is the probability that candidate X will fail to endorse the Act if 52% of *all* voters in his district *are* in favor of it?

Solution (a):   The random selection of 200 voters from a district in which 45% of all voters are in favor of the Educational Opportunity Act can be regarded as a binomial experiment with $n = 200$ and $\pi = .45$. The mean μ and the standard deviation σ of this binomial experiment are computed below.

$$\mu = n\pi = 200(.45) = 90 \text{ voters};$$
$$\sigma = \sqrt{n\pi(1 - \pi)} = \sqrt{200(.45)(.55)} = \sqrt{49.5} = 7.04 \text{ voters}.$$

Candidate X will endorse the Act only when 100 or more (100, or 101, or 102, ..., or 200) voters are in favor of it. Using the normal distribution as an approximation to the binomial distribution, the probability that the Act will be endorsed by 100 or more voters is approximated by the area under the normal curve lying to the right of $x = 99.5$ (see shaded area in Fig. 4–25). We calculate

$$z = \frac{x - \mu}{\sigma} = \frac{99.5 - 90}{7.04}$$

$$= \frac{9.5}{7.04} = 1.35;$$

$$A = .4115.$$

Therefore

$$P(X \geq 99.5) = .50 - .4115$$
$$= .0885.$$

Thus the probability that candidate X will endorse the Educational Opportunity Act is .0885.

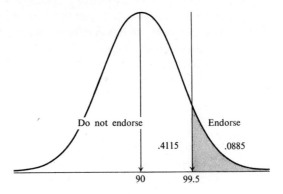

Figure 4–25

Solution (b): The random selection of 200 voters from a district in which 52% of all voters endorse the Educational Opportunity Act can be regarded as a binomial experiment with $n = 200$ and $\pi = .52$. The mean and the standard deviation of this binomial experiment are computed below.

$$\mu = n\pi = 200(.52) = 104 \text{ voters};$$
$$\sigma = \sqrt{n\pi(1 - \pi)} = \sqrt{200(.52)(.48)} = \sqrt{49.92} = 7.07 \text{ voters}.$$

Candidate X will fail to endorse the Act if less than 100 (99, or 98, or 97, ..., or 0) voters are in favor of it. The probability that the Act is endorsed by 99 voters (or less) is approximated by the area under the normal curve lying to the left of $x = 99.5$ (see shaded area in Fig. 4–26). We compute z and A:

$$z = \frac{x - \mu}{\sigma}$$
$$= \frac{99.5 - 104}{7.06}$$
$$= \frac{-4.5}{7.07} = -.64;$$
$$A = .2389.$$

Therefore

$$P(X \leq 99.5) = .50 - .2389$$
$$= .2611.$$

Thus the probability that candidate X will fail to endorse the Educational Opportunity Act is .2611.

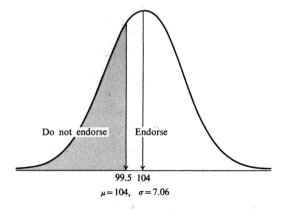

99.5 104

$\mu = 104, \quad \sigma = 7.06$

Figure 4–26

EXERCISES, SET 7

1. A balanced coin is tossed 36 times. What is the probability
   a) of getting 15 heads?
   b) of getting less than 20 heads?

2. Determine the probability of getting 110 or more "sixes" when a fair die is rolled 720 times.

3. If 64% of all students attending this college are males, what is the probability that a random sample of 400 students will contain 250 or more male students?

4. A fair coin is to be tossed 100 times. The player receives $10 if the number of heads is 57 or more, and loses $5 if the number of heads is 47 or less; no gain or loss is incurred otherwise. Determine the expected value of this game to the player.

5. The following rule is used in controlling the operation of a machine that produces certain parts. Select a random sample of 400 parts every hour. If the number of defective parts is 12 or more, then stop the machine. If the number of defective parts is less than 12, continue the operation of the machine. What is the probability
   a) of stopping the machine when it produces, on the average, 2% defectives?
   b) of continuing the machine operation when it produces, on the average, 4% defectives?

6. A pair of dice consists of a red die and a blue die. The pair is to be rolled 180 times. In order to win, the player must get at least 37 sixes with the red die and no more than 27 sixes with the blue die. If the player wins, he receives $21. Determine the expected value of this game if it costs the player $1 to play.

7. It is claimed that 10% of all students attending this college have an IQ score of 125 or more. What can you conclude about this claim if a random sample of 900 students shows less than 50 students having an IQ score of 125 or more?

REVIEW EXERCISES I

1. Given the following information concerning a population of 100 accounts receivable.

$$x = \text{the amount of any individual account,}$$
$$\sum x = \$13,000,$$
$$\sum (x - \mu)^2 = \$62,500.$$

Determine the population mean and its standard deviation.

2. Given the sets

$$A = \{x \mid x^2 - 2x - 24 = 0\} \quad \text{and} \quad B = \{x \mid x^2 + 2x = 8\}.$$

Find

$$P(A \cap B).$$

3. An experiment consists of drawing three cards from an ordinary deck of playing cards and observing the color of each card.

a) Denoting a red card by R and a black card by B, list all the elements of the sample space of this experiment.

b) List the elements of S contained in the event that there is exactly one red card.

c) How do you define event A where

$$A = \{B, B, B\}?$$

4. A population consists of the following 10 accounts receivable: $26, $20, $18, $22, $20, $12, $24, $24, $20, $14.

a) Determine the arithmetic mean and the standard deviation.

b) One account receivable is to be selected at random. What is the probability that it would fall in the range $\mu \pm 1\sigma$?

5. A committee consists of 4 Republicans and 6 Democrats. Two persons are to be selected at random, the first to be the committee chairman and the second to be vice chairman.

a) What is the probability that the committee chairman is a Republican?

b) What is the probability that the chairman and vice chairman are Republicans?

c) What is the probability that only the vice chairman is a Republican?

6. The probability that Oscar would marry a blonde is 0.20. The probability that he would marry an intelligent girl is 0.40, and the probability that he would marry a rich girl is 0.05. Assuming that the three characteristics are independent, what is the probability that Oscar would marry a rich, dumb blonde?

7. The probability of an engine failure of an aircraft is 0.10. With how many engines should the aircraft be equipped to be 0.999 sure against an engine failure (assume that only one engine is needed for a successful operation of the aircraft)?

8. The probability that a rifleman hits the target with any single shot is 0.4.

   a) What is the probability of missing the target in four (4) consecutive shots?
   b) What is the probability of hitting the target at least once in four (4) consecutive shots?
   c) How many shots should the rifleman fire in order to be approximately 0.95 sure that the target is hit at least once?

9. A system consists of four independent components: A, B, C_1, C_2. The probability of failure is 0.01 for A, 0.02 for B, 0.10 for C_1, and 0.10 for C_2. If component A, component B, and at least one of components C are needed for a successful operation of the system, what is the probability that the system would succeed?

10. Miss Smith leaves work promptly at 4:45 p.m. The probability is 0.7 that she goes directly home; the probability is 0.3 that she stops at the Rendezvous Room before going home.

    If Miss Smith goes directly home, the probability is 0.90 that she arrives home by 6 p.m. If she visits the Rendezvous Room after work the probability is only 0.05 that she will arrive home by 6 p.m.

    Given that Miss Smith arrived at home after 6 p.m., what is the probability that her delay was due to a visit to the Rendezvous Room?

11. Professor X teaches statistics and would like to give a quiz every class meeting. Being aware that he sometimes forgets to come to class, Professor X has arranged for his Graduate Assistant to take over the class in his absence.

    If Professor X appears in the class, the probability is 70 percent that he would give a quiz. In his absence, the probability is only 10 percent that a quiz is given by his Graduate Assistant.

    If Professor X is known to be absent 80 percent of the time:

    a) What is the probability that there will be a quiz in any given class meeting?
    b) Assume that there was a quiz in a given class meeting, what is the probability that Professor X was present?

12. Urn A contains 3 white, 5 black, and 2 green marbles; urn B contains 4 white, 3 black, and 3 green marbles. One marble is to be selected at random from each urn. You will win $4.00 when the two marbles selected are white, $3.00 when both are black, and $5.00 when both are green. If it costs you $1.50 to play the game, what is the expected value of this game to you?

13. A true/false quiz consists of 5 questions. Assuming that you answer each question by sheer guess,

    a) what is the probability that you answer only the first two questions correctly?
    b) what is the probability that you answer exactly two questions correctly?

14. Mr. Monday is an amateur thief.   The probability that he opens the jewel safe on any single trial is 0.2.

    a) What is the probability that it would take Mr. Monday exactly two trials to open the safe?

    b) What is the probability that Mr. Monday would open the safe by the third trial?

15. It is known that 80 percent of all students at Valley State College are under-graduates.

    a) Four students are selected at random.   What is the probability that exactly three (3) students are undergraduate?

    b) Calculate the mean number of undergraduate students in a random sample of four (4) students from Valley State College.   Calculate also the standard deviation.

16. It is known that 60 percent of all registered voters in the 27th Congressional District are Democrats.   Four registered voters are selected at random from this district.

    a) What is the probability that there will be exactly two Democrats in the sample?

    b) What is the probability that the sample would contain at least one Democrat?

    c) What is the probability that only the first two voters selected are Democrats?

17. Fifty percent of all students at Bradford College are freshmen, 25 percent are sophomores, 15 percent are juniors, and 10 percent are seniors.   Five students are selected at random.

    a) What is the probability that exactly two students are freshmen?

    b) What is the probability that *none* of the students selected are juniors or seniors?

18. The probability of a success in a single trial is $\frac{1}{2}$.

    a) What is the probability of getting *exactly* two successes in three consecutive trials?

    b) What is the probability of having *at least* one success in three consecutive trials?

    c) How many consecutive trials should be performed in order to be 0.9375 sure of obtaining *at least* one success?

19. The heights of orange trees are normally distributed.   2.28 percent are taller than 14 feet and 84.13 percent are shorter than 12 feet.   Find the mean height of an orange tree.   Find the standard deviation.

20. Beta Airlines has introduced the Red Eye Special, a nonstop midnight flight between Los Angeles and Atlanta.   Flight time is normally distributed with a mean of 4 hours and a standard deviation of 1 hour.

    If an aircraft left Los Angeles at 12:00 a.m., what is the probability that it will arrive in Atlanta by 6:00 a.m. (Los Angeles time, of course)?

21. The Hob-Nob Bar has installed an automatic draft beer machine.   The machine can be regulated so that the average fill per glass can be set at any desired level.

At any given set level, however, the amounts of beer discharged are normally distributed with a standard deviation of 0.2 oz.

a) If the filling level is set at 10.3 oz. per glass, what percentage of glasses will contain less than 10 oz.?

b) At what filling level should the machine be set, if it is desired that only 2.28 percent of the glasses contain less than 10 oz.?

c) At what filling level should the machine be set, if it is desired that only 84.13 percent of the glasses contain less than 10.6 oz.?

22. The lives of a certain brand of batteries are normally distributed with a mean of 80 hours and a standard deviation of 10 hours. The manufacturer guarantees to replace any battery which fails prior to a guaranteed time. How long a guarantee should the manufacturer give so that no more than five percent (5%) of the batteries fail prior to the guaranteed time?

23. The ABC Tire Company produces a radial tire which has a mean lifetime of 50,000 miles and a standard deviation of 5,000 miles. Assuming that the lifetime is normally distributed:

a) What is the probability that any one tire would last longer than 60,000 miles?

b) Fifty percent of the tires would last between x_1 and x_2. Find the values of x_1 and x_2 using symmetrical limits around the mean.

c) The manufacturer guarantees to replace free any tire that lasts less than x. Determine the value of x so that the manufacturer would have to replace only one percent of his tires.

24. The machine time required to produce a unit of product A is normally distributed with a mean of 50 minutes and a standard deviation of 5 minutes. A lot of 4,000 units of product A is to be produced.

a) How many units of the lot are expected to require more than 53 minutes of machine time?

b) How many units of the lot are expected to require no less than 48 minutes and no more than 53 minutes of machine time?

c) Fifty percent of the lot requires machine time that is greater than X_1 minutes but less than X_2 minutes. Determine X_1 and X_2 using symmetrical limits around the mean.

25. In deciding on whether to undertake a public housing project or not in East Los Angeles, the Public Housing Administration adopted the following rule: select 100 housing units from East Los Angeles at random. If 40 percent or more of the units selected are dilapidated, undertake the project. If less than 40 percent are dilapidated, do not undertake the project.

a) What is the probability of undertaking the public housing project when only 36 percent of *all* housing units in East Los Angeles are actually dilapidated?

b) What is the probability of not undertaking the public housing project when 50 percent of all housing units in East Los Angeles are actually dilapidated?

26. A firm markets its product exclusively by mail to a list of 100,000 customers. The firm is considering to market a new gadget. In deciding whether to market

the new gadget or not, the firm adopted the following rule: Select and contact 100 persons from the customers list at random.   If 30 or more of the customers contacted wished to purchase the gadget, market the gadget.   If less than 30 of those contacted desire to purchase the gadget, do not market it.

a) What is the probability of marketing the gadget when *actually* 20 percent of *all* customers on the list would purchase the gadget?

b) What is the probability of not marketing the gadget when *actually* 36 percent of all customers would purchase the gadget?

27. An electronic tube is used in the operation of an aircraft.   The life of the tube is normally distributed with a mean of 100 hours and a standard deviation of 10 hours.

a) What is the probability that a tube will last longer than 112 hours?

b) After how many hours of usage should a tube be replaced if the probability of a tube failure is to be approximately .001?

28. Urn X contains four balls numbered 1 through 4; urn Y contains three balls numbered 6, 7, and 8.   One ball is selected from each urn.

a) What is the probability that the sum of the two numbers selected is 11?

b) What is the probability that the sum of the two numbers selected is 10?

29. Arithmetic test scores are normally distributed with a mean of 60 and a standard deviation of 10.   English test scores are normally distributed with a mean of 70, and a standard deviation of 15.   Tom's test score in arithmetic is 72. How high should Tom score in English in order to attain an equivalent level of achievement in both subjects?

30. A manufacturer estimates the daily demand for his product as follows:

Number of units sold per day	Probability
5	.2
6	.5
7	.3

a) What is the probability that the manufacturer sells six units in two successive days?

b) Determine the *expected* number of units sold per day.

31. Smith is an insurance salesman who makes several personal sales contacts during the day.   It is known that 50% of all these contacts result in signed contracts.   Assume that Mr. Smith makes 5 sales contacts on a given day.

a) What is the probability that Mr. Smith signs *exactly* one contract on that day?

b) What is the probability that he signs *at least* one contract?

32. An automobile agency sells either two, three, or five cars per day, with probability of .2, .3, and .5, respectively. The selling price of an automobile is either $4000, $5000, or $8000, with probability of .6, .2, and .2, respectively.

   a) What is the expected number of automobiles sold per day?

   b) What is the expected price per automobile sold?

   c) Determine the expected sales per day for the agency.

33. A certain type of an electronic tube has an average life of 100 hours with a standard deviation of 10 hours.

   a) What is the probability that a single tube will fail before 95 hours of operation time?

   b) What is the probability that in a random sample of five tubes, three tubes will fail before 95 hours of operation time?

34. The owner of a newspaper stand estimates his daily sales for a certain newspaper as follows:

Number of newspapers sold	Probability
30	1/4
40	1/2
50	1/4

   a) What is the probability that the owner will sell 30 newspapers in two consecutive days?

   b) What is the probability of selling 30 papers in two days out of five days?

   c) What is the expected number of newspapers sold per day?

5
Sampling

Sampling

In many problems it is either impossible or unnecessary to have complete population data. Only a part of the population data can provide the information necessary to make a decision or to test a hypothesis concerning the entire population. It is not practical, for example, to burn all electric bulbs of a certain brand in order to determine the bulb's average life. The average life may be estimated by testing only 100 bulbs.

Any part of a population is called a sample. The objective of sampling is to select that part which is representative of the entire population. Various sampling techniques and designs have been developed in attempts to improve the representation. Each design is suitable for certain types of problems.

Sample designs are classified into *probability samples* and *nonprobability samples*. A sample is a probability sample if each unit in the population is given *some* chance of being selected in the sample. Furthermore, the probability of selecting each unit must be known. Contrary to the general belief, therefore, it is not necessary that each unit in the population be given an *equal* chance of being selected in the sample, in order to obtain a probability sample. However, each unit in the population must be given a chance of being selected, and this chance must be known to the investigator.

Since the probability of selecting each population unit into the sample is known, a statistician can then utilize the various rules and laws of probability to evaluate the reliability of conclusions derived from probability samples. In other words, when a sample is a probability sample, the risk of incorrect decisions and conclusions can be measured using the theory of probability.

A sample is a nonprobability sample when some units in the population are not given any chance of being selected and if the probability of selecting any unit into the sample cannot be determined or is not known.

Once again, when a decision is made on the basis of sample information, there is always the risk of error. Thus, if 1000 items selected from the output of machines *A* and *B* showed 20% and 24% defectives, respectively, there is always a risk in concluding that machine *A* is better than machine *B*. It is always possible that the two machines are of the same efficiency, and the observed difference is due to chance or to sampling fluctuations. The risk of arriving at the wrong conclusion can be measured only when the sample is a probability sample; it cannot be determined for a nonprobability sample. For this reason, only probability samples are subject to statistical treatment and analysis. The simple random sample, the most common type of probability sample, is treated in great detail in this chapter. Other sample designs are surveyed briefly.

SIMPLE RANDOM SAMPLE

In order to illustrate the procedure used in selecting a simple random sample, let our population of interest consist of five people: Adams, Bakers, Camel, Daniels, and Edwards. Furthermore, let us assume that we desire to select a committee (a sample) of three members from this population, using a simple random selection.

How many possible samples of *three* can we actually select from this population? Referring to the names by their initials, we have 10 possible samples:

ABC	*ACD*	*BCD*
ABD	*ACE*	*BCE*
ABE	*ADE*	*BDE*
	CDE	

Our method of selection would result in a simple random sample if each of these 10 samples were given an equal chance of being selected. One way of accomplishing this is to write each of the 10 combinations on a slip of paper, thoroughly mix the slips, and draw one slip from the bowl. If the combination drawn is *CDE*, our sample consists of Camel, Daniels, and Edwards. Although this method of selecting a simple random sample is straightforward, statisticians prefer an alternative procedure that is equally sound but easier to apply.

It has been found that each of the 10 possible samples is given an equal chance of being selected if each of the five names is written on a slip of paper, the slips thoroughly mixed, and three slips drawn in succession from that bowl. This alternative method eliminates the task of enumerating all possible samples of a given size that can be selected from a population—an enormous task when the number of elementary units in the population is very large.

It may be interesting to prove that both methods of selecting a simple random sample give each possible sample an equal chance of being selected. Under the first method, each of the 10 possible samples is given the same chance of being selected. The probability of selecting any of these samples, say *ABC* (Adams, Bakers, and Camel), is $\frac{1}{10}$. To prove our proposition, therefore, we must show that the probability of selecting a committee consisting of Adams, Bakers, and Camel under the second method of selection is also $\frac{1}{10}$.

Using the second method of sample selection, there are six ways in which one can obtain a sample consisting of Adams, Bakers, and Camel (not necessarily in that order). The resulting sample is Adams, Bakers, and Camel if Adams is selected on the first draw, Bakers is selected on the second draw, and Camel is selected on the third draw. The resulting sample is also Adams, Bakers, and Camel if Adams is selected on the first draw, Camel is selected on the second draw, and Bakers is selected on the third draw. The six possible ways of selecting a committee consisting of Adams, Bakers, and Camels are listed below.

1. Adams, then Bakers, then Camel (*ABC*);
2. Adams, then Camel, then Bakers (*ACB*);
3. Bakers, then Adams, then Camel (*BAC*);
4. Bakers, then Camel, then Adams (*BCA*);
5. Camel, then Adams, then Bakers (*CAB*);
6. Camel, then Bakers, then Adams (*CBA*).

Thus each of the above six events represents the selection of a committee (a simple random sample) consisting of Adams, Bakers, and Camel. Since the six events are mutually exclusive, the probability of selecting a committee consisting of Adams, Bakers, and Camel is equal to the sum of the probabilities of these six events. Let us determine the probability of each event:

$$P(A \cap B \cap C) = \tfrac{1}{5} \times \tfrac{1}{4} \times \tfrac{1}{3} = \tfrac{1}{60}$$
$$P(A \cap C \cap B) = \tfrac{1}{5} \times \tfrac{1}{4} \times \tfrac{1}{3} = \tfrac{1}{60}$$
$$P(B \cap A \cap C) = \tfrac{1}{5} \times \tfrac{1}{4} \times \tfrac{1}{3} = \tfrac{1}{60}$$
$$P(B \cap C \cap A) = \tfrac{1}{5} \times \tfrac{1}{4} \times \tfrac{1}{3} = \tfrac{1}{60}$$
$$P(C \cap A \cap B) = \tfrac{1}{5} \times \tfrac{1}{4} \times \tfrac{1}{3} = \tfrac{1}{60}$$
$$P(C \cap B \cap A) = \tfrac{1}{5} \times \tfrac{1}{4} \times \tfrac{1}{3} = \tfrac{1}{60}$$

Probability of selecting the sample *ABC* = $\frac{6}{60} = \frac{1}{10}$.

The Use of Random Numbers

Using slips of paper and a bowl in the selection of a simple random sample is a rather crude approach, which becomes even more cumbersome and unmanageable as the number of elementary units in the population gets larger and larger. In practice, therefore, a table of random numbers is used instead of the "fish bowl."

What are *random numbers*? They are the digits 0, 1, 2, 3, 4, 5, 6, 7, 8, 9 scrambled in a particular fashion. Table 5–1 is an example of a table of random numbers. The various digits are arranged in such a way that all digits appear with approximately the same frequency in the table. Furthermore any given digit has no relationship to the digit above, below, to the right, or to the left of it. In other words, the various digits are scattered at random. However, for convenience, 40 digits, in groups of five, are placed on each line.

A practical sampling problem will best illustrate the use of the table of random numbers in selecting a simple random sample. Let us assume that we wish to select 10 accounts receivable from a ledger containing 6532 accounts.

As a first step, we must assign a serial number to each account in the ledger—in our case, a four-digit serial number. (If the number of accounts in the ledger had been 623, we would have assigned a three-digit serial number to each account. Thus the number of digits of which a serial number consists depends on the total number of accounts in the ledger.) We now assign 0001 to the first account, 0002 to the second account, 0123 to the 123rd account, and 6532 to the last account. Once all accounts in the ledger are serialized, we can begin to select our sample by reading the first four digits

Table 5–1

An illustrative page of random numbers

Line	(1)	(2)	(3)	(4)	(5)	(6)	(7)	(8)
1	78994	36244	02673	25475	84953	61793	50243	63423
2	04909	58485	70686	93930	34880	73059	06823	80257
3	46582	73570	33004	51795	86477	46736	60460	70345
4	29242	89792	88634	60285	07190	07795	27011	85941
5	68104	81339	97090	20601	78940	20228	22803	96070
6	17156	02182	82504	19880	93747	80910	78260	25136
7	50711	94789	07171	02103	99057	98775	37997	18325
8	39449	52409	75095	77720	39729	03205	09313	43545
9	75629	82729	76916	72657	58992	32756	01154	84890
10	01020	55151	36132	51971	32155	60735	64867	35424
11	08337	89989	24260	08618	66798	25889	52860	57375
12	76829	47229	19706	30094	69430	92399	98749	22081
13	39708	30641	21267	56501	95182	72442	21445	17276
14	89836	55817	56747	75195	06818	83043	47403	58266
15	25903	61370	66081	54076	67442	52964	23823	02718

(continued)

Table 5-1 (*Continued*)

Line	(1)	(2)	(3)	(4)	(5)	(6)	(7)	(8)
16	71345	03422	01015	68025	19703	77313	04555	83425
17	61454	92263	14647	08473	34124	10740	40839	05620
18	80376	08909	30470	40200	46558	61742	11643	92121
19	45144	54373	05505	90074	24783	86299	20900	15144
20	12191	88527	58852	51175	11534	87218	04876	85584
21	62936	59120	73957	35969	21598	47287	39394	08778
22	31588	96798	43668	12611	01714	77266	55079	24690
23	20787	96048	84726	17512	39450	43618	30629	24356
24	45603	00745	84635	43079	52724	14262	05750	89373
25	31606	64782	34027	56734	09365	20008	93559	78384
26	10452	33074	76718	99556	16026	00013	78411	95107
27	37016	64633	67301	50949	91298	74968	73631	57397
28	66725	97865	25409	37498	00816	99262	14471	10232
29	07380	74438	82120	17890	40963	55757	13492	68294
30	71621	57688	58256	47702	74724	89419	08025	68519
31	03466	13263	23917	20417	11315	52805	33073	07723
32	12692	32931	97387	34822	53775	91674	76549	37635
33	52192	30941	44998	17833	94563	23062	95725	38463
34	56691	72529	66063	73570	86860	68125	40436	31303
35	74952	43041	58869	15677	78598	43520	97521	83248
36	18752	43693	32867	53017	22661	39610	03796	02622
37	61691	04944	43111	28325	82319	65589	66048	98498
38	49197	63948	38947	60207	70667	39843	60607	15328
39	19436	87291	71684	74850	76501	93456	95714	92518
40	39143	64893	14606	13543	09621	68301	69817	52140
41	82244	67549	76491	09761	74494	91307	64222	66592
42	55847	56155	42878	23708	97999	40131	52360	90390
43	94095	95970	07826	25991	37584	56966	68623	83454
44	11751	69469	25521	44097	07511	88976	30122	67542
45	69902	08995	27821	11758	64989	61902	32121	28165
46	21850	25352	25556	92161	23592	43294	10479	37879
47	75850	46992	25165	55906	62339	58958	91717	15756
48	29648	22086	42581	85677	20251	39641	65786	80689
49	82740	28443	42734	25518	82827	35825	90288	32911
50	36842	42092	52075	83926	42875	71500	69216	01350

A portion of page 5 of *Table of 105,000 Random Decimal Digits*, constructed by H. Burke Horton and R. Tynes Smith, III, for the Bureau of Transport Economics and Statistics, Interstate Commerce Commission. Reproduced by permission.

from the first line of the table of random digits. The number is 7899. Since no account in the ledger corresponds to this number, we move down and read the first four digits of the second line. The number 0490 corresponds to the 490th account in the ledger, which becomes our first account in the sample. Continuing in the same fashion, we obtain the other nine accounts:

4658, 2924, 1715, 5071, 3944, 0102, 0833, 3970, 2590.

(Note that when a serial number appears twice, it must be ignored on the second appearance.)

The table of random numbers could have been used in some other manner. We could have read our first four digits from any place in the table. Furthermore, in selecting our second four digits, we could have moved sideways or crosswise, so long as our movement remained systematic.

Before concluding this section, we must point out that in order to select a simple random sample, one must obtain a list of all elementary units in the population. Such a list is called a *population frame*. Ideally, a population frame should contain each elementary unit in the population, while excluding duplication of any unit.

In many market research surveys, it is somewhat difficult to obtain a good population frame. Researchers have often used various "workable" frames such as city directories, tax assessors' lists, telephone directories, voters' registration records, auto registrations, or lists prepared by name-gathering agencies. Although some of these frames such as city directories and tax assessors' lists are better than others, all such frames are subject to considerable imperfections which limit their use and introduce some hazards.

Our discussion has dealt so far with the selection of a simple random sample from a finite population for which it was possible to prepare a list of all elementary units. But how can we select a simple random sample from an infinite population for which such a list cannot be obtained? How can we select, for example, a simple random sample of 100 electric bulbs from a production process that produces thousands day after day? The answer to this question is simple: a production process is a random process, and any 100 bulbs are a simple random sample of the entire output. Fortunately, most infinite populations are generated by, more or less, random processes, and any number of elementary units is a simple random sample of all elementary units that can be produced by a process.

A Statistic

A *statistic* is a number that describes a certain aspect of a sample. The arithmetic mean, the median, the mode, the proportion, and the standard deviation are all statistics if they are computed from a sample (only part of the population).

A clear distinction must be made between a parameter and statistic. While a parameter describes a certain aspect of a population, a statistic describes a certain aspect of a sample. The arithmetic mean is a parameter if it was computed from complete population data. However, if the arithmetic mean was computed from a sample, it is called a statistic.

Since a parameter is computed from complete population data, its value is constant. The value of a statistic, however, varies from sample to sample. Thus a statistic is a *variable*. To illustrate this point, let our population of interest consist of the ages of the following five houses:

$$x_1 = 2 \text{ years}, \quad x_3 = 4 \text{ years},$$
$$x_2 = 2 \text{ years}, \quad x_4 = 5 \text{ years},$$
$$x_5 = 2 \text{ years}.$$

The arithmetic mean of the population is constant and is equal to 3:

$$\mu = \frac{2 + 2 + 4 + 5 + 2}{5} = \frac{15}{5} = 3 \text{ years}.$$

If we select two houses as a simple random sample, and these two happen to be the first and the second house, the arithmetic mean of this sample is 2 years. If our sample consists of the third and the fourth house, the arithmetic mean is equal to 4.5 years. Thus the arithmetic mean differs from sample to sample.

The most popular statistics are the arithmetic mean, the proportion, and the standard deviation. The arithmetic mean of a *simple random sample*, denoted by $\bar{x}$, is calculated in the same manner as the arithmetic mean of a population. Symbolically,

$$\bar{x} = \frac{\sum\limits_{i=1}^{n} x_i}{n},$$

where x_i indicates the ith member of the sample and n refers to the sample size. The proportion of a *simple random sample*, denoted by p, is also calculated in the same manner as the population proportion.

The standard deviation of a *simple random sample*, on the other hand, is calculated in a slightly different way from the standard deviation of a population. Denoted by s, the standard deviation of a sample can be written symbolically as

$$s = \sqrt{\frac{\sum\limits_{i=1}^{n} (x_i - \bar{x})^2}{n - 1}} \, * \, .$$

---

* It is beyond the scope of this text to explain why the sum of squared deviations is divided by $n - 1$ instead of by n.

OTHER PROBABILITY SAMPLES

Stratified Sample

In a stratified sample the population is divided into a number of groups or strata. The objective of such stratification is to obtain a relatively more homogeneous group with respect to the characteristic under study. A simple random sample is selected from each group or stratum, and these subsamples are combined into one grand sample.*

Thus to estimate the annual family expenditure on rent in Los Angeles, for example, it might be desirable to divide Los Angeles families into three groups: those earning less than $5000 a year, those with annual incomes ranging from $5000 to $10,000, and those whose annual earnings exceed $10,000. After the population is divided into these three strata, a simple random sample is selected from each group, and these three subsamples are combined into one grand sample.

Provided our stratification is really successful in dividing the population into more homogeneous groups with respect to annual rent expenditures, we will obtain a more accurate estimate of average annual rent using a stratified sample than a simple random sample selected from the population at large, assuming that the two samples are of the same size. Stratification has another related advantage: information is obtained concerning the individual strata as well as the entire population. It must be mentioned, however, that the cost of obtaining a stratified sample is higher than that of obtaining a simple random sample of the same size, since stratification entails the additional work of classifying the population into the various groups.

NONPROBABILITY SAMPLES

Nonprobability samples are characterized by the fact that the chance of including any elementary unit of the population in the sample cannot be determined. For this reason, there is no means of measuring the risk of making erroneous conclusions derived from nonprobability samples. Since the reliability of the results of nonprobability samples cannot be measured, such samples do not lend themselves to statistical treatment and analysis. Convenience and judgment samples are the most common types of nonprobability samples.

---

* The formulas used in computing the various statistics (mean, standard deviation, etc.) from a stratified sample are entirely different from those used to compute the same statistics for a simple random sample.

Convenience Sample

The elementary units to be included in a convenience sample are chosen because they can be reached more easily and conveniently than other units. A manufacturer of a certain brand of food who wants to get some idea about consumer preference for his product may use a convenience sample to obtain such information. He may insert a refund coupon (requiring the customer to provide the needed information) in a number of packages of his product. Returned coupons provide the manufacturer with a convenience sample. Samples of this sort are popular among manufacturers of cigars and pipe tobacco.

Judgment Sample

The elementary units to be included in a judgment sample are chosen by an expert on the basis that these units are representative of the majority of elementary units in the population. Judgment samples are used whenever such limitations as time and cost make it necessary to choose a very small sample.

EXERCISES

1. The final grades of 112 students enrolled in Elementary Statistics are listed in Table 5–2. Use the table of random digits (Table 5–1) to select a random sample of 10 grades and calculate the sample mean.

2. A random sample of 100 accounts receivable is selected from a retail store ledger. The sample results are summarized below.

$$x = \text{the amount of an account receivable in dollars,}$$
$$\sum x = \$12,000,$$
$$\sum (x - \bar{x})^2 = \$89,100.$$

Compute the mean and the standard deviation of the sample.

3. The nicotine contents of five cigarettes of a certain brand, measured in milligrams, are as follows:

$$21, \quad 19, \quad 23, \quad 19, \quad 23.$$

Determine the mean nicotine content in the sample. Determine the standard deviation.

4. Under what circumstances is a stratified sample more desirable than a simple random sample?

5. What is the difference between a statistic and a parameter?

6. J. Paul Betty, an investor, wishes to invest a sum of $100,000 in either of two stocks: A or B. Mr. Betty is a prudent man. Before investing this large sum of money, he decides to engage the services of an investment counselor, Mr.

Table 5–2

Student	Grade	Student	Grade	Student	Grade	Student	Grade
1	78	29	89	57	67	85	58
2	46	30	81	58	56	86	32
3	68	31	94	59	95	87	43
4	50	32	52	60	69	88	71
5	39	33	82	61	46	89	14
6	75	34	55	62	22	90	76
7	76	35	89	63	42	91	42
8	39	36	47	64	70	92	93
9	89	37	55	65	88	93	51
10	71	38	61	66	97	94	60
11	61	39	92	67	82	95	77
12	80	40	88	68	75	96	72
13	45	41	59	69	76	97	51
14	62	42	96	70	36	98	56
15	45	43	96	71	56	99	75
16	66	44	64	72	66	100	54
17	71	45	33	73	58	101	68
18	52	46	64	74	73	102	90
19	56	47	97	75	43	103	51
20	74	48	74	76	84	104	43
21	61	49	57	77	84	105	56
22	82	50	30	78	76	106	99
23	94	51	72	79	67	107	53
24	69	52	43	80	82	108	92
25	74	53	43	81	58	109	55
26	82	54	63	82	97	110	85
27	58	55	87	83	44	111	25
28	73	56	64	84	66	112	83

Harry Goldstein. Mr. Goldstein consequently decides to gather pertinent information concerning the performance of the two stocks over the last two years, and the following information is obtained:

Stock A	Stock B
$\bar{x} = \$80$	$\bar{x} = \$80$
$s = \$30$	$s = \$5$

Assuming that the current prices of the two stocks are the same, which stock should Mr. Goldstein recommend and why?

7. A random sample of 9 one-pound packages of Oscar Mire's hot dogs is taken to estimate the amount of meat contained in his products. The sample revealed the following figures (in ounces):

$$14, 5, 6, 11, 13, 14, 14, 5, 8.$$

a) Find the sample mean.

b) Find the sample standard deviation.

c) Determine the proportion of packages, p, that contain at least 10 ounces of meat.

6
Estimating the
Population
Mean
from a
Sample

Estimating the Population Mean from a Sample

Introduction

The principal objective in many statistics problems is to determine the value of a population parameter such as the arithmetic mean μ. To achieve this objective, the statistician rarely calculates the value of the parameter using the complete population data, but instead, he tries to estimate its value from a carefully selected sample.

The statistician thus uses the sample mean $\bar{x}$ as an estimate of the population mean μ. He also uses the sample proportion p as an estimate of the population proportion π, and the sample standard deviation s as an estimate of the population standard deviation σ.

But how reliable is $\bar{x}$ as an estimate of μ? The value of μ is constant. On the other hand, the value of $\bar{x}$ varies from sample to sample. The reliability of $\bar{x}$ as an estimate of μ cannot be determined unless we can precisely describe the behavior of $\bar{x}$ or its pattern of variability from sample to sample. If μ is known, what are the possible values that $\bar{x}$ can assume in different samples? If the last question can be answered satisfactorily, the reliability of $\bar{x}$ as an estimate of μ can easily be assessed.

The first part of this chapter describes the behavior of $\bar{x}$ in relation to μ. The latter part deals with the problem of estimating μ from sample data.

PATTERN OF BEHAVIOR OF SAMPLE MEANS

The relationship between the population mean and the means of the various samples that can be selected from such a population can best be illustrated by an actual sampling operation from a known population. Let the known population consist of the heights of five different plants. The individual heights of these plants are

$$x_1 = 2 \text{ feet}, \qquad x_3 = 6 \text{ feet},$$
$$x_2 = 4 \text{ feet}, \qquad x_4 = 8 \text{ feet},$$
$$x_5 = 10 \text{ feet},$$

where x_1 is the height of the first plant, x_2 that of the second plant, etc. The population mean μ and the population standard deviation σ are calculated below.

x_i	$(x_i - \mu)$ $(x_i - 6)$	$(x_i - \mu)^2$
2	-4	16
4	-2	4
6	0	0
8	$+2$	4
10	$+4$	16
$\sum x = 30$		$\sum (x - \mu)^2 = 40$

$$\mu = \frac{\sum x}{N} = \frac{30}{5} \qquad\qquad \sigma = \sqrt{\frac{\sum (x - \mu)^2}{N}}$$

$$= 6 \text{ feet} \qquad\qquad = \sqrt{\frac{40}{5}} = \sqrt{8}$$

$$= 2.83 \text{ feet}$$

Thus the mean of our population is 6 feet and the standard deviation is 2.83 feet.

Now let us select *all* possible samples of size *two* from this population, and compute the mean of each sample.

Possible samples	Sample means, feet
x_1 and x_2 (2, 4)	$\bar{x}_1 = 3$
x_1 and x_3 (2, 6)	$\bar{x}_2 = 4$
x_1 and x_4 (2, 8)	$\bar{x}_3 = 5$
x_1 and x_5 (2, 10)	$\bar{x}_4 = 6$
x_2 and x_3 (4, 6)	$\bar{x}_5 = 5$
x_2 and x_4 (4, 8)	$\bar{x}_6 = 6$
x_2 and x_5 (4, 10)	$\bar{x}_7 = 7$
x_3 and x_4 (6, 8)	$\bar{x}_8 = 7$
x_3 and x_5 (6, 10)	$\bar{x}_9 = 8$
x_4 and x_5 (8, 10)	$\bar{x}_{10} = 9$

There are ten possible samples of size two that can be selected from our population of five plants. Consequently, there are ten possible sample means. How representative are these sample means of the true population mean? While the means of the first and the tenth samples ($\bar{x}_1 = 3$, $\bar{x}_{10} = 9$) differ from the mean of our population by 3 feet, the means of the fourth and sixth samples ($\bar{x}_4 = 6$, $\bar{x}_6 = 6$, $\mu = 6$) are exactly equal to the population mean. The means of the remaining six samples are either above or below the true mean by 1 or 2 feet.

To examine these ten sample means in a greater detail, let us first calculate their arithmetic mean or their average. The average of all sample means is noted by $E(\bar{x})$:

$$E(\bar{x}) = \frac{\bar{x}_1 + \bar{x}_2 + \bar{x}_3 + \bar{x}_4 + \bar{x}_5 + \bar{x}_6 + \bar{x}_7 + \bar{x}_8 + \bar{x}_9 + \bar{x}_{10}}{10}$$

$$= \frac{3 + 4 + 5 + 6 + 5 + 6 + 7 + 7 + 8 + 9}{10}$$

$$= \frac{60}{10} = 6 = \mu.$$

We can thus say that the average of all possible sample means is equal to the population mean, or $E(\bar{x}) = \mu$. However, individual sample means may either overstate or understate the population mean, but their arithmetic mean is always equal to the population mean.

Continuing our examination of the properties of the sample means, let us calculate the standard deviation of the ten possible sample means. Their standard deviation, called the *standard error of the mean* (denoted by $\sigma_{\bar{x}}$) is computed below.

$\bar{x}$'s	$(\bar{x} - \mu)$	$(\bar{x} - \mu)^2$
3	-3	9
4	-2	4
5	-1	1
6	0	0
5	-1	1
6	0	0
7	$+1$	1
7	$+1$	1
8	$+2$	4
9	$+3$	9

$E(\bar{x}) = \mu = 6$ 　　 $\sigma_{\bar{x}} = \sqrt{\dfrac{\Sigma(\bar{x} - \mu)^2}{\text{No. of samples}}}$ 　　 $\Sigma(\bar{x} - \mu)^2 = 30$

$$= \sqrt{\tfrac{30}{10}} = \sqrt{3}$$

$$= 1.73 \text{ feet}$$

The standard error of the mean indicates the "average" difference between the various $\bar{x}$'s and μ.* The first sample mean differs from the population mean by 3 feet; the second sample mean by 2 feet; the third by 1 foot, etc. On the average, each sample mean differs from the population mean by 1.73 feet.

A small value for $\sigma_{\bar{x}}$ indicates two facts: first, that the various values of $\bar{x}$ are close to each other; second, that the average difference between these $\bar{x}$'s and μ is small. Consequently, any one $\bar{x}$ is a good estimate of μ.

Although the standard error of the mean measures the average difference between all possible sample means and the population mean, it is not necessary to consider all the samples in order to determine the error. Fortunately, the standard error of the mean can be determined if the standard deviation of the original population is known. It has been found that

$$\sigma_{\bar{x}} = \frac{\sigma}{\sqrt{n}} \frac{\sqrt{N - n}}{\sqrt{N - 1}},$$

where σ is the standard deviation in the population, N is the population size, and n is the sample size.

In reference to our population of five plants with $\mu = 6$ feet and $\sigma = 2.83$ feet, the standard error of the mean for all possible samples of *size two* is

$$\sigma_{\bar{x}} = \frac{\sigma}{\sqrt{n}} \sqrt{\frac{N - n}{N - 1}}$$

$$= \frac{2.83}{\sqrt{2}} \sqrt{\frac{5 - 2}{5 - 1}} = \frac{2.83}{\sqrt{2}} \sqrt{\frac{3}{4}}$$

$$= \frac{2.83}{1.414} \frac{1.73}{2} = \frac{(2.83)(1.73)}{(2.83)}$$

$$= 1.73 \text{ feet.}$$

The above value for $\sigma_{\bar{x}}$ is identical to the one previously obtained.

Thus we can say that if all possible samples of size n are selected from a given population, then

$$E(\bar{x}) = \mu, \quad \text{and} \quad \sigma_{\bar{x}} = \frac{\sigma}{\sqrt{n}} \sqrt{\frac{N - n}{N - 1}}.$$

The term $\sqrt{(N - n)/(N - 1)}$, called the *finite population correction*, is approximately equal to 1 whenever we are sampling from an infinite population. If our population is finite but the sample size is small in relation to the

---

*   To be specific, $\sigma_{\bar{x}}$ is the square root of the average square difference.

population (the sample size is 10% or less of the population size), the finite population correction will also approach unity. If in a certain problem, for example,

$$N = 1001, \qquad n = 49,$$

then the finite population correction is

$$\sqrt{\frac{N-n}{N-1}} = \sqrt{\frac{1001-49}{1001-1}} = \sqrt{\frac{952}{1000}}$$

$$= \sqrt{.952} = .9757.$$

In the following discussion, whenever a population is finite, it is assumed that our sample is less than 10% of the population (a very realistic assumption indeed). Consequently, the finite population correction approaches 1, and it can, therefore, be completely ignored. Under such circumstances, the standard error of the mean is equal to the population standard deviation divided by the square root of the sample size. Symbolically,

$$\sigma_{\bar{x}} = \frac{\sigma}{\sqrt{n}}.$$

It is very interesting to observe that the standard error of the mean gets smaller and smaller as the sample size gets larger and larger. As the sample size is increased, the various sample means become more and more uniform in value and, consequently, any one sample mean is a good estimate of the population mean. In other words, a large sample is more reliable than a small one.

Concluding our examination of the properties of the sample mean, we may add a final but important property. The sample means are approximately normally distributed whenever the sample size is 30 or more. It must be emphasized that this statement is true even when the samples are obtained from a parent population which is not normally distributed. For example, the first diagram in Fig. 6–1 represents a parent population which looks like a rectangle (called rectangular distribution). Now, if repeated samples of size two are drawn from this population, the resulting sample means will have a distribution which looks like a triangle (see second diagram in Fig. 6–1). However, as the sample size increases, the distribution of the sample means becomes more and more normal. The third and fourth diagrams of Fig. 6–1 represent the distributions of sample means when $n=5$ and $n=30$, respectively. Hence, as shown in the fourth diagram of Fig. 6–1, when repeated samples of size 30 are drawn from our rectangular population, the resulting sample means are approximately normally distributed.

The parent populations in Fig. 6–2 and Fig. 6–3 are not in fact normally distributed. However, the distributions of the sample means are approximately normally distributed when $n = 30$. When the parent population is normally distributed, on the other hand (Fig. 6–4), the sample means are normally distributed no matter *how* small the sample size is.

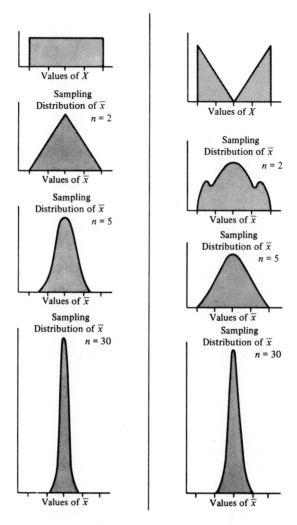

Fig. 6–1   Parent population*   Fig. 6–2   Parent population.*

In conclusion we can therefore state: If all possible samples of size n ($n \geq 30$) are selected from a given population, then the various sample means are approximately normally distributed, have an average equal to the population mean, and have a standard error which is equal to the popu-

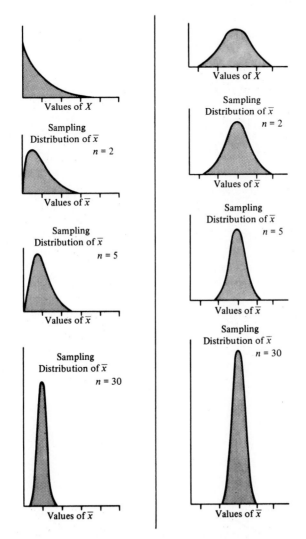

Fig. 6–3   Parent population.*   **Fig. 6–4**   Parent population.*

*   Reproduced with permission from Kurnow, Glasser, Ottman, *Statistics for Business Decisions* (Homewood, Ill.: Richard D. Irwin, Inc., © 1959), pp. 182–183.

lation standard deviation divided by the square root of the sample size (Fig. 6–5).

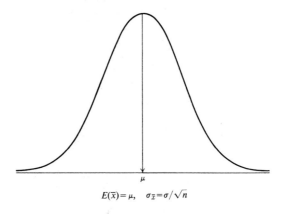

$$E(\bar{x}) = \mu, \quad \sigma_{\bar{x}} = \sigma/\sqrt{n}$$

Figure 6–5

Example 1

Two thousand different simple random samples of 100 students each are selected from a very large university, where the average age of the student, μ, is 20 years and the standard deviation σ is 2 years. There will be 2000 sample means. The mean of the first sample may be 22 years; the mean of the second sample, 19; the third sample, 20, etc. However, the 2000 sample means will be approximately normally distributed, have an average of 20 years, and a standard deviation or a standard error of $2/\sqrt{100}$ or 0.2 years (Fig. 6–6).

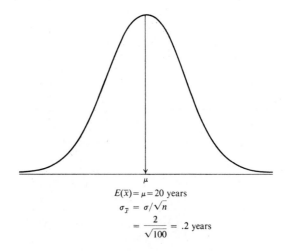

$$E(\bar{x}) = \mu = 20 \text{ years}$$
$$\sigma_{\bar{x}} = \sigma/\sqrt{n}$$
$$= \frac{2}{\sqrt{100}} = .2 \text{ years}$$

Figure 6–6

The mean of each of these 2000 samples may or may not be 20 years. However, according to Table A, 68% of these sample means will be in the range

$$\mu \pm 1\sigma_{\bar{x}}, \qquad 20 \pm .2,$$

or between 19.8 years and 20.2 years (Fig. 6–7).

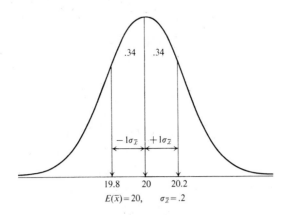

Figure 6–7

In the same manner, according to Table A, 95% of all sample means will be in the range

$$\mu \pm 1.96\sigma_{\bar{x}}, \qquad 20 \pm 1.96(.2), \qquad 20 \pm .4,$$

or between 19.6 years and 20.4 years (Fig. 6–8).

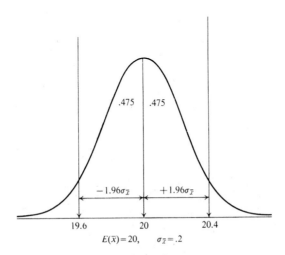

Figure 6–8

Example 2

The average account receivable in a ledger is $125 with a standard deviation of $24 ($\mu = 125$, $\sigma = 24$). What is the probability that a simple random sample of 36 accounts selected from this ledger will show an average of $115 or less per account?

If all possible samples of size 36 are selected from this ledger, then all $\bar{x}$'s are normally distributed, have an average of $125, and a standard error of $24/\sqrt{36}$, or $4 (Fig. 6–9).

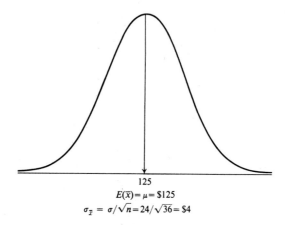

$$E(\bar{x}) = \mu = \$125$$
$$\sigma_{\bar{x}} = \sigma/\sqrt{n} = 24/\sqrt{36} = \$4$$

Figure 6–9

The probability that the mean of a sample of size 36 is $115 or less can be calculated as follows:

$$z = \frac{\bar{x} - \mu}{\sigma_{\bar{x}}} = \frac{115 - 125}{4} = -2.5.$$

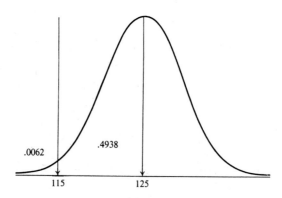

Figure 6–10

Therefore

$$P(\bar{x} \leq 115) = .5000 - .4938 = .0062.$$

See Fig. 6–10.

Example 3

The mean IQ scores of all students attending this college is 110 with a standard deviation of 10.

a) If the IQ scores are normally distributed, what is the probability that the score of any one student is greater than 112?

b) What is the probability that the mean score in a random sample of 36 students is greater than 112?

c) What is the probability that the mean score in a random sample of 100 students is greater than 112?

Solution (a): The probability that the IQ score of any one student is greater than 112 is calculated as follows:

$$z = \frac{x - \mu}{\sigma}$$

$$= \frac{112 - 110}{10} = .2.$$

Therefore

$$P(x > 112) = .5000 - .0793$$
$$= .4207.$$

We must note here that the normal distribution in Fig. 6–11 describes a *real* population. This real population consists of the IQ scores of all students

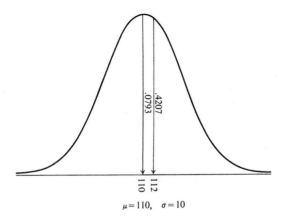

$\mu = 110, \quad \sigma = 10$

Figure 6–11

attending this college. The mean of this population is 110 and its standard
deviation is 10 ($\mu = 110$ and $\sigma = 10$).

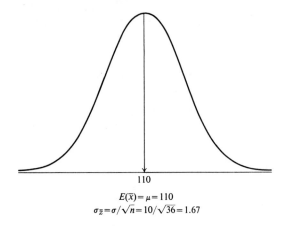

$$E(\bar{x}) = \mu = 110$$
$$\sigma_{\bar{x}} = \sigma/\sqrt{n} = 10/\sqrt{36} = 1.67$$

Figure 6–12

Solution (b):   The *means* of the various samples of 36 students are described
by the normal distribution shown in Fig. 6–12.   The probability that the mean
IQ score in a random sample of 36 students is greater than 112 can now be
calculated as follows (Fig. 6–13):

$$z = \frac{\bar{x} - \mu}{\sigma_{\bar{x}}} = \frac{112 - 110}{1.67} = 1.2.$$

Therefore

$$P(\bar{x} > 112) = .5000 - .3849$$
$$= .1151.$$

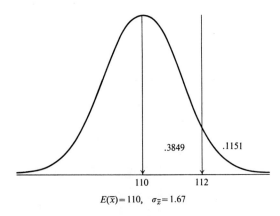

$$E(\bar{x}) = 110, \quad \sigma_{\bar{x}} = 1.67$$

Figure 6–13

It may be useful to compare the two normal distributions used in parts (a) and (b) of this example. Once again, the normal distribution in (a) describes a real population, consisting of the IQ scores of all students. The mean of this population is 110 and the standard deviation is 10 ($\mu = 110$ and $\sigma = 10$).

The normal distribution in (b), on the other hand, describes a *theoretical population*, consisting of the means of all possible random samples of 36 students that can be selected from this college. The mean of this population is also 110, but its standard deviation, called standard error, is only 1.67 [$E(\bar{x}) = 110$ and $\sigma_{\bar{x}} = 1.67$].

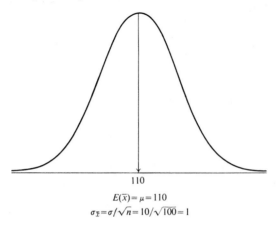

$$E(\bar{x}) = \mu = 110$$
$$\sigma_{\bar{x}} = \sigma/\sqrt{n} = 10/\sqrt{100} = 1$$

Figure 6–14

Solution (c): The means of the various samples of 100 students are described by the normal distribution shown in Fig. 6–14. The probability that the

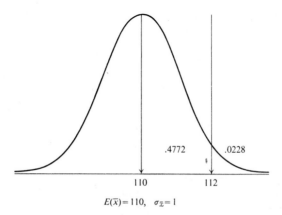

.4772   .0228

110   112

$$E(\bar{x}) = 110, \quad \sigma_{\bar{x}} = 1$$

Figure 6–15

mean IQ score in a random sample of 100 students is greater than 112 can now be determined as follows (Fig. 6–15):

$$z = \frac{\bar{x} - \mu}{\sigma_{\bar{x}}} = \frac{112 - 110}{1} = 2.0.$$

Therefore

$$P(\bar{x} > 112) = .5000 - .4772 = .0228.$$

We may finally note that the normal distribution in Fig. 6–15 also describes a theoretical population. This population, however, consists of the *means* of all possible random samples of 100 students that can be selected from this college. The mean of this population is 110 and its standard deviation is 1 [$E(\bar{x}) = 110$ and $\sigma_{\bar{x}} = 1$].

EXERCISES, SET 1

1. A population consists of children's ages in a family of four children. These ages are

$$X_1 = 2 \text{ years}$$
$$X_2 = 4 \text{ years}$$
$$X_3 = 6 \text{ years}$$
$$X_4 = 8 \text{ years}$$

   a) Determine the mean μ and the standard deviation σ of the population of the four children.
   b) List *all* possible samples of two children that can be selected from this family, and determine the arithmetic mean $\bar{x}$ for each sample.
   c) Determine the mean $E(\bar{x})$ and the standard deviation $\sigma_{\bar{x}}$ of the six sample means.
   d) Demonstrate that

$$\sigma_{\bar{x}} = \frac{\sigma}{\sqrt{n}} \sqrt{\frac{N - n}{N - 1}}.$$

2. The mean height of all soldiers in the U.S. Army is 69 inches with a standard deviation of 2 inches.

   a) Assume that very many random samples of 100 soldiers are selected and the mean height in each sample is computed. Determine an interval within which 90% of all sample means are expected to fall.
   b) Assume that very many random samples of 400 soldiers are selected and the mean height in each sample is computed. Determine an interval within which 90% of all sample means are expected to fall.
   c) Explain why the interval in (a) is larger than that in (b).

3. A radio manufacturer receives a shipment of 100,000 nine-volt batteries each week from a supplier. The following sampling rule is used in deciding whether to accept or reject the shipment: Measure the life of 36 batteries from each shipment received. If the mean life of the sample is 50 hours or more, accept the shipment; otherwise reject the shipment.

a) What is the probability of accepting a shipment that has a mean life of 49 hours with a standard deviation of 3 hours?

b) What is the probability of rejecting a shipment that has a mean life of 50.5 hours with a standard deviation of 3 hours?

c) What is the probability of rejecting a shipment that has a mean life of 50 hours? of accepting such a shipment?

4. A certain brand of tire has a mean life of 21,000 miles with a standard deviation of 800 miles.

   a) Assuming that the lives of the tires are normally distributed, what is the probability that any one tire would last less than 20,900 miles?

   b) What is the probability that the mean life of 64 tires is less than 20,900 miles?

   c) What is the probability that the mean life of 256 tires is less than 20,900 miles?

5. A food processor packages his coffee in one-pound cans. He uses the following sampling rule in deciding whether or not to stop the filling process: Select a sample of 64 cans every hour. If the mean weight of the sample is less than a critical level, L, stop and readjust the process. If the sample mean weight is equal to or greater than L, continue the operation. Determine the value of L so that the probability is only .05 of stopping a process which turns out, on the average, 16.3-ounce cans with a standard deviation of .1 ounce ($\mu = 16.3$ and $\sigma = .1$).

6. The life of a certain type of electronic tube is normally distributed with $\mu = 50$ hours and $\sigma = 5$ hours.

   a) What is the probability that a *single* tube would last longer than 55 hours?

   b) What is the probability that the mean, $\bar{x}$, of a simple random sample of 100 tubes would last longer than 51 hours?

   c) The probability is .90 that the mean of a random sample of 100 tubes is between L_1 and L_2. Find L_1 and L_2 using symmetrical limits about μ.

Estimating the Population Mean from a Large Sample*

In discussing the pattern of behavior of the sample means, we have learned that when all possible random samples of size n are selected from a population that has a mean of μ and a standard deviation of σ, then these sample means are approximately normally distributed. Furthermore, the various sample means have an arithmetic mean which is equal to μ and a standard error which is equal to $\sigma/\sqrt{n}$ (Fig. 6–16).

---

* A sample is regarded large when $n \geq 30$. Estimating the population mean from a sample of small size is discussed in a later section of this chapter.

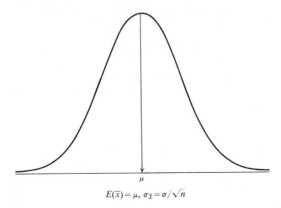

$$E(\bar{x}) = \mu, \ \sigma_{\bar{x}} = \sigma/\sqrt{n}$$

Figure 6–16

Since these sample means are normally distributed, we expect 95% of the $\bar{x}$'s to fall in the range defined by

$$\mu \pm 1.96\sigma_{\bar{x}} \quad \text{or} \quad \mu \pm 1.96 \frac{\sigma}{\sqrt{n}}.$$

(See Fig. 6–17). We can also conclude that if only one sample of size n is selected, there is a 0.95 chance that its mean $\bar{x}$ falls within the interval $\mu \pm 1.96\sigma_{\bar{x}}$. The probability that it falls outside this particular interval is .05.

Now let us assume that a simple random sample of size n is actually selected and its arithmetic mean calculated. The interval $\bar{x} \pm 1.96\sigma_{\bar{x}}$ *may* or *may not* contain μ. If the mean of the sample happens to be from within any part of the shaded area under the normal curve shown in Fig. 6–18 (selected from within the interval $\mu \pm 1.96\sigma_{\bar{x}}$), the interval $\bar{x} \pm 1.96\sigma_{\bar{x}}$ will certainly contain μ. (Figs. 6–18 and 6–19).

If on the other hand, $\bar{x}$ happened to be from the unshaded area under the

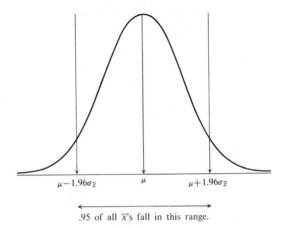

$$\mu - 1.96\sigma_{\bar{x}} \qquad \mu \qquad \mu + 1.96\sigma_{\bar{x}}$$

.95 of all $\bar{x}$'s fall in this range.

Figure 6–17

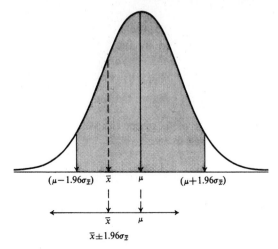

Figure 6–18

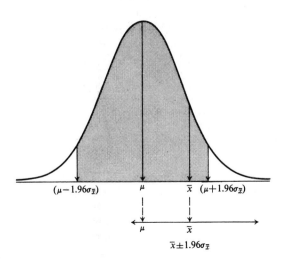

Figure 6–19

normal curve shown in Fig. 6–20(selected from outside the range $u \pm 1.96\sigma_{\bar{x}}$), the interval $\bar{x} \pm 1.96\sigma_{\bar{x}}$ would fail to contain μ.

We may, therefore, conclude that if all possible samples of size n are selected and the interval $\bar{x} \pm 1.96\sigma_{\bar{x}}$ is established for each sample, then .95 of all such intervals are expected to contain μ. The interval $\bar{x} \pm 1.96\sigma_{\bar{x}}$ or $\bar{x} \pm 1.96(\sigma/\sqrt{n})$ is called a *.95 confidence interval estimate of* μ.

In order to determine the above interval estimate of μ, we must know the standard deviation of the population. But the standard deviation of the population is not known. However, when the sample size is 30 or more, the standard deviation in the population, σ, may be approximated by the

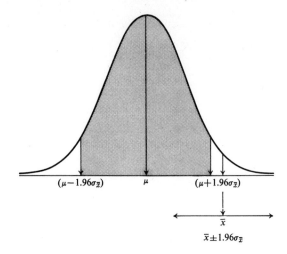

Figure 6–20

sample standard deviation s, and the exact confidence interval estimate $(\bar{x} \pm 1.96(\sigma/\sqrt{n}))$ is approximated by the interval $(\bar{x} \pm 1.96(s/\sqrt{n}))$. This interval is called an *approximate .95 confidence interval estimate of μ*. In the same manner, the range $\bar{x} \pm 2.58(s/\sqrt{n})$ is called an *approximate .99 confidence interval estimate of μ*. In general, the range $\bar{x} \pm z\dfrac{s}{\sqrt{n}}$ is an approximate confidence interval estimate of μ, where the value of z is determined by the degree of confidence required in estimating μ.

Example 4

A random sample of 100 students from this college showed an average IQ score of 112 with a standard deviation of 10.

a) Establish a .95 confidence interval estimate of the mean IQ score of all students attending this college.

b) Establish a .99 confidence interval estimate of the mean IQ score of all students in this college.

Solution (a): The .95 confidence interval estimate of μ is

$$\bar{x} \pm 1.96\ \frac{s}{\sqrt{n}};$$

$$112 \pm 1.96\ \frac{10}{\sqrt{100}}, \qquad 112 \pm 1.96\ (1), \qquad 112 \pm 1.96,$$

or between 110.04 and 113.96.

Solution (b): The .99 confidence interval estimate of μ is

$$\bar{x} \pm 2.58 \frac{s}{\sqrt{n}};$$

$$112 \pm 2.58 \frac{10}{\sqrt{100}}, \qquad 112 \pm 2.58\,(1), \qquad 112 \pm 2.58,$$

or between 109.42 and 114.58.

Example 5

A sample survey of 400 families in Reseda, California, shows an average annual expenditure of \$74 per family on shoes. The standard deviation of the sample is \$40.

a) Establish a .95 confidence interval estimate of the average annual expenditure on shoes by a family residing in Reseda, California.

b) What can we conclude with .99 confidence about the maximum error in our estimate if the average annual expenditure on shoes by a family in Reseda is estimated as \$74?

c) With what degree of confidence can we assert that the average annual family expenditure on shoes in Reseda is somewhere between \$71 and \$77?

Solution (a): The .95 confidence interval estimate of μ is

$$\bar{x} \pm 1.96 \frac{s}{\sqrt{n}};$$

$$74 \pm 1.96 \frac{40}{\sqrt{400}}, \qquad 74 \pm 1.96(2), \qquad 74 \pm 3.92.$$

Therefore, we can say with .95 confidence that the average annual family expenditure on shoes is somewhere between \$70.08 and \$77.92.

Solution (b): The .99 confidence interval estimate of μ is

$$\bar{x} \pm 2.58 \frac{s}{\sqrt{n}};$$

$$74 \pm 2.58 \frac{40}{\sqrt{400}}, \qquad 74 \pm 2.58(2), \qquad 74 \pm 5.16,$$

or between \$68.84 and \$79.16.

    Since we have added and subtracted the expression $2.58(40/\sqrt{400})$, or 5.16, from the sample mean, \$74, in order to establish a .99 confidence interval estimate of μ, we can assert with a probability of .99 that our maximum

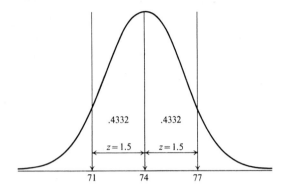

.4332   .4332

$z = 1.5$ $z = 1.5$

71   74   77

Figure 6–21

error is $5.16 if μ is estimated as $74. In general, the maximum expected error in a point sample estimate of μ is determined by the expression $z(s/\sqrt{n})$.

Solution (c): If μ is estimated to be between $71 and $77, then we must have added and subtracted $3 from the sample mean, $74, in order to obtain this interval estimate of μ. In other words, our interval estimate was obtained as follows:

$$\bar{x} \pm z \frac{s}{\sqrt{n}}, \qquad 74 \pm 3.$$

Therefore,

$$z \frac{s}{\sqrt{n}} = 3;$$

$$z \frac{40}{\sqrt{400}} = 3, \qquad z(2) = 3, \qquad z = 1.5.$$

Using the area under the normal curve in Table A we conclude that the range $71 to $77 is the 0.8664 confidence interval estimate of μ (see Fig. 6–21).

Example 6

A survey was conducted in 1980 to determine the average hourly earnings of a female sales clerk employed by a department store in metropolitan Los Angeles. A simple random sample of 225 female clerks was selected and the following information obtained:

$$X = \text{hourly wage rate earned by a female sales clerk,}$$

$$\Sigma x = \$450.00, \qquad \Sigma (x - \bar{x})^2 = \$2016.00.$$

What is the .99 confidence interval estimate of the average hourly wage rate?

Solution:

Step 1. $\bar{x} = \dfrac{\Sigma x}{n}$

$\quad = \dfrac{\$450.00}{225}$

$\quad = \$2.00$ (the sample mean).

Step 2. $s = \sqrt{\dfrac{\Sigma (x - \bar{x})^2}{n - 1}}$

$\quad = \sqrt{\dfrac{2016.00}{224}}$

$\quad = \sqrt{9.00} = 3.00$ (the standard deviation of the sample).

Step 3. $\bar{x} \pm 2.58 \dfrac{s}{\sqrt{n}}$;

$2.00 \pm 2.58 \dfrac{3.00}{\sqrt{225}}, \qquad 2.00 \pm 2.58\left(\dfrac{3.00}{15}\right), \qquad 2.00 \pm 2.58(.20), \qquad 2.00 \pm .52.$

Thus the average hourly earnings of a female sales clerk is estimated to be between $1.48 and $2.52.

Example 7

A department store has 10,000 customers' charge accounts. To estimate the total amount owed by its customers, it selected a random sample of 36 charge accounts which showed a mean amount of $150 per account and a standard deviation of $60 ($\bar{x} = 150$, $s = 60$). Establish a 0.95 confidence interval estimate of the *total amount* owed by customers to the department store.

Solution: This problem is solved in two steps. In the first step we will establish a 0.95 confidence interval estimate of the average amount owed per customer (a 0.95 confidence interval estimate of μ). In the second step we will multiply the range established for μ by the number of customers to obtain a range for the total amount owed by customers.

Step 1. The 0.95 confidence-interval estimate of μ is

$$\bar{x} \pm 1.96\left(\frac{s}{\sqrt{n}}\right) = \$150 \pm 1.96\left(\frac{\$60}{\sqrt{36}}\right)$$
$$= \$150 \pm 1.96\,(\$10)$$
$$= \$150 \pm \$19.60,$$

or between $130.40 and $169.60.   This is a 0.95 confidence-interval estimate
of the average amount owed per customer.

Step 2. Since the department store has 10,000 customers, the 0.95 confidence-
interval estimate for the total amount owed by customers can be obtained as

$$10,000 \ (\$130.40 \leq \mu \leq \$169.60),$$

or between $1,304,000 and $1,696.000.

It may be useful now to summarize the solution to our problem.   A 0.95
confidence-interval estimate of the total amount owed by customers to the
department store can be determined as

$$N\left(\bar{x} \pm z\left(\frac{s}{\sqrt{n}}\right)\right) = 10,000\left(\$150 \pm 1.96 \ \frac{\$60}{\sqrt{36}}\right)$$

or between $1,304,000 and $1,696, 000.

Determining the Sample Size

The question of sample size must be resolved before a sample is selected.
How large should the sample be?   Generally speaking, the sample size is
determined by the desired degree of accuracy required in estimating the
population mean from a sample.

There are two aspects to the degree of accuracy required:

1. The magnitude of the maximum allowable error.

2. The degree of confidence that the error in the estimate will not exceed
the maximum allowable error.

If we are asked by the Los Angeles City Council to estimate the average
annual family earnings in Los Angeles within $500, the maximum allowable
error is $500. However, regardless of the size of the sample selected, there is
always a chance that the error in our estimate will exceed $500. The larger
the sample size, however, the smaller is the risk that the allowable error will
be exceeded, or the greater the confidence that our estimate is within the
allowable error. If the City Council states that there should be .95 confidence
that the error in the estimate will not exceed $500, the two aspects of accuracy
become

1. maximum allowable error, denoted by e, is $500,

2. the degree of confidence that the maximum allowable error will not be
exceeded is .95.

Having determined the accuracy requirements of our estimate, we must
determine the sample size that satisfies these two requirements. Specifically,

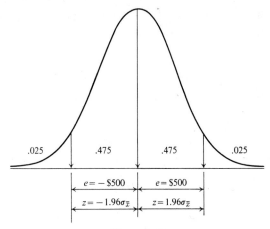

Figure 6–22

we must have a sample size n that will ensure

$$1.96\sigma_{\bar{x}} = \$500.$$

(See Fig. 6–22)

In general, the sample size is manipulated until

$$z\sigma_{\bar{x}} = e,$$

where z is determined by the degree of confidence. Since

$$\sigma_{\bar{x}} = \sigma/\sqrt{n},$$

it follows that

$$z\,\frac{\sigma}{\sqrt{n}} = e,$$

or

$$\sqrt{n} = \frac{z\sigma}{e}, \qquad n = \frac{z^2\sigma^2}{e^2}.$$

Thus, the sample size is determined by e, z, and σ, where e is equal to the maximum allowable error, z is determined by the degree of confidence, and σ is the standard deviation in the population whose mean we are trying to estimate.

The value of z is determined by the degree of confidence. Thus, if the degree of confidence is .95, z is equal to 1.96. If, on the other hand, it is required that the degree of confidence be .99, z is equal to 2.58. Finally, the value of z is 1.64 if the degree of confidence is .90. All these values are obtained from the area under the normal curve.

In addition to e and z, the standard deviation of the population from which the sample is selected, σ, must also be known. We can obtain σ from previous surveys or census records if such records exist. Otherwise, the statistician may have to select a preliminary sample for the purpose of estimating the standard deviation in the population.

It is interesting to note that as the value of σ (the variability of the population we are sampling) increases, we must increase the size of the sample. The conclusion is logical indeed: for example, if each family in Los Angeles had the same annual income, a sample of size one would be sufficient; however, the greater the disparity in family earnings, the larger the required sample size.

To summarize: the sample size must be increased as

1. the allowable error becomes smaller,

2. the degree of confidence increases, and

3. the variability in the population from which the sample is chosen increases.

Example 8

What sample size is required to estimate the average family income in Los Angeles if it is specified that there should be .95 confidence that the error in the estimate will not exceed $500? The 1980 Census shows an average family income of $15,000 with a standard deviation of $4000.

Solution:

$$e = \$500, \qquad z = 1.96, \qquad \sigma = \$4000,$$

but

$$n = \frac{z^2\sigma^2}{e^2} \quad \text{or} \quad n = \frac{(1.96)^2(4000)^2}{(500)^2}$$

so $n = 245.8,$ or $n = 246$ families.

Estimating the Population Mean From a Small Sample

We have already learned that when the sample size is 30 or more, the confidence interval estimate of the population mean which is $(\bar{x} \pm z(\sigma/\sqrt{n}))$ is approximated by the interval $(\bar{x} \pm z(s/\sqrt{n}))$. But when the sample size is small, say 10, such an approximation is no longer appropriate. Consequently, we must develop an alternative approach for estimating the population mean from a small sample.

Let us begin this alternative approach by assuming that there is a *normally distributed* population with a mean of μ. The standard deviation of this population is unknown. Let us further assume that all possible simple random samples of size n are selected from this population and the

mean $\bar{x}$ and the standard deviation s of each of these samples is computed.
Now, if the variable t, where

$$t = \frac{\bar{x} - \mu}{s/\sqrt{n}},$$

is computed for each of these samples, then according to statistical theory,
the various values of t are distributed according to what is called the Student
t-distribution.   A Student t-distribution, together with a standard normal
distribution, are shown in Fig. 6–23 (in a standard normal distribution,
$\mu = 0$ and $\sigma = 1$).

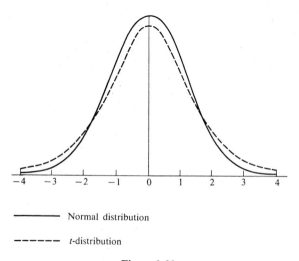

————— Normal distribution

------ t-distribution

Figure 6–23

In studying the t-distribution, we must recognize that there is not only
one but that there are several t-distributions.   Each of these is associated
with what is called *the number of degrees of freedom* (df), which, in this case,
is equal to the sample size minus one.
Thus the distribution of the t-values of samples of size 6 can be described
by a t-distribution of 5 degrees of freedom.   Similarly, a t-distribution of 10
degrees of freedom describes the distribution of the t-values of samples of
size 11.   Two distinct t-distributions (one with 5 degrees of freedom and the
other with 10 degrees of freedom), together with the *normal distribution*,
are shown in Fig. 6–24.   As shown by the diagram, a t-distribution, like the
normal distribution, is bell-shaped and symmetrical.   However, unlike the
normal distribution, a t-distribution has a greater dispersion which becomes
larger as the number of degrees of freedom decreases.

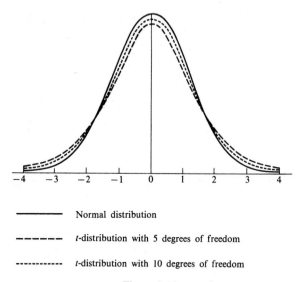

—————— Normal distribution

— — — — *t*-distribution with 5 degrees of freedom

- - - - - - - *t*-distribution with 10 degrees of freedom

Figure 6–24

Information concerning *t*-distributions is obtained from Table E. Each *row* in Table E corresponds to *one particular t*-distribution. The fifth row in the table, for example, corresponds to a *t*-distribution with 5 degrees of freedom. Similarly, the tenth row of the table deals with a *t*-distribution having 10 degrees of freedom. Confining our discussion for the moment to the *t*-distribution with 5 degrees of freedom (row 5), we explain the various values recorded in this row as follows: The value 2.571, located in the column labeled $t_{.025}$, indicates that 2.5% of the area under this particular *t*-curve is to the right of $+2.571$ or to the left of -2.571 (Fig. 6–25). Consequently, 95% of the area under this *t*-curve is enclosed in the interval ±2.571, or the probability is 0.95 that the variable *t* is in the interval ±2.571.

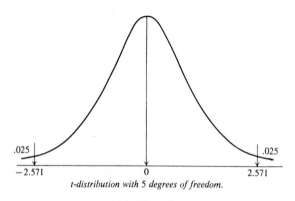

t-distribution with 5 degrees of freedom.

Figure 6–25

Similarly, the value 4.032, located in the column labeled $t_{.005}$, indicates that .005 of the area under the t-curve is to the right of $+4.032$ or to the left of -4.032 (Fig. 6–26). Consequently, 0.99 of the area under the t-curve is enclosed in the interval ± 4.032, or the probability is 0.99 that the variable t is in the interval ± 4.032.

Other entries in Table E can be interpreted in a similar fashion. The value 1.812, for example, is located in the 10th row and in the column labeled $t_{.05}$, indicating that in a t-distribution with 10 degrees of freedom, 5% of the area under the t-curve is to the right of $+1.812$ or to the left of -1.812. Consequently, 90% of the area under the t-curve is enclosed in the interval ± 1.812, or the probability is .90 that the variable t is in the interval ± 1.812.

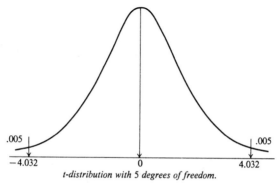

.005   .005

-4.032 0   4.032

t-distribution with 5 degrees of freedom.

Figure 6–26

Again confining our discussion to a t-distribution with 5 degrees of freedom, we find from Table E that 95% of the area under this particular t-curve is in the interval ± 2.571, or the probability is .95 that

$$-2.571 \le t \le +2.571.$$

Since

$$t = \frac{\bar{x} - \mu}{s/\sqrt{n}},$$

the inequality becomes

$$-2.571 \le \frac{\bar{x} - \mu}{s/\sqrt{n}} \le +2.571,$$

which, finally, can be written as

$$\bar{x} - 2.571 \frac{s}{\sqrt{n}} \le \mu \le \bar{x} + 2.571 \frac{s}{\sqrt{n}}.$$

Thus the probability is .95 that μ is in the interval $\bar{x} \pm 2.571(s/\sqrt{n})$.

We can, therefore, state that when the sample size is 6 (df $= 5$), then there is .95 confidence that μ is in the interval

$$\bar{x} \pm 2.571 \frac{s}{\sqrt{n}}.$$

Similarly, we can show that when the sample size is 6, there is .99 confidence that μ is in the interval

$$\bar{x} \pm t_{.005} \frac{s}{\sqrt{n}} \quad \text{or} \quad \bar{x} \pm 4.032 \frac{s}{\sqrt{n}}.$$

To summarize, we can construct a confidence interval estimate of the population mean based on a small-size sample by choosing the appropriate value of t in the formula $\bar{x} \pm t(s/\sqrt{n})$. The appropriate value of t is determined by the number of degrees of freedom (df $= n - 1$) and by the degree of confidence required in estimating the population mean. As an illustration, several confidence interval estimates of the population mean based on a random sample of size 23 (df $= 22$) are shown below.

$$\bar{x} \pm t_{.05} \frac{s}{\sqrt{n}}, \qquad \text{(.90 confidence interval estimate of } \mu\text{)};$$

$$\bar{x} \pm 1.717 \frac{s}{\sqrt{n}},$$

$$\bar{x} \pm t_{.025} \frac{s}{\sqrt{n}}, \qquad \text{(.95 confidence interval estimate of } \mu\text{)};$$

$$\bar{x} \pm 2.074 \frac{s}{\sqrt{n}},$$

$$\bar{x} \pm t_{.005} \frac{s}{\sqrt{n}}, \qquad \text{(.99 confidence interval estimate of } \mu\text{)}.$$

$$\bar{x} \pm 2.819 \frac{s}{\sqrt{n}},$$

Before concluding this discussion, we present a few illustrative problems.

Example 9

Nine cars of the same model were driven in an identical manner, using one gallon of regular gasoline. The mean distance traveled by those cars was 19 miles with a standard deviation of 2.7 miles. Establish a .95 confidence interval estimate of the average mileage per gallon for this car model.

Solution:

1. Determine the number of degrees of freedom:

$$\text{df} = n - 1$$
$$= 8.$$

2. Determine the appropriate value of t: With 8 degrees of freedom and a desired .95 confidence interval estimate of μ, the appropriate value of t is located at the intersection of row 8 and the column labeled $t_{.025}$ in Table E. This t-value is 2.306.

3. Construct the interval estimate of μ:

$$\bar{x} \pm t_{.025}\frac{s}{\sqrt{n}};$$

$$19 \pm 2.306\frac{2.7}{\sqrt{9}}, \qquad 19 \pm (2.306)(.9), \qquad 19 \pm 2.0754;$$

the interval estimate is 16.9246–21.0754 miles.

Example 10

The nicotine contents of five cigarettes of a certain brand, measured in milligrams, are: 21, 19, 23, 19, 23. Establish a .99 confidence interval estimate of the average nicotine content of this brand of cigarettes.

Solution:

1. Determine the mean of the sample:

$$\bar{x} = \frac{21 + 19 + 23 + 19 + 23}{5} = \frac{105}{5} = 21 \text{ milligrams.}$$

2. Determine the standard deviation of the sample:

x	$(x - \bar{x})$	$(x - \bar{x})^2$
21	0	0
19	−2	4
23	+2	4
19	−2	4
23	+2	4
	$\Sigma(x - \bar{x})^2 = 16$	

$$s = \sqrt{\frac{\Sigma(x - \bar{x})^2}{n - 1}} = \sqrt{\frac{16}{5 - 1}} = 2 \text{ milligrams.}$$

3. Determine the number of degrees of freedom:

$$df = n - 1 = 5 - 1 = 4.$$

4. Construct the .99 confidence interval estimate of μ:

$$\bar{x} \pm t_{.005}\frac{s}{\sqrt{n}};$$

$$21 \pm 4.604\frac{2}{\sqrt{5}}, \qquad 21 \pm 4.604(.8944), \qquad 21 \pm 4.1178;$$

the interval is 16.8822–25.1178.

Before we conclude this chapter, it may be useful to compare the two methods used in estimating the population mean (small vs. large samples). In the small-sample method, the range

$$\bar{x} \pm t \frac{s}{\sqrt{n}}$$

is an *exact* confidence interval estimate of the population mean. In the large-sample method, the range

$$\bar{x} \pm z \frac{s}{\sqrt{n}}$$

is only an approximate confidence interval estimate of the population mean. If this is the case, then why not use the small-sample method all the time?

In the small-sample method, the range $\bar{x} \pm t(s/\sqrt{n})$ is an exact confidence interval estimate of the population mean only when such a population is *normally distributed*. Thus, if a population is not normally distributed, the use of the small sample can no longer be justified.

In comparison, the large-sample method of estimating the population mean does not require any assumption about the nature of the population distribution. The only requirement is that the sample standard deviation s be a good approximation of the population standard deviation σ. Thus we conclude that the small-sample method of estimating the population mean should not supplant the large-sample method unless the assumption of normality of the population is justified.

EXERCISES, SET 2

1. A random sample of 30 members was selected from the faculty of a large university in order to estimate the average length of teaching experience of its faculty members. The records of teaching experience (measured in years) for the 30 members are

$$\begin{array}{cccccccccc}
3, & 4, & 4, & 6, & 2, & 3, & 4, & 6, & 2, & 4 \\
6, & 4, & 3, & 4, & 4, & 7, & 3, & 4, & 5, & 6 \\
1, & 6, & 4, & 5, & 4, & 3, & 2, & 4, & 3, & 4
\end{array}$$

Using the above data, establish a .99 confidence interval estimate of the average length of teaching experience of a faculty member.

2. The management of a manufacturing concern wishes to determine the average time required to complete a certain manual operation. There should be .95 confidence that the error in the estimate will not exceed 2 minutes.

   a) What sample size is required if the standard deviation of the time needed to complete the manual operation is estimated by a time and motion study expert as 10 minutes?

b) What sample size is required if the standard deviation of the time needed to complete the manual operation is estimated by a time and motion study expert as 16 minutes?

c) Explain intuitively (without referring to the formula) why the required sample size is larger in (b) than in (a).

3. The nicotine content of 36 cigarettes of a certain brand is measured. The results are summarized below.

$$x = \text{nicotine content of a cigarette measured in milligrams,}$$
$$\Sigma x = 756 \text{ milligrams,}$$
$$\Sigma (x - \bar{x})^2 = 315 \text{ milligrams.}$$

Establish a .95 confidence interval estimate of the average nicotine content of this brand of cigarettes.

4. The Miser Saving and Loan Association wishes to determine the mean amount of its customers' savings accounts. The standard deviation of all savings accounts is estimated by the manager as $400.

a) What sample size is required to ensure with .95 confidence that the error in the estimate will not exceed $20?

b) What sample size is required to ensure with .95 confidence that the error in the estimate will not exceed $40?

c) Compare the sample size and the maximum allowable error in (a) and (b). What happens to the sample size when the maximum allowable error is doubled?

5. The systolic blood pressure of 100 patients taking 20 milligrams of a certain drug exhibits a mean increase of 18 with a standard deviation of 6.

a) Establish a .99 confidence interval estimate of the increase in blood pressure caused by 20 milligrams of this drug.

b) What can we assert with probability .95 about the maximum error in our estimate if the average increase in blood pressure is estimated as 18?

6. A constructing firm wishes to estimate the average sheering strength of steel bars used in the construction of several "high-rise" apartment buildings. What sample size is required to ensure that there will be only a .001 risk of exceeding an error of 10 pounds or more in the estimate? The standard deviation of the sheering strength of this type of steel bars is estimated as 50 pounds.

7. Forty-nine pigs were fed a special feed for a period of three months. The mean weight gain in these three months was 120 pounds with a standard deviation of 14 pounds. With what degree of confidence can we assert that this feed will cause an average weight gain of 118.5 to 121.5 pounds in a period of three months?

8. Construct a .98 confidence interval estimate of the population mean based on each of the following samples:

a) $n = 9$, $\bar{x} = 40$, $s = 6$ b) $n = 16$, $\bar{x} = 20$, $s = 2$

c) $n = 25$, $\bar{x} = 70$, $s = 10$

9. A random sample of 64 accounts receivable was selected from a ledger containing 10,000 accounts. The sample showed a mean amount of $120 per account with a standard deviation of $40.

   a) Establish a .90 confidence interval estimate of the average account receivable in the ledger.

   b) What can we assert with a probability of .95 about the maximum error in our estimate if the average account receivable in the ledger is estimated as $120?

10. A sample of 5 jars of instant coffee is selected from a production process. The contents of these jars, measured in ounces, are as follows: 10.5, 10.7, 10.3, 10.6, and 10.4. Establish a .95 confidence interval estimate of the average net weight of a jar turned out by this production process.

11. To estimate the total amount of customers' demand deposits, a commercial bank selected a random sample of 400 demand deposit accounts. The sample showed a mean amount of $500 per account and a standard deviation of $100. Assuming that the bank has 12,000 demand deposits accounts, establish a 0.99 confidence interval estimate of the aggregate amount of demand deposits held by the commercial bank.

12. A random sample of 100 families was selected from a community of 5000 families. The sample showed an average annual family income of $15,000 and a standard deviation of $2,000. Establish 0.90 confidence interval estimate of aggregate annual income in that community.

13. A random sample of 100 families in Boston shows an average family income of $12,000 with a standard deviation of $4000.

   a) Establish a .99 confidence interval estimate of the average family income in Boston.

   b) With what degree of confidence can we assert that average family income in Boston is somewhere between $11,400 and $12,600?

14. A random sample of five batteries of a certain brand is tested in order to determine the average battery life. The lives of the batteries tested measured in hours are as follows:

$$52, 48, 50, 44, 56$$

Establish a .95 confidence interval estimate of the average battery life.

7
Test of Hypothesis Concerning the Mean of a Population

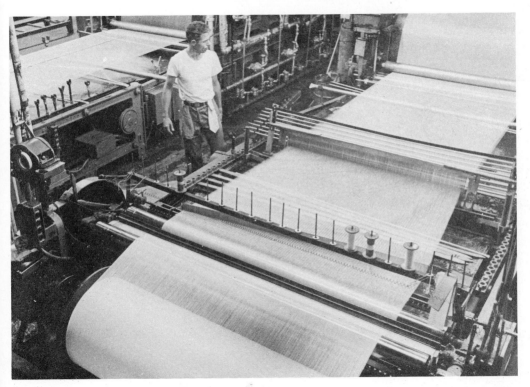

Test of
Hypothesis
Concerning
the Mean
of a Population

The preceding chapter dealt with the problem of estimating the population mean μ from a simple random sample. In other statistical investigations, however, the objective may be to test a hypothesis concerning the mean of a population. We may evaluate, for example, a manufacturer's claim that his tire has an average life of 21,000 miles, or that his brand of battery will, on the average, last 27 months. In the same manner, we may test the hypothesis that the average IQ of a college student is 115.

When a manufacturer claims that his brand of tires has an average life of 21,000 miles, the statistician must formulate this claim in statistical language. To the statistician, this claim means that: If *all* the tires produced by this manufacturer are tested, and the life of each tire is recorded, and the arithmetic mean of all tires is calculated, the value of the mean is 21,000 miles. Stated briefly, the manufacturer claims that μ is equal to 21,000 miles, where μ is the arithmetic mean of an infinite population (all tires made and to be made by the manufacturer).

In order to test the claim that μ is equal to 21,000 miles, only a sample of the tires can be tested. The sample results may then either support or refute the manufacturer's claim. To be specific, let us illustrate the procedure with several examples:

Example 1

A manufacturer of a certain brand of nine-volt batteries claims that the average life of his battery is 50 hours with a standard deviation of 5 hours. A simple random sample of 100 batteries shows an average of 40 hours. What do you conclude about the manufacturer's claim?

If the manufacturer's claim is true, and an infinite number of samples of size 100 are selected from the claimed population, then we expect all

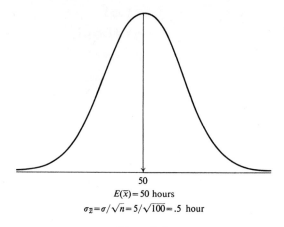

$$E(\bar{x}) = 50 \text{ hours}$$

$$\sigma_{\bar{x}} = \sigma/\sqrt{n} = 5/\sqrt{100} = .5 \text{ hour}$$

Figure 7–1

possible $\bar{x}$'s to be normally distributed, have a mean of 50 hours, and a standard error of $5/\sqrt{100}$ or .5 hour (Fig. 7–1).

To evaluate the manufacturer's claim, let us calculate the probability that any one sample mean is 40 hours or less.

$$z = \frac{\bar{x} - \mu}{\sigma_{\bar{x}}}$$

$$= \frac{40 - 50}{.5} = -20.$$

Therefore $P(\bar{x} \le 40)$ is null and the manufacturer's claim is totally rejected.

Example 2

Suppose that the sample in the previous example has an average of 49 hours. Do you still reject the manufacturer's claim?

To evaluate the claim, let us calculate the probability of any one sample mean being 49 hours or less.

$$z = \frac{\bar{x} - \mu}{\sigma_{\bar{x}}}$$

$$= \frac{49 - 50}{.5} = -2.0.$$

Therefore

$$P(\bar{x} \le 49) = .5000 - .4772$$

$$= .0228.$$

Now, if the manufacturer's claim is rejected, there is a risk of .0228 that we are rejecting a true claim. Are we willing to take such a risk? The answer to this question depends on the consequences of rejecting the claim when it is true.

Example 3

A manufacturer of synthetic fiber advertised that his fiber has an average tensile strength of 30 pounds. A simple random sample of 100 fibers was tested for breaking strength. It showed an average of 28 pounds and a standard deviation of 12 pounds. Test the manufacturer's claim against the hypothesis that the average tensile strength of his fiber is less than 30 pounds.

If the manufacturer's claim is true, and an infinite number of samples of 100 fibers are selected from his product, then the various sample means $\bar{x}$ are normally distributed, have an average of 30 pounds, and a standard error of $\sigma/\sqrt{100}$. But since the standard deviation in the population, σ, is not known, we use the standard deviation of the sample, s, as an estimate of σ. The estimated standard error of the mean, denoted by $\hat{\sigma}_{\bar{x}}$, will be $s/\sqrt{n}$. (This is in contrast to the true standard error of the mean which is $\sigma/\sqrt{n}$.) See Fig. 7–2.

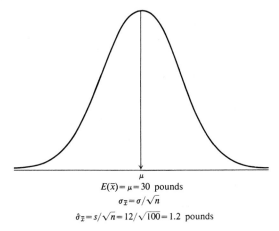

$$E(\bar{x}) = \mu = 30 \text{ pounds}$$
$$\sigma_{\bar{x}} = \sigma/\sqrt{n}$$
$$\hat{\sigma}_{\bar{x}} = s/\sqrt{n} = 12/\sqrt{100} = 1.2 \text{ pounds}$$

Figure 7–2

To test the manufacturer's claim, let us calculate the probability that a sample mean of 100 fibers will be 28 pounds or less.

$$z = \frac{\bar{x} - \mu}{\hat{\sigma}_{\bar{x}}}$$

$$= \frac{28 - 30}{1.2} = 1.67.$$

Therefore

$$P(\bar{x} \leq 28) = .5000 - .4525$$
$$= .0475.$$

If the manufacturer's claim is rejected, there is only a risk of .0475 that we are rejecting a true claim.

The above example is typical of a frequent problem in statistics, that of testing a hypothesis about a population. In our case, the hypothesis is made by the manufacturer who claims that the average tensile strength of all his fibers, or μ, is 30 pounds. This hypothesis is tested against another hypothesis which states that μ is less than 30 pounds.

In problems of this sort the hypothesis to be tested is generally referred to as the *null hypothesis* (denoted by H_0). The null hypothesis is tested against some other hypothesis, generally known as the *alternative* hypothesis. The alternative hypothesis is denoted by H_1. Referring to the fiber problem, we find that our two hypotheses are

H_0: μ = 30 pounds (null hypothesis),
H_1: μ < 30 pounds (alternative hypothesis).

In order to test a hypothesis (the null hypothesis) about a population, a sample is selected from that population. Depending on the findings in the sample, the investigator will either *reject the null hypothesis* or *will not reject it*. The null hypothesis is rejected if the sample findings are inconsistent with it. The null hypothesis is not rejected if the sample findings do not contradict it.

When the null hypothesis is not rejected, a statistician prefers to say that he *reserves judgment* (in contrast to accepting the null hypothesis). What he means by "reserving judgment" is that, on the basis of the gathered sample information, he does not have enough evidence to reject the null hypothesis. For all practical purposes, however, a statistician reserving judgment is in a way accepting the null hypothesis without really saying so and without commitment. Thus, when a statistician reserves judgment, he absolves himself, at least theoretically, from the sin of committing any error.

If, on the other hand, the null hypothesis is rejected, there is always a risk that we are rejecting a true hypothesis. The risk of rejecting the null hypothesis when it is true is called the α-*risk*, and the error committed is called a *type-1 error*.

How great a risk should the statistician take in rejecting a true hypothesis? Generally speaking, the greater the consequence of rejecting the null hypothesis when it is true, the smaller the risk he should take. Statisticians, however, have set arbitrary limits of .05 or .01. Each limit is called a *level of significance*.

The .05 level of significance is used whenever it is desired that the α-risk (the risk of rejecting the null hypothesis when it is true) should not exceed .05. The .01 level of significance is used when the risk tolerated must not exceed .01.

Returning to our fiber problem, we have found that the probability of obtaining a sample mean of 28 pounds or less is only .0475 if the null hypothesis is true (the null hypothesis is that the population mean is 30 pounds). Hence, if the null hypothesis is rejected, there is a .0475 risk that we are rejecting a true hypothesis.

In our example, the decision of whether to reject the null hypothesis or reserve judgment depends on the maximum risk we are willing to take. Using the .05 level of significance, we reject the null hypothesis since the maximum allowable risk is .05 and the actual risk is .0475. Using the .01 level of significance, we must reserve judgment (maximum allowable risk is .01 and actual risk is .0475).

Example 4

In justifying their demand for higher wages, the employees in the shipping department of a large mail order house report that, on the average, the department completes an order in 13 minutes. As a general manager for this firm, what can you conclude if a sample of 400 orders shows an average completion time of 14 minutes with a standard deviation of 10 minutes? Use a .05 level of significance.

Denoting the null hypothesis by H_0, and the alternative hypothesis by H_1, we can summarize the problem as follows:

$$H_0: \mu = 13 \text{ minutes,}$$
$$H_1: \mu > 13 \text{ minutes,}$$
$$\text{Level of significance} = .05.$$

Note that in this example the null hypothesis is tested against an alternative hypothesis which states that the average completion time is *greater* than 13 minutes. This is because management feels that the demand for a wage increase is unjustified if the true mean is greater than 13 minutes.

To test the null hypothesis, let us calculate the probability of obtaining a sample mean of 14 minutes or more from a population for which the true mean is 13 minutes (Fig. 7–3).

$$z = \frac{\bar{x} - \mu}{\hat{\sigma}_{\bar{x}}}$$

$$= \frac{14 - 13}{10/\sqrt{400}} = \frac{1}{.5} = 2.0.$$

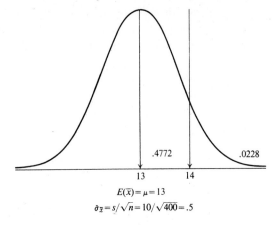

$$E(\bar{x}) = \mu = 13$$
$$\hat{\sigma}_{\bar{x}} = s/\sqrt{n} = 10/\sqrt{400} = .5$$

Figure 7–3

Therefore

$$P(\bar{x} \geq 14) = .5000 - .4772$$
$$= .0228.$$

Using the .05 level of significance, we reject the null hypothesis.

The reader may have already observed that the decision whether to reject the null hypothesis or to reserve judgment becomes obvious once the z-value is calculated. Using the .05 level of significance, we reject the null hypothesis if the *absolute* value of z exceeds 1.64. (If the z-value is greater than 1.64, the risk of rejecting the null hypothesis when it is true is less than .05.) This is shown in Fig. 7–4. If the .01 level of significance is used instead,

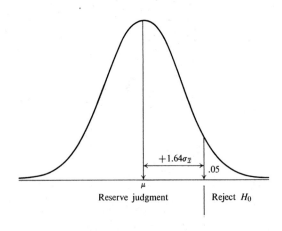

Figure 7–4

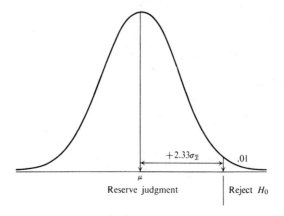

Figure 7–5

the null hypothesis is rejected only when the absolute z-value is greater than 2.33. (If the z-value is greater than 2.33 the risk is less than .01.) (See Fig. 7–5.)

Example 5

In investigating several complaints concerning the weight of the "NET WT. 12 OZ." jar of a local brand of peanut butter, the Better Business Bureau selected a sample of 36 jars. The sample showed an average net weight of 11.92 ounces and a standard deviation of .3 ounce. Using the .01 level of significance, what would the Bureau conclude about the operation of the local firm? The problem is formulated below:

$$H_0: \mu = 12 \text{ ounces,}$$
$$H_1: \mu < 12 \text{ ounces,}$$
$$\text{Level of significance} = .01.$$

To test the null hypothesis, let us calculate the deviation of a sample mean of 11.92 ounces from a population mean of 12 ounces (the deviation is measured in units of z).

$$z = \frac{\bar{x} - \mu}{\hat{\sigma}_{\bar{x}}} = \frac{11.92 - 12.00}{.3/\sqrt{36}} = \frac{-.08}{.05} = -1.60.$$

Using the .01 level of significance, we must reserve judgment (z is less than 2.33). In other words, the sample evidence is not strong enough to indicate that the local firm is not maintaining the advertised weight.

Two-Sided Test

The alternative hypotheses in all the problems discussed so far were essentially of two kinds. In some, the alternative hypothesis stated that the mean of the population is less than a given number. In others, the alternative hypothesis stated that the mean of the population is greater than a given number.

When the alternative hypothesis is formulated as "less than," we obviously do not mind when the mean of the population is actually above the figure stated by the null hypothesis. In the fiber problem, for example, the null hypothesis states that the average tensile strength of the fiber is 30 pounds. The alternative hypothesis is formulated to state that the average is less than 30 pounds, simply because we are not concerned, and may even be delighted, if the average tensile strength is actually greater than 30 pounds. For exactly the same reason, the null hypothesis in the peanut butter problem which states that the average weight of all jars is 12 ounces is tested against the alternative hypothesis that the average weight of all jars is less than 12 ounces. Customers will be very happy indeed if the average weight is greater than 12 ounces.

At other times, the investigator is concerned only when the mean of the population is greater than the value stipulated by the null hypothesis. Under such circumstances, the alternative hypothesis is formulated in "greater than" form. The null hypothesis in the mail order house problem, for example, was tested against the alternative hypothesis that the average completion time is greater than 13 minutes. An alternative hypothesis of the "greater than" type is certainly appropriate here, since the shipping department would be operating more efficiently than stated in the report if the average completion time turned out to be less than 13 minutes. Management will be concerned and feel that the demand for wage increase is unjustified only if the average completion time is greater than 13 minutes.

When the null hypothesis is tested against an alternative hypothesis of a "greater than" or a "less than" type, the test is called a *one-sided* test. As we have already learned, in such problems the null hypothesis is rejected only if the absolute value of z exceeds 1.64 or 2.33, depending on whether we are using the .05 or the .01 level of significance.

In contrast to the one-sided test problems, there are some problems which will cause the investigator to be equally concerned whether the mean of the population is understated or overstated by the null hypothesis. Under such circumstances, the null hypothesis is rejected whenever the sample data strongly suggest that the population mean is greater or less than the value stipulated in the null hypothesis. To clarify this point, let us assume that a manufacturing firm places an order for 10,000 ball bearings of two-inch diameter. Having received the shipment, the firm would be equally alarmed whether the actual average diameter was smaller or greater than two inches.

In other words, the consequences are very grave when the average diameter is not equal to two inches.

If the null hypothesis is rejected whenever sample evidence indicates that the population mean is overstated or understated by the null hypothesis, the test is called a *two-sided* or *two-tailed* test. In a two-sided test problem, therefore, the null hypothesis which states that the mean of the population is equal to a given value is tested against the alternative hypothesis that the mean is not equal to that value ("not equal" means "greater" or "smaller").

Using the .05 level of significance, the null hypothesis is rejected in a two-sided test whenever the *total* risk of rejecting the null hypothesis when it is true does not exceed .05. And since the null hypothesis is rejected whether the mean of the population is greater or less than the value stipulated by the null hypothesis, the total .05 risk must be divided equally on both sides of the normal curve as shown in Fig. 7–6. Hence, the null hypothesis can be rejected only when the absolute value of z exceeds 1.96.

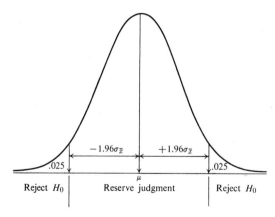

$-1.96\sigma_{\bar{x}}$ $+1.96\sigma_{\bar{x}}$

.025 .025

μ

Reject H_0 Reserve judgment   Reject H_0

Figure 7–6

Using the .01 level of significance, on the other hand, we must reject the null hypothesis in a two-sided test whenever the absolute value of z exceeds 2.58 (see Fig. 7–7).

Example 6

A circuit fuse is designed to burn out as the electric current reaches 20 amperes. From a lot of 10,000 fuses, 36 are selected and tested for their breaking point. What do you conclude about the amperage specification of the lot if the sample reveals a mean of 20.9 amperes and a standard deviation of 1.5 amperes? Use a level of significance of .01.

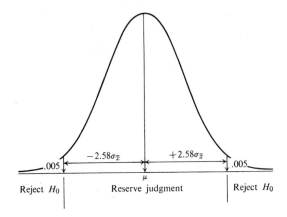

Figure 7–7

It is apparent here that the fuse should withstand a current of up to 20 amperes and should break the circuit once the amperage exceeded 20 amperes. The null hypothesis stating that the average breaking point $\mu = 20$ amperes should, therefore, be tested against the alternative hypothesis that $\mu \neq 20$ amperes. The problem can be summarized as follows:

$$H_0: \mu = 20 \text{ amperes,}$$
$$H_1: \mu \neq 20 \text{ amperes,}$$
$$\text{Level of significance} = .01.$$

To test the validity of the null hypothesis, let us calculate the deviation, measured in z-units, of a sample mean of 20.9 amperes from a population mean of 20 amperes.

$$z = \frac{\bar{x} - \mu}{\hat{\sigma}_{\bar{x}}}$$

$$= \frac{20.9 - 20}{1.5/\sqrt{36}} = \frac{.9}{.25} = 3.60.$$

Since the actual z-value, 3.60, exceeds 2.58, the null hypothesis is rejected, using the .01 level of significance.

Example 7

In a report prepared by the Economic Research Department of a major bank, the department manager maintains that the average annual family income in Los Angeles is $8432. What do you conclude about the validity of the report if a simple random sample of 400 families shows an average income of $8574 with a standard deviation of $2000? Use a .05 level of significance. The problem is formulated as shown below:

$$H_0: \mu = \$8432,$$
$$H_1: \mu \neq \$8432,$$
$$\text{Level of significance} = .05.$$

$$z = \frac{\bar{x} - \mu}{\hat{\sigma}_{\bar{x}}}$$

$$= \frac{8574 - 8432}{2000/\sqrt{400}} = \frac{142}{100} = 1.42.$$

Since the actual z-value, 1.42, is less than 1.96, we must reserve judgment, using the .05 level of significance.

Before concluding this chapter, we must recognize that our discussion, so far, has dealt with large-size samples. The means of large-size samples are approximately normally distributed. Consequently, the normal distribution is used in all the preceding testing problems. Using the normal distribution, we reject the null hypothesis whenever the *absolute* value of z exceeds some critical level. The various critical levels are summarized below.

	One-sided test	Two-sided test
.05 level of significance	1.64	1.96
.01 level of significance	2.33	2.58

Now, what happens when the samples size is small? A hypothesis concerning a population mean can be tested using a small-size sample only if it can be assumed that the sample is selected from a *normally distributed population* with unknown standard deviation (see the section on estimating the population mean from a small sample in Chapter 6). If the assumption regarding the normality of the population can be justified, a hypothesis concerning the mean of that population can be tested by means of a small-size sample utilizing a t-distribution with $n - 1$ degrees of freedom. In order to test the null hypothesis in this case, a t-value is computed as

$$t = \frac{\bar{x} - \mu}{s/\sqrt{n}}.$$

The null hypothesis is then rejected only when the absolute value of t exceeds some critical level. The critical level is determined by the nature of the test (whether it is one-sided or two-sided), by the level of significance, and by the degrees of freedom.

The various critical levels of t are shown below.

	One-sided test	Two-sided test
.05 level of significance	$t_{.05}$	$t_{.025}$
.01 level of significance	$t_{.01}$	$t_{.005}$

The following problem illustrates the use of the t-distribution in testing a hypothesis concerning the mean of a population.

Example 8

The maker of a certain car model claimed that his car averaged 31 miles per gallon of regular gasoline. A random sample of 9 cars was selected, and each car was driven with one gallon of regular gasoline. The sample showed a mean of 29.43 miles with a standard deviation of 3 miles. Using the .05 level of significance, what do you conclude about the auto manufacturer's claim?

Solution:

$$H_0: \mu = 31,$$
$$H_1: \mu < 31,$$
$$\text{Level of significance} = .05.$$

To test the null hypothesis, we compute the *actual* value of t as

$$t = \frac{\bar{x} - \mu}{s/\sqrt{n}}$$

$$= \frac{29.43 - 31}{3/\sqrt{9}} = -1.57.$$

From Table E, we find that the critical t-value is 1.860 (row 8, column $t_{.05}$). Since the actual t-value is -1.57 and the critical level is 1.860, we reserve judgment about the manufacturer's claim.

EXERCISES, SET 1

1. The Retail Clerk Union claims that the average hourly earnings of its members amount to $2.10. The Retail Trade Owners' Association suspects that the union understates the average hourly wage. In a random sample of 100 sales-clerks, the Retail Trade Owners' Association finds that the mean hourly earnings are $2.17 with a standard deviation of $1.20. If the Association is willing to

reject a true claim no more than 5 in 100, will the Association reject the union claim?

2. The Retail Trade Owners' Association claims that the average hourly wage of its salesclerks is $2.20. The Retail Clerk Union suspects that the Association overstates the average hourly wage. In a random sample of 400 salesclerks, the Retail Clerk Union finds that the mean hourly wage is $2.11, with a standard deviation of $0.80. If the Union is willing to reject a true claim no more than once in 100, will the union reject the Association's claim?

3. A government testing agency tests a sample of 36 packages of ground beef sold by the Shop & Save Super Market. The label on each package reads: "contains no more than 25% fat." Can the testing agency conclude that the ground beef sold by the supermarket contains more than 25% fat, if the sample shows a mean fat content of 0.265 and a standard deviation of 0.030? Use a 0.05 level of significance.

4. An ice cream producer desires the fat content of his ice cream to remain at 10% in order to maintain a uniform quality of his product. An analysis of a sample of 49 cartons of his ice cream shows a mean fat content of 0.103 with a standard deviation of 0.014. Using a 0.05 level of significance, will the producer reject the hypothesis that the fat content of his ice cream is properly maintained?

5. The mean weight of a tablet of a certain drug is to be 50 milligrams. A sample of 64 tablets shows a mean weight of 50.15 milligrams with a standard deviation of 0.4 milligrams. Using a 0.01 level of significance can we conclude that the desired weight of the tablet is not properly maintained?

6. A manufacturer of a certain brand of cigarettes claims that his cigarettes contain, on the average, 18 milligrams of nicotine per cigarette. An independent testing agency examines a sample of 100 cigarettes. Using a level of significance of 0.01, can the agency conclude that the manufacturer understates the average nicotine content of his brand of cigarettes, if the mean nicotine content in the sample is 19.2 milligrams with a standard deviation of 2 milligrams?

7. A manufacturer of electric ovens purchases from a supplier glass components with a specified mean heat resistance of 800°. A sample of 25 components selected from a very large shipment shows an average heat resistance of 792° with a standard deviation of 20°. Using the 0.01 level of significance, can the oven manufacturer conclude that the heat resistance of the glass components received is less than 800°?

8. It has been theorized that the melting point of a certain metal is 1657°. A metallurgist repeats the same experiment four times and observes a mean melting point of 1655° with a standard deviation of 2°. Using a level of significance of 0.01, what can the metallurgist conclude about the validity of the hypothesis that the melting point of the metal is 1657°?

9. A manufacturer of multiple vitamin tablets asserts that a tablet contains, on the average, 2.50 milligrams of riboflavin (vitamin B-2). A representative of the Food and Drug Administration tests 64 tablets. He finds a mean riboflavin content of 2.48 milligrams with a standard deviation of 0.16 milligrams. If the

Food and Drug Administration representative is willing to reject a true assertion no more than 5 times in 100, will he reject the manufacturer's assertion?

10. A manufacturer of a certain car model claims that the average mileage of this model is 30 miles per gallon of regular gasoline.   A consumer protection agency believes that the average mileage of the car is exaggerated by the manufacturer. Nine cars of this particular model are driven in the same manner with one gallon of regular gasoline.   The distances traveled by the different cars are

$$30, \quad 28, \quad 26, \quad 27, \quad 29, \quad 28, \quad 31, \quad 26, \quad 27 \quad \text{miles.}$$

If the agency is willing to reject a true claim no more than once in 100, would the agency reject the manufacturer's claim?

11. The Department of Health, Education, and Welfare reports that average annual family expenditure on health is $980.   If a random sample of 100 families shows an average annual expenditure of $920 with a standard deviation of $200, would you conclude that the annual average family expenditure on health reported by H.E.W. is exaggerated?

12. A manufacturer of a certain average brand of cigarettes claims that the nicotine content of his cigarettes is 18 milligrams per cigarette.   A government testing agency measured the nicotine content of 16 cigarettes, and the following information was obtained:

$$X = \text{nicotine content of a cigarette measured in milligrams}$$
$$\sum X = 336 \text{ milligrams}$$
$$\sum (X - \bar{X})^2 = 240 \text{ milligrams}$$

a) Using a .05 level of significance, can the agency conclude that the nicotine content of this brand of cigarettes is underestimated by its manufacturer?

b) With what degree of confidence can we assert that the average nicotine content of this brand of cigarettes is between 18.869 and 23.131 milligrams?

13. The Accidental Automobile Insurance Company paid an average settlement of $512 per claim last year ($\mu = 512$).   A random sample of 100 claims selected from settlements paid this year shows an average settlement cost of $556 per claim with a standard deviation of $200.   Using a level of significance of .05, can the company conclude that average settlement cost per claim has risen since last year?

14. A pharmaceutical laboratory advertises that the mean time required for its brand of aspirin to dissolve is 10 minutes.   The Food and Drug Administration tests five tablets.   The times, in minutes, required to dissolve the tablets are

$$9, 12, 13, 11, 10.$$

Using a level of significance of .05, can the FDA conclude that the drug manufacturer exaggerates the speed with which his brand of aspirin dissolves?

15. The ABC Company's balance sheet shows an aggregate accounts receivable of 1.2 million dollars.   Suspecting that this amount is overstated, an auditor selects a random sample of 150 accounts from all company accounts receivable (10,000 accounts).   The sample shows a mean amount of $113 per account with

a standard deviation of $40. Using a .05 level of significance can the auditor conclude that the ABC Company has overstated its aggregate accounts receivable?

TYPE I AND TYPE II ERROR

The concepts of type I and type II error can best be explained with reference to a concrete example. Hence, let us consider the case of a radio manufacturer who receives a very large lot of batteries from a supplier each week. Since the manufacturer cannot test all batteries in a lot, he has adopted the following rule in deciding on whether to accept or to reject the weekly lot: select a random sample of 36 batteries from each lot received. If the average life of the batteries selected is 48 hours or more, accept the entire lot. If the average battery life in the sample is less than 48 hours, reject the lot. Briefly stated, the decision rule is

1. If $\bar{x} \geq 48$ hours, accept the lot.
2. If $\bar{x} < 48$ hours, reject the lot.

Now how much protection does the use of this decision rule provide against accepting a bad lot and against rejecting a good lot? Before we can answer this question, we must first describe what the manufacturer considers to be a good lot and a bad lot.

The manufacturer considers a lot to be of a good quality when the average battery life of the entire lot, μ, is at least 50 hours. And a lot is considered to be of a bad quality when its average battery life is 47 hours or less.

$$\text{Good lot:} \quad \mu \geq 50 \text{ hours.}$$
$$\text{Bad lot:} \quad \mu \leq 47 \text{ hours.}$$

The reader is quite justified to ask now: how does the manufacturer classify a lot whose average battery life is greater than 47 hours but less than 50 hours? For example, does he consider a lot with $\mu = 49$ hours to be a good lot or a bad lot? We will provide a satisfactory answer to this question next.

First, a lot is certainly considered to be a good lot when $\mu \geq 50$ hours. And in this case, the manufacturer wishes to take very small risk of rejecting such a lot. Second, a lot is certainly considered to be a bad lot when $\mu \leq 47$ hours. And similarly, the manufacturer wishes to take very small risk of accepting such a lot. Finally, the manufacturer is indifferent about a lot when its average battery life is between 47 and 50 hours. In other words, the manufacturer is not much alarmed here whether such a lot is accepted or rejected. Having defined a good lot and a bad lot, we are ready to deal now with measuring the risks of rejecting a good lot and accepting a bad lot.

Since the decision on whether to accept or to reject a lot depends on the sample outcome, and because of the nature of sample fluctuations, the use of the sample decision rule could result in any of four situations:

1. The lot, received is a good lot ($\mu \geq 50$ hours) and the sample outcome indicates that the lot should be accepted ($\bar{x} \geq 48$ hours). Hence, the shipment is accepted, and the decision made is a correct decision.

2. The lot received is again a good lot ($\mu \geq 50$ hours) but the sample outcome indicates that this lot should be rejected ($\bar{x} < 48$ hours). Hence, this lot is erroneously rejected, and the error committed is called type I error.

3. The lot received is a bad lot ($\mu \leq 47$ hours), and the sample outcome indicates that the lot should be rejected ($\bar{x} < 48$ hours). Hence, the lot is correctly rejected.

4. Finally, the lot received is a bad lot ($\mu \leq 47$ hours), but the sample outcome suggests that the lot should be accepted ($\bar{x} \geq 48$ hours). Hence, the lot is erroneously accepted, and the error committed is called type II error. Table 7-1 is a summary of these four situations.

Table 7–1

Summary of Decisions

Decision dictated by sample outcome	Quality of the lot	
	Good ($\mu \geq 50$)	Bad ($\mu \leq 47$)
Reject lot ($\bar{x} < 48$)	Type I error	Correct decision
Accept lot ($\bar{x} \geq 48$)	Correct decision	Type II error

In order to determine the probability of committing the type I error (rejecting a good lot), let us assume now that the manufacturer receives a lot of batteries with an *actual* mean battery life of 50 hours and a standard deviation of 6 hours ($\mu = 50$ hours and $\sigma = 6$ hours). Although this lot is a good lot, the manufacturer would *reject* such a lot whenever the selected sample shows an average of less than 48 hours. The means of the various samples of 36 batteries that can be selected from this lot are described by the normal distribution shown in Fig. 7–8.

The probability that the mean battery life in a random sample of 36 batteries is less than 48 can now be calculated as follows:

$$z = \frac{\bar{x} - \mu}{\sigma_{\bar{x}}} = \frac{48 - 50}{1} = -2.$$

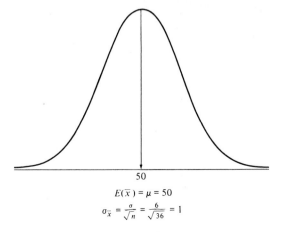

$$E(\bar{x}) = \mu = 50$$

$$\sigma_{\bar{x}} = \frac{\sigma}{\sqrt{n}} = \frac{6}{\sqrt{36}} = 1$$

Figure 7–8

Therefore

$$P(\bar{x} < 48) = 0.5000 - 0.4772 = 0.0228.$$

(See Fig. 7–9.) Thus, using the sampling decision rule, we find that the probability of rejecting a lot with $\mu = 50$ hours and $\sigma = 6$ hours is 0.0228. This is the probability of a type I error, which is denoted by the symbol α.

Similarly, in order to determine the probability of committing a type II error (accepting a bad lot), let us now assume that the manufacturer receives a lot with an actual mean battery life of 47 hours and a standard deviation of

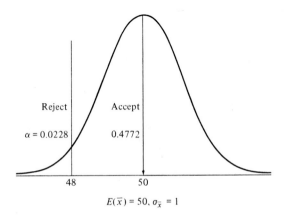

$$E(\bar{x}) = 50, \sigma_{\bar{x}} = 1$$

Figure 7–9

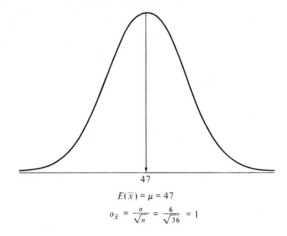

$$E(\bar{x}) = \mu = 47$$

$$\sigma_{\bar{x}} = \frac{\sigma}{\sqrt{n}} = \frac{6}{\sqrt{36}} = 1$$

Figure 7–10

6 hours ($\mu = 47$ hours and $\sigma = 6$ hours).   Although this lot is a bad lot, the manufacturer would *accept* such a lot whenever the selected sample shows an average life of 48 hours or more.   The means of the various samples of 36 batteries that can be selected from this lot are described by the normal distribution shown in Fig. 7–10.   The probability that the mean battery life in random samples of 36 batteries is 48 hours or more can now be calculated as follows (Fig. 7–11):

$$z = \frac{\bar{x} - \mu}{\sigma_{\bar{x}}} = \frac{48 - 47}{1} = 1.$$

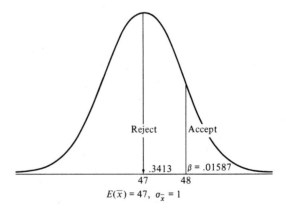

$$E(\bar{x}) = 47, \ \sigma_{\bar{x}} = 1$$

Figure 7–11

Therefore,

$$P(\bar{x} \geq 48) = 0.5000 - 0.3413$$
$$= 0.1587.$$

Thus, using the sampling decision rule, we find that the probability of accepting a lot with $\mu = 47$ hours and $\sigma = 6$ hours is 0.1587. This is the probability of type II error, which is denoted by the symbol β.

The probabilities of making type I error and type II error are displayed in Table 7–2.

Table 7–2

Summary of decisions

Decision dictated by sample outcome	Quality of lot	
	Good lot: $\mu = 50$	Bad lot: $\mu = 47$
Reject lot ($\bar{x} < 48$)	Type I error ($\alpha = 0.0228$)	Correct decision (Probability $= 0.8413$)
Accept lot ($\bar{x} \geq 48$)	Correct decision (Probability $= 0.9772$)	Type II error ($\beta = 0.1587$)

In order to generalize our discussion concerning type I and type II errors, let the null hypothesis, H_0, stand for the assumption that the lot received is a good lot; and the alternative hypothesis, H_1, for the assumption that the lot received is a bad lot.

Null hypothesis (H_0): $\mu \geq 50$ hours.

Alternative hypothesis (H_1): $\mu \leq 47$ hours.

Now, if the manufacturer accepts a lot, the null hypothesis is accepted and the alternative hypothesis is rejected. On the other hand, if the manufacturer rejects a lot, the null hypothesis is rejected and the alternative hypothesis is accepted. Hence, we can now define type I error as the error committed by rejecting H_0 when it is true (error incurred by rejecting a good lot). And we define type II error as the error incurred by accepting H_0 when H_0 is false (error incurred by accepting a bad lot).* Once again, the probability of making a type I error is denoted by the symbol α, and the probability of

---

*   It is useful to point out that type II error is not committed when the decision maker *reserves judgment* instead of accepting H_0.

making a type II error is denoted by the symbol β.   These generalizations are summarized in Table 7–3.

Table 7–3

Type 1 and Type II Errors

Decision dictated by sample outcome	H_0 is true	H_1 is true
Reject H_0	Type I error	Correct decision
Accept H_0	Correct decision	Type II error

Effect of Changing Sample Size on Type I and Type II Errors

We begin our discussion by stating the decision rule adopted by the radio manufacturer.   The decision rule is: select a random sample of 36 batteries from each lot received.   If the average life of the batteries selected is 48 hours or more, accept the entire lot; if the observed sample mean is less than 48 hours, reject the lot.   Briefly stated, the decision rule is

1. if $\bar{x} \geq 48$ hours, accept the lot,
2. if $\bar{x} < 48$ hours, reject the lot.

This decision rule can now be generalized as follows: select a random sample of size n from each lot received.   If the observed sample mean, $\bar{x}$, is equal to or greater than some critical value (this critical value is denoted by c^*), accept the lot; if $\bar{x}$ is less than this critical value, reject the lot.   Once again, briefly stated, the general decision rule is

1. if $\bar{x} \geq c^*$, accept the lot,
2. if $\bar{x} < c^*$, reject the lot.

In formulating his decision rule, the manufacturer has adopted a sample size of 36 and a critical value of 48 hours ($n = 36$, $c^* = 48$).

We must emphasize now, that in formulating a decision rule of the kind we are dealing with, two elements are of utmost importance:

1. the sample size (n),
2. the critical value (c^*).

In fact, the probabilities of type I and type II errors depend entirely on the values assigned to n and c^*.

Now, although we are mainly interested in examining the effect of changing the sample size on type I and type II errors, it is useful to examine the effect of modifying the value of c^* first.

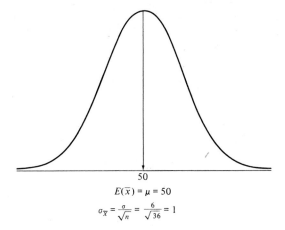

$$E(\bar{x}) = \mu = 50$$

$$\sigma_{\bar{x}} = \frac{\sigma}{\sqrt{n}} = \frac{6}{\sqrt{36}} = 1$$

Figure 7–12

Modifying the Value of c*. Let us now assume that the manufacturer modifies his original decision rule. The modified rule is: select a random sample of 36 batteries from each shipment received. If the observed sample mean, $\bar{x}$, is equal to or greater than 48.4 hours, accept the lot; if the sample mean is less than 48.4 hours, reject the lot (the critical value c^* is changed from 48 to 48.4 and the sample size is kept the same). We will examine the effect of modifying the value of c^* on type I error and type II in turn.

In order to determine the effect of modifying the value of c^* on type I error, let us once again assume that the manufacturer receives a lot with $\mu = 50$ hours and $\sigma = 6$ hours. Although this lot is a good lot, the manufacturer would reject such a lot (commit type I error) whenever the selected sample shows an average of less than 48.4 hours. The means of the various samples of 36 batteries that can be selected from this lot are described by the normal distribution shown in Fig. 7–12. The probability that the mean battery life in a random sample of 36 is less than 48.4 hours can now be calculated as follows:

$$z = \frac{\bar{x} - \mu}{\sigma_{\bar{x}}} = \frac{48.4 - 50}{1} = -1.6.$$

Therefore,

$$P(\bar{x} < 48.4) = 0.5000 - 0.4452 = 0.0548.$$

(See Fig. 7–13). Thus, using the modified decision rule ($n = 36$, $c^* = 48.4$), the probability of making type I error is 0.0548. The reader may recall, however, that in using the original decision rule ($n = 36$, $c^* = 48$), the probability of making type I error is 0.0228. Hence, by changing the critical value,

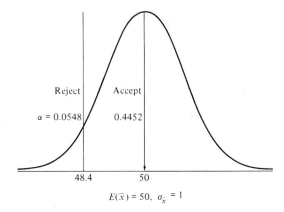

$$E(\bar{x}) = 50, \quad \sigma_{\bar{x}} = 1$$

Figure 7–13

c^*, from 48 hours to 48.4 hours, the probability of making type I error has increased from 0.0228 to 0.0548.

Similarly, in order to determine the effect of modifying the value of c^* on type II error, let us assume now that the manufacturer receives a lot with $\mu = 47$ hours and $\sigma = 6$ hours. Although this lot is a bad lot, the manufacturer would accept such a lot (commit type II error), whenever the observed sample mean is equal to or greater than 48.4 hours. The means of various battery samples of size 36 that can be selected from this lot are described by the normal distribution shown in Fig. 7-14. The probability that the mean

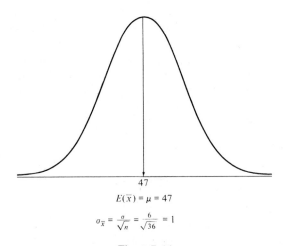

$$E(\bar{x}) = \mu = 47$$

$$\sigma_{\bar{x}} = \frac{\sigma}{\sqrt{n}} = \frac{6}{\sqrt{36}} = 1$$

Figure 7–14

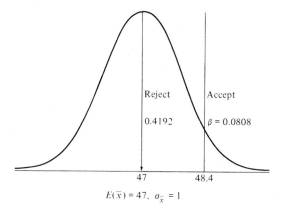

$E(\bar{x}) = 47, \ \sigma_{\bar{x}} = 1$

Figure 7–15

battery life in a random sample of 36 batteries is 48.4 hours or more can now
be calculated as follows (Fig. 7–15):

$$z = \frac{\bar{x} - \mu}{\sigma_{\bar{x}}} = \frac{48.4 - 47}{1} = 1.4.$$

Therefore,

$$P(\bar{x} \geq 48.4) = 0.5000 - 0.4192 = 0.0808.$$

Thus, using the modified decision rule ($n = 36$, $c^* = 48.4$), the probability
of making type II error is 0.0808. Using the original decision rule ($n = 36$,
$c^* = 48$), on the other hand, the probability of making type II error is 0.1587.
Hence, the probability of type II error is reduced from 0.1587 to 0.0808 as a
result of changing the critical value, c^*, from 48 to 48.4.

The effect of modifying the critical value of a decision rule on the probability of making type I and type II errors may now be summarized as follows:
as a result of changing the critical value, c^*, from 48 to 48.4, the probability
of making type I error is increased from 0.0228 to 0.0548; but the probability
of making type II error is decreased from 0.1587 to 0.0808. We may now,
therefore, generalize these results and say, that as the critical value, c^*, is
modified, the probabilities of making type I error and type II error change
in opposite directions. While the probability of making one type of error
decreases, the probability of making the other type of error increases.

Effect of Increasing Sample Size. We have already learned that changing the
critical value, c^*, would result in reducing the probability of making one type
of error only at the expense of increasing the probability of making the other
type of error. By contrast, we will now demonstrate that the probabilities

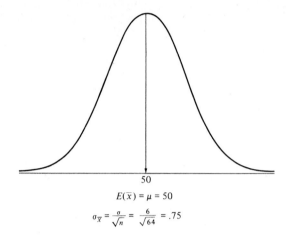

$$E(\bar{x}) = \mu = 50$$

$$\sigma_{\bar{x}} = \frac{\sigma}{\sqrt{n}} = \frac{6}{\sqrt{64}} = .75$$

Figure 7–16

of making both type of errors can *simultaneously* be reduced by increasing the *sample size* in the decision rule.

In order to show that enlarging the sample size, n, reduces the probability of making both types or errors, let us modify the original decision rule to read as follows: select a random batch of 64 batteries from each lot received. If average battery life in the observed sample is equal to or greater than 48 hours, accept the lot; if the sample mean is less than 48 hours, reject the lot (compared to the original decision rule, n is increased from 36 to 64 and c^* is kept the same at 48). Now, to measure the probability of type I error under this modified rule, let us once again assume that the manufacturer receives a lot with $\mu = 50$ hours and $\sigma = 6$ hours. Although the received lot is a good lot, the manufacturer would reject such a lot (commit type I error) whenever the selected sample reveals an average of less than 48 hours. The means of the various samples of 64 batteries that can be selected from this lot are described by the normal distribution shown in Fig. 7–16. The probability that the mean battery life in a random sample of 64 is less than 48 hours can now be calculated as follows (Fig. 7–17):

$$z = \frac{\bar{x} - \mu}{\sigma_{\bar{x}}} = \frac{48 - 50}{0.75} = -2.67$$

Therefore,

$$P(\bar{x} < 48) = 0.5000 - 0.4962 = 0.0038.$$

Thus, using the modified decision rule ($n = 64$, $c^* = 48$), the probability of

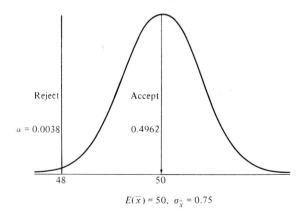

$$E(\bar{x}) = 50, \quad \sigma_{\bar{x}} = 0.75$$

Figure 7–17

making type I error is 0.0038. Using the original decision rule ($n = 36$, $c^* = 48$), on the other hand, the probability of making type I error is 0.0228. Hence, the probability of making type I error is reduced from 0.0228 to 0.0038 as a result of changing the sample size from 36 to 64 batteries.

Similarly, in order to measure the probability of making type II error using the modified rule ($n = 64$, $c^* = 48$) let us now assume again that the manufacturer receives a lot with $\mu = 47$ hours and $\sigma = 6$ hours. Although the lot received is a bad lot, the manufacturer would accept such a lot (commit type II error), whenever the sample selected from the lot shows an average equal to or greater than 48 hours. The means of the various samples of 64 batteries that can be selected from this lot are described by the normal distribution shown in Fig. 7–18. The probability that the mean battery life in a

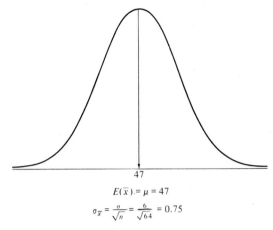

$$E(\bar{x}) = \mu = 47$$

$$\sigma_{\bar{x}} = \frac{\sigma}{\sqrt{n}} = \frac{6}{\sqrt{64}} = 0.75$$

Figure 7–18

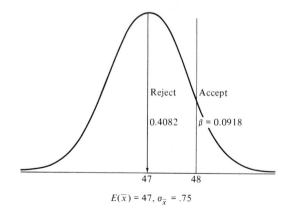

$E(\bar{x}) = 47,\ \sigma_{\bar{x}} = .75$

Figure 7–19

random sample of 64 is greater than or equal to 48 hours can now be calculated as follows (Fig. 7–19):

$$z = \frac{\bar{x} - \mu}{\sigma_{\bar{x}}} = \frac{48 - 47}{0.75} = 1.33.$$

Therefore

$$P(\bar{x} \geq 48) = 0.5000 - 0.4082 = 0.0918.$$

Thus, using the modified decision rule ($n = 64$, $c^* = 48$), we find that the probability of making type II error is 0.0918. Using the original decision rule ($n = 36$, $c^* = 48$), on the other hand, the probability of making type II error is 0.1587. Hence the probability of type II error is reduced from 0.1587 to 0.0918 as a result of enlarging the sample size from 36 to 64 batteries. In conclusion, we may now generalize and say that as the sample size, n is increased, the probabilities of making both type I and type II errors are reduced.

Example 9

A soft-drink bottling process is considered to be operating satisfactorily when the mean fill per bottle is 355 ml ($\mu = 355$). The following decision rule is adopted in order to ascertain whether the process continues to operate in a satisfactory manner: Select a random sample of 40 bottles each hour and determine their average fill. If this average, $\bar{x}$, is between 350 ml and

360 ml, let the process continue; otherwise stop and readjust the process. (Assume that the standard deviation of the process always remains 15 ml.)

a) Formulate the null hypothesis and the alternative hypothesis.

b) What is the risk of type I error?

c) What is the risk of type II error when $\mu = 365$ (excessively overfilling process)?

Solution (a): The null hypothesis is that the process is operating satisfactorily, or $\mu = 355$ ml. The alternative hypothesis is that the value of μ is no longer 355 ml and the process is either overfilling or underfilling. Stated briefly the two hypotheses are

$$H_0: \mu = 355,$$

$$H_1: \mu \neq 355.$$

Solution (b): Type I error is committed when the process is operating satisfactorily, i.e., $\mu = 355$, but the process is erroneously stopped because the mean of the sample selected does not fall between 350 and 360. Referring to the normal curve in Fig. 7–20, we can determine the probability of type I error (stopping the process erroneously) as follows: We denote the area under the normal curve between 350 and 355 by A_1 and the area between 355 and 360 by A_2; then

$$P(\text{Stop}) = P(\bar{x} \leq 350) + P(\bar{x} \geq 360),$$

$$= (.5 - A_1) + (.5 - A_2),$$

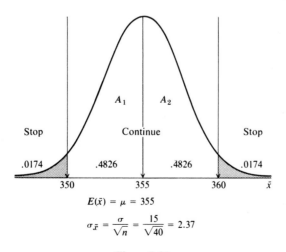

$$E(\bar{x}) = \mu = 355$$

$$\sigma_{\bar{x}} = \frac{\sigma}{\sqrt{n}} = \frac{15}{\sqrt{40}} = 2.37$$

Figure 7–20

which is the shaded area in Fig. 7–20. We now proceed to find $Area_1$ and $Area_2$:

$$z_1 = \frac{\bar{x} - \mu}{\sigma_{\bar{x}}} \qquad\qquad z_2 = \frac{\bar{x} - \mu}{\sigma_{\bar{x}}}$$

$$= \frac{350 - 355}{2.37} \qquad\qquad = \frac{360 - 355}{2.37}$$

$$= -2.11; \qquad\qquad = 2.11;$$

$$Area_1 = .4826. \qquad\qquad Area_2 = .4826.$$

Therefore,

$$P(\text{Stop}) = (.5 - .4826) + (.5 - .4826)$$

$$= .0174 + .0174$$

$$= .0348.$$

Thus, using the sampling decision rule, we find the probability of stopping a satisfactory process is .0348. This is the probability of type I error.

Solution (c): When $\mu = 365$ the process is overfilling and, hence, should be stopped. However, the process is allowed to continue (type II error is committed) whenever the mean of the sample selected falls between 350 and 360. We refer to Fig. 7–21 to calculate the probability of committing this error as follows: We denote the area under the normal curve between 350 and

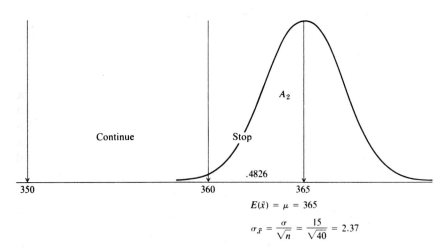

$$E(\bar{x}) = \mu = 365$$

$$\sigma_{\bar{x}} = \frac{\sigma}{\sqrt{n}} = \frac{15}{\sqrt{40}} = 2.37$$

Figure 7–21

365 by A_1 and the area between 360 and 365 by A_2; then

$$P(\text{Continue}) = A_1 - A_2.$$

We proceed to find Area$_1$ and Area$_2$:

$$z_1 = \frac{\bar{x} - \mu}{\sigma_{\bar{x}}} \qquad\qquad z_2 = \frac{\bar{x} - \mu}{\sigma_{\bar{x}}}$$

$$= \frac{350 - 365}{2.37} \qquad\qquad = \frac{360 - 365}{2.37}$$

$$= -6.33; \qquad\qquad = -2.11;$$

$$\text{Area}_1 = .5000. \qquad\qquad \text{Area}_2 = .4826.$$

Therefore,

$$P(\text{Continue}) = A_1 - A_2$$

$$= .5000 - .4826$$

$$= .0174.$$

Thus, although the process is overfilling ($\mu = 365$), this process is allowed to continue with a probability of .0174.

Example 10

A governmental housing agency undertakes public housing projects in various communities. The housing agency considers a project essential to a community when the average family income, μ, in that community is $5000 or less. The project is deemed unjustified when average income is $6000 or more. The agency wishes to develop a sampling decision rule to be used in deciding whether to undertake any given project. The decision rule is, select a number of n families from any given community and determine their average income. If this average, $\bar{x}$, is equal to c^* or less, undertake the project. If it is greater than c^*, do not undertake the project. Determine the values of n and c^* so that the probablity of not undertaking the project when it is justified (the probability of type I error) is $\alpha = .05$; and the probability of undertaking the project when it is not justified (probability of type II error) is $\beta = .10$. Assume that for every community considered by the agency that the standard deviation is $2000 ($\sigma = 2000).

Solution: We need to develop a sampling rule (determine n and c^*) that will satisfy the conditions stipulated by the agency. We summarize these conditions in Table 7–4. According to Table 7–4 when $\mu = $5000, then application of the sampling rule should result in undertaking a project with a probability of .95 and not undertaking it with probability of .05. When $\mu =$

Table 7–4

Summary of Decisions

Decision dictated by sample outcome	Average family income	
	$H_0: \mu = \$5000$	$H_1: \mu = \$6000$
Undertake project $(\bar{x} \leq c^*)$	Correct decision (probability $= .95$)	Type II error $(\beta = .10)$
Do not undertake $(\bar{x} > c^*)$	Type I error $(\alpha = .05)$	Correct decision (probability $= .90$)

$6000, on the other hand, application of the sampling rule should result in undertaking a project with a probability of .10 and not undertaking with a probability of .90. In summary, we must find the values for n and c^* that will render the risk of type I error .05 and the risk of type II error .10. We consider the implications of these two conditions in turns.

Type I Error

Type I error is committed when a project is justified, i.e., $\mu = \$5000$, but the project is erroneously not undertaken because the mean of the sample selected is greater than c^*. The probability of committing this error is set as .05.

The means of the various samples of size n that can be selected from a population with $\mu = 5000$ and $\sigma = 2000$ are described by the normal distribution shown in Fig. 7–22. Since the probability of committing type I

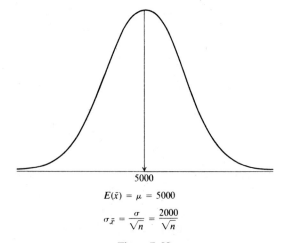

$$E(\bar{x}) = \mu = 5000$$

$$\sigma_{\bar{x}} = \frac{\sigma}{\sqrt{n}} = \frac{2000}{\sqrt{n}}$$

Figure 7–22

error is set as .05, we specify the value of c^* so that .05 of the area under the normal curve is to the right of c^*, as in Fig. 7–23.

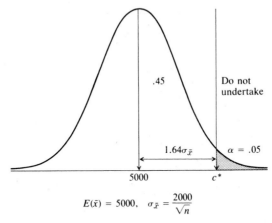

$$E(\bar{x}) = 5000, \quad \sigma_{\bar{x}} = \frac{2000}{\sqrt{n}}$$

Figure 7–23

With .45 of the area under the normal curve between 5000 and c^*, Table A indicates

$$c^* = 5000 + 1.64\sigma_{\bar{x}}$$

$$= 5000 + 1.64\frac{2000}{\sqrt{n}},$$

or,

$$c^* = 5000 + \frac{3280}{\sqrt{n}}. \qquad (\text{I})$$

Equation (I) contains the unknowns n and c^*. By considering type II error next, we will obtain another equation containing the same unknowns. The two equations are then simultaneously solved to find c^* and n.

Type II Error

Type II error is committed when a project is not justified, i.e., $\mu = \$6000$ but the project is undertaken because the mean of the sample selected is below c^*. The probability of committing this error is set as .10.

The means of the various samples of size n that can be selected from a population with $\mu = 6000$ and $\sigma = 2000$ are described by the normal curve shown in Fig. 7–24. Since the probability of type II error is set as .10, we specify the value of c^* so that .10 of the area under the normal curve

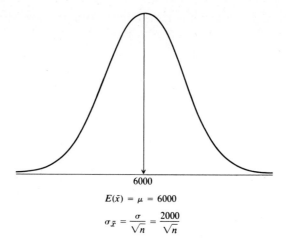

$$E(\bar{x}) = \mu = 6000$$

$$\sigma_{\bar{x}} = \frac{\sigma}{\sqrt{n}} = \frac{2000}{\sqrt{n}}$$

Figure 7–24

is to the left of c^* as in Fig. 7–25. With .40 of the area under the normal curve between c^* and 6000, Table A indicates that

$$c^* = 6000 - 1.28\sigma_{\bar{x}}$$

$$= 6000 - 1.28\frac{2000}{\sqrt{n}},$$

or

$$c^* = 6000 - \frac{2560}{\sqrt{n}}. \tag{II}$$

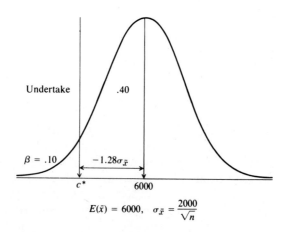

$$E(\bar{x}) = 6000, \quad \sigma_{\bar{x}} = \frac{2000}{\sqrt{n}}$$

Figure 7–25

Equation (II) can be combined with equation (I), developed in the preceding section, to solve for c^* and n. We proceed by subtracting equation (II) from equation (I) to obtain equation (III):

$$c^* = 5000 + \frac{3280}{\sqrt{n}} \tag{I}$$

$$c^* = 6000 - \frac{2560}{\sqrt{n}} \tag{II}$$

$$0 = -1000 + \frac{5840}{\sqrt{n}} \tag{III}$$

We solve for n in equation (III) as follows:

$$\frac{5840}{\sqrt{n}} = 1000$$

$$\sqrt{n} = \frac{5840}{1000} = 5.84.$$

Hence,

$$n = (5.84)^2 = 34.11.$$

Substituting $\sqrt{n} = 5.84$ in equation (I) gives

$$c^* = 5000 + \frac{3280}{5.84} = 5561.64.$$

With $n = 34.11$ (say 35 for convenience) and with $c = 5561.64$, the sampling decision rule becomes, select a random sample of 35 families from any given community and determine their average annual income. If this average is equal to or less than \$5561.64, undertake the project. If it is greater than \$5561.64, do not undertake the project.

Example 11

Given the two hypotheses

$$H_0: \mu = 80 \quad \text{and} \quad H_1: \mu \neq 80,$$

use a sample of 40 observations ($n = 40$) to develop a sampling decision rule so that the probability of committing type I error is .05. Assume that the population standard deviation, σ, is 12.

Solution:

This is a two-sided test. The null hypothesis is rejected when $\mu < 80$ as well as when $\mu > 80$. Consequently, the appropriate decision rule is, select a random sample of 40 observations and determine the mean of the sample

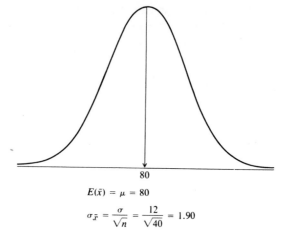

$$E(\bar{x}) = \mu = 80$$

$$\sigma_{\bar{x}} = \frac{\sigma}{\sqrt{n}} = \frac{12}{\sqrt{40}} = 1.90$$

Figure 7–26

($\bar{x}$). If $\bar{x}$ does not fall between c_1^* and c_2^*, reject H_0. We must now find the values of c_1^* and c_2^* so that the probability of committing type I error is .05.

Type I error is committed when H_0 is true ($\mu = 80$), but this hypothesis is erroneously rejected because the mean of the sample selected does not fall between c_1^* and c_2^*. Once again, the probability of committing such error is set as .05.

The means of the various samples of size 40 that can be selected from a population with $\mu = 80$ and $\sigma = 12$ are described by the normal curve shown in Fig. 7–26.

Since the probability of committing type I error is set as .05, we specify the values of c_1^* and c_2^* so that .025 of the area under the normal curve is to the left of c_1^*, and .025 of the area is to the right of c_2^*, as shown in Fig. 7–27.

With .475 of the area under the normal curve between c_1^* and 80, Table A indicates that

$$c_1^* = 80 - 1.96\sigma_{\bar{x}}$$

$$= 80 - 1.96(1.90)$$

$$= 76.276.$$

In a similar fashion we find

$$c_2^* = 80 + 1.96\sigma_{\bar{x}}$$

$$= 80 + 1.96(1.90)$$

$$= 83.724.$$

Hence, our sampling decision rule is, select a random sample of 40 observa-

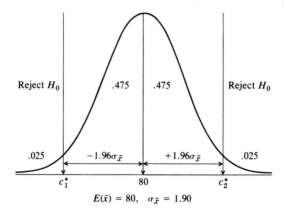

$$E(\bar{x}) = 80, \quad \sigma_{\bar{x}} = 1.90$$

Figure 7–27

tions and determine the mean of the sample. If this mean does not fall between 76.276 and 83.724, reject H_0. Using this decision rule, the probability of committing type I error is .05.

Example 12

Given the two hypotheses

$$H_0: \mu = 80 \quad \text{and} \quad H_1: \mu < 80,$$

use a sample of 40 observations to develop a sampling decision rule so that the probability of type I error is .05. Assume that the population standard deviation is 12.

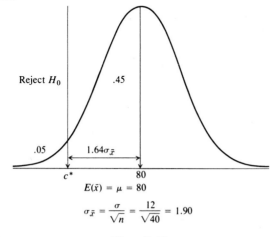

$$E(\bar{x}) = \mu = 80$$

$$\sigma_{\bar{x}} = \frac{\sigma}{\sqrt{n}} = \frac{12}{\sqrt{40}} = 1.90$$

Figure 7–28

Solution:

In contrast to example 11, this is a one-sided test. The null hypothesis is rejected only when $\mu < 80$. Consequently, the appropriate decision rule is, select a random sample of 40 observations and determine the mean of the sample ($\bar{x}$). If $\bar{x}$ is less than c^*, reject H_0. We must now find the value of c^* so that the probability of committing type I error is .05.

Type I error is committed when H_0 is true ($\mu = 80$), but this hypothesis is erroneously rejected because the mean of the sample selected is less than c^*. Since the probability of committing such error is set as .05, we must specify the value of c^* so that .95 of the area under the normal curve is to the left of c^*, as indicated in Fig. 7–28.

With .45 of the area under the normal curve between c^* and 80, Table A indicates that

$$c^* = 80 - 1.64\sigma_{\bar{x}}$$

$$= 80 - 1.64(1.90)$$

$$= 76.865$$

Hence, our sampling decision rule is, select a random sample of 40 observations and determine the mean of the sample. If this mean is less than 76.865, reject H_0. Using this decision rule, the probability of committing type I error is .05.

EXERCISES, SET 2

1. A foundry desires to produce iron castings with an average weight of 20 pounds and a standard deviation of 2 pounds ($\mu = 20$, $\sigma = 2$). To decide whether the manufacturing process is operating satisfactorily a sample of 40 castings is selected each hour from the output of the process and their average weight is determined. If this average is between 19.5 and 20.5 pounds, no action is taken; otherwise, stop and correct the process. Assume that the standard deviation of the process is always two pounds.

   a) What is the probability of type I error?

   b) What is the probability of type II error when $\mu = 19.0$ pounds?

   c) What is the probability of type II error when $\mu = 21.1$ pounds?

2. The Standard Shoe Company operates a growing chain of retail shoe stores in various communities throughout California. The company wishes to have a branch in any community with an average annual family expenditure on shoes of $150 or more. In contrast, the company considers it unprofitable to operate a store in a community where an average expenditure on shoes is $125 or less. The company wishes to develop a decision rule to be used in deciding whether to open a new branch. The decision rule is, select a number of n families from any given community and determine their average annual expenditure on shoes. If this average is equal to c^* or more, establish a store in that community. If it is less than c^*, do not open a store. Determine the

values of n and c^* so that the probability of not establishing a store when $\mu = \$150$ is .01; and the probability of establishing a store when $\mu = \$125$ is .05. (Assume that the standard deviation of annual family expenditure on shoes, σ, is $50.)

3. A manufacturer of oleomargarine uses national television advertising to promote their brand of oleomargarine. In addition they undertake local advertising campaigns to supplement national advertising in some areas. The manufacturer believes that local advertising is not necessary in areas where the average amount of oleomargarine used by a family in a week is 2 pounds or more. Local advertising is needed however when average consumption is 1.5 pounds or less. In order to decide whether to undertake a local advertising campaign in any given area, the Market Research Department has adapted the following decision rule: Select a random sample of 36 families from the area and determine their average consumption of oleomargarine per week. If this average is 1.8 pounds or more, do not advertise locally. If it is less than 1.8 pounds, advertise. (Assume that the standard deviation of the amount of oleomargarine consumed per family in a week is .6 pounds).

a) Formulate the null hypothesis and the alternative hypothesis.

b) What is the risk of type I error?

c) What is the risk of type II error when $\mu = 1.5$ pounds?

4. Automobile insurance companies operating in the State of California have petitioned the State Insurance Commission for an increase in automobile insurance rates, claiming rising accident settlement costs. Last year, the average settlement cost was $502 per claim with a standard deviation of $182 ($\mu = 502$ and $\sigma = 182$). In order to decide whether to approve the increase in premium rate, the State Commission adopted the following decision rule: Select a random sample of 50 claims from settlements paid this year and determine the average amount paid per claim. If it is less than $550, do not approve the increase. Assume that the population standard deviation, σ, remains $182.

a) Formulate the null hypothesis and alternative hypothesis.

b) What is the probability of type I error?

c) What is the probability of type II error when $\mu = \$600$?

5. The Department of Health, Education, and Welfare has adopted the following decision rule in deciding on whether to introduce a comprehensive national health insurance plan: Select a random sample of n families and determine the average annual medical expenses. If this average is equal to c^* or more, introduce the insurance plan. If it is less than c^*, do not introduce the plan. Determine the value of n and c^* so that the probability of introducing the plan is .05 when average family medical expenses is $800 per year ($\mu = \$800$), and the probability of not introducing the plan is .01 when $\mu = \$1000$. Assume that the population standard deviation for medical expenses is $300.

6. Given the two hypothesis

$$H_0: \mu = 150 \qquad \text{and} \qquad H_1: \mu > 150,$$

use a sample of 64 observations to develop a sampling decision rule so that the probability of type I error is .05. Assume that the population standard deviation is 16.

7. Given the two hypothesis

$$H_0: \mu = 65 \quad \text{and} \quad H_1: \mu \neq 65,$$

use a sample of 49 observations to develop a sampling decision rule so that the probability of type I error is .01. Assume that the population standard deviation is 14.

8. Given the two hypothesis

$$H_0: \mu = 90 \quad \text{and} \quad H_1: \mu < 90,$$

use a sample of 40 observations to develop a sampling decision rule so that the probability of type II error is .10 when $\mu = 83$.

Assume that the population standard deviation is 15.

8
Tests of
Significance:
Differences
between
Means of
Various Samples

Tests of
Significance:
Differences
between
Means of
Various Samples

When a random sample of students from San Francisco State College shows a mean age of 23 years and a random sample from Los Angeles State College shows a mean age of 24 years, can we safely conclude that the average ages in the two colleges are really different? In other words, can we use the difference between the means of the two samples as conclusive evidence that the means of the two populations are not the same?

Similarly, when three samples from three different brands of 9-volt batteries show average lives of 40, 50, and 60 hours respectively, are we justified in concluding that these three brands are really different in regard to their average lives?

The theoretical foundations and techniques needed in order to provide a satisfactory answer to these important questions are the subject matter of this chapter. Specifically, the first two sections deal with the theory and procedure used in comparing the means of *two* samples. The technique of comparing more *than two* sample means is dealt with in the third section.

THE SAMPLING DISTRIBUTION OF THE DIFFERENCE BETWEEN TWO SAMPLE MEANS

Let us assume that we have two distinct populations: the first consists of the ages of all students attending Harvard University, and the second consists of the ages of all students attending Columbia University. The mean ages and the standard deviation for the two populations are recorded below.

	Harvard	Columbia
Population mean	$\mu_1 = 24$ years	$\mu_2 = 22$ years
Population standard deviation	$\sigma_1 = 7$ years	$\sigma_2 = 6$ years

235

If we select *many* random samples of 100 students from the Harvard population and record the mean of each sample, the various sample means might be as shown below.

Harvard sample means, years
24
23
25
23
26
24
22
⋮
25

Thus the mean of the first sample of 100 students is 24 years, the mean of the second sample is 23 years, and the mean of the last sample is 25 years.

Similarly, let us assume that we have selected an equal number of random samples of 50 students from Columbia University. The means of these samples might be as given below.

Columbia sample means, years
23
21
22
20
24
22
21
⋮
22

In comparing the results of the two sets of sample means, let us randomly pair these samples and compute the difference between the means of each pair. The difference is denoted by d.

Table 8–1

$\bar{x}_1$'s (Harvard)	$\bar{x}_2$'s (Columbia)	$d = (\bar{x}_1 - \bar{x}_2)$
24	23	1
23	21	2
25	22	3
23	20	3
26	24	2
24	22	2
22	21	1
$\vdots$	$\vdots$	$\vdots$
25	22	3

Remembering that $\bar{x}_1$'s are sample means drawn from a population with $\mu_1 = 24$, and that $\bar{x}_2$'s are sample means drawn from a population with $\mu_2 = 22$, let us examine some of the properties of the differences recorded in the last column of Table 8–1.

In the first place, although the difference between the means of the first pair of samples is 1 year, that between the second pair is 2 years, and that between the last pair is 3 years, we would expect the average of these differences to be 2 years (that is, $24 - 22$). The first property of the differences between two sample means is therefore

$$E(d) = \mu_1 - \mu_2,$$

where $E(d)$ is the average of *all* differences, μ_1 is the mean of the population from which the first set of samples is selected (Harvard), and μ_2 is the mean of the population from which the second set of samples is selected (Columbia).

Statisticians have studied these differences in great detail. They have discovered that these differences are approximately normally distributed and have a standard deviation or standard error equal to

$$\sqrt{\frac{\sigma_1^2}{n_1} + \frac{\sigma_2^2}{n_2}},$$

where σ_1 is the standard deviation of the population from which the set of samples is drawn (Harvard), n_1 is the size of each sample of the first set of samples (100 students), σ_2 is the standard deviation of the population from which the second set of samples is drawn (Columbia), and n_2 is the size of each sample of the second set (50 students).

Denoting the standard deviation or the standard error of the differences by σ_d, we summarize the properties of these differences as follows (see Fig. 8–1).

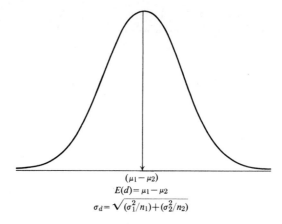

$$(\mu_1 - \mu_2)$$
$$E(d) = \mu_1 - \mu_2$$
$$\sigma_d = \sqrt{(\sigma_1^2/n_1) + (\sigma_2^2/n_2)}$$

Figure 8–1

For our samples from Harvard and Columbia, these properties imply that the average of all differences between the means of paired samples is 2 years and their standard error is 1.1 years (see Fig. 8–2 and the computation below).

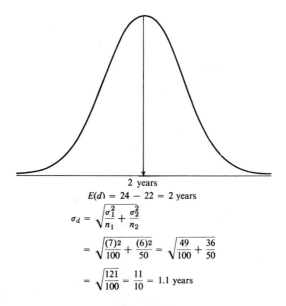

2 years
$$E(d) = 24 - 22 = 2 \text{ years}$$
$$\sigma_d = \sqrt{\frac{\sigma_1^2}{n_1} + \frac{\sigma_2^2}{n_2}}$$
$$= \sqrt{\frac{(7)^2}{100} + \frac{(6)^2}{50}} = \sqrt{\frac{49}{100} + \frac{36}{50}}$$
$$= \sqrt{\frac{121}{100}} = \frac{11}{10} = 1.1 \text{ years}$$

Figure 8–2

Having determined the mean and the standard error of the differences between the means of the paired samples, we can draw some useful conclusions.   Since these differences are approximately normally distributed, we expect 68% of the differences to fall within the range

$$E(d) \pm 1\sigma_d \quad \text{that is} \quad 2 \pm 1.1,$$

or between .9 and 3.1 years (Fig. 8–3).

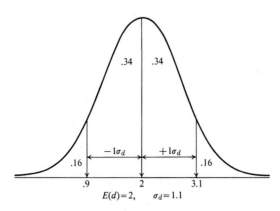

$E(d)=2, \qquad \sigma_d=1.1$

Figure 8–3

In the same manner, we expect 95% of all these differences to fall within the range

$$E(d) \pm 1.96\sigma_d,$$
$$2 \pm 1.96(1.1), \quad \text{or} \quad 2 \pm 2.16,$$

or between −.16 and 4.16 years (Fig. 8–4).

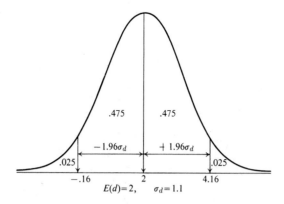

$E(d)=2, \qquad \sigma_d=1.1$

Figure 8–4

Let us now summarize the implications of the preceding discussion. Let us assume that the mean ages at Harvard and Columbia are 24 and 22 years respectively, and that their respective standard deviations are 7 and 6 years. Then, if a random sample of 100 students is selected from Harvard and another sample of 50 students from Columbia, one expects the mean of the Harvard sample to be, *on the average,* higher than the mean of the Columbia sample by 2 years. However, the difference between the means of the two samples may very well be smaller or greater than 2 years. For example, the Harvard sample mean could *exceed* the Columbia mean by 4.16 years or more or it could be *lower* than the Columbia mean by .16 year or more. But the probability of obtaining either of these extreme results is only .025 (see Fig. 8–4).

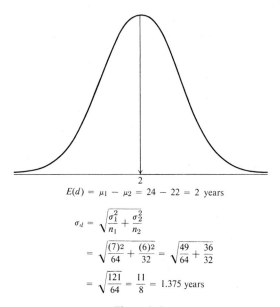

$$E(d) = \mu_1 - \mu_2 = 24 - 22 = 2 \text{ years}$$

$$\sigma_d = \sqrt{\frac{\sigma_1^2}{n_1} + \frac{\sigma_2^2}{n_2}}$$

$$= \sqrt{\frac{(7)^2}{64} + \frac{(6)^2}{32}} = \sqrt{\frac{49}{64} + \frac{36}{32}}$$

$$= \sqrt{\frac{121}{64}} = \frac{11}{8} = 1.375 \text{ years}$$

Figure 8–5

As a final remark, let us assume that we have selected a sample of 64 students from Harvard and a sample of 32 students from Columbia. What is the probability that the Harvard sample mean exceeds that of Columbia by 3 years or more? We have already learned that if *many* samples of 64 students are selected from Harvard and many samples of 32 students from Columbia, then the differences between the means of paired samples are normally distributed with an average equal to $(\mu_1 - \mu_2)$ and a standard deviation or standard error equal to

$$\sqrt{\frac{\sigma_1^2}{n_1} + \frac{\sigma_2^2}{n_2}}$$

(see Fig. 8–5). Having determined the mean and the standard error of the differences between the sample means, we can now calculate the probability that the difference between any two sample means is 3 years or more (see Fig. 8–6):

$$z = \frac{d - E(d)}{\sigma_d}$$

$$= \frac{3 - 2}{1.375} = \frac{1}{1.375} = .73.$$

Therefore

$$P(d \geq 3) = .5000 - .2673 = .2327.$$

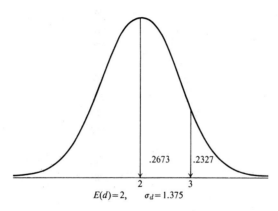

.2673   .2327

2   3

$E(d) = 2,$ $\sigma_d = 1.375$

Figure 8–6

TEST OF HYPOTHESIS CONCERNING THE DIFFERENCE BETWEEN TWO SAMPLE MEANS

A simple random sample of 100 students attending San Francisco State College showed an average age of 23 years and a standard deviation of 4 years while a simple random sample of 50 Los Angeles State College students revealed an average age of 21 years and a standard deviation of 5 years. From the results of these two samples, can we safely conclude that the average age in the two colleges is not the same?

The mere fact that the mean of a sample of San Francisco State College is higher than the mean of a sample from Los Angeles State College is not conclusive evidence that the average age in the two schools is not the same. Remembering that two samples drawn from the *same* population rarely show the same arithmetic mean, we must be very careful in interpreting the difference between the means of two samples.

Generally speaking, when two sample means are different, we have two hypotheses to explore. First, there is the null hypothesis that the two populations from which the two samples originate have the same mean ($\mu_1 = \mu_2$). If this is the case, then the observed difference between the two sample means is *not significant* and is attributed to chance or random sampling fluctuations. The alternative hypothesis to be explored is that the two samples are drawn from populations which have different means ($\mu_1 \neq \mu_2$). If this alternative hypothesis is true, the observed difference between the two sample means is deemed *significant*.

When two sample means are different, how can we decide whether or not the difference between the two means is significant? The standard procedure is to test the validity of the null hypothesis, which states that $\mu_1 = \mu_2$, utilizing the information from the two samples. On the basis of the evidence produced by the two samples, we will either reject the null hypothesis or reserve judgment. If the null hypothesis is rejected, the observed difference between the two sample means is significant. However, the observed difference is not significant whenever judgment is reserved.

Having outlined the general nature of our test, we can now proceed with the solution to our problem. For clarity, the problem is divided into four sections.

1. Data

San Francisco State	Los Angeles State
$\bar{x}_1 = 23$ years	$\bar{x}_2 = 21$ years
$s_1 = 4$ years	$s_2 = 5$ years
$n_1 = 100$ students	$n_2 = 50$ students

2. Question: Is the observed difference between the two sample means significant or not?

3. Hypothesis

$$H_0: \mu_1 = \mu_2,$$
$$H_1: \mu_1 \neq \mu_2.$$

4. Solution: Let us assume that we have selected many samples of size 100 from San Francisco State and an equal number of samples of size 50 from Los Angeles State. When these two sets of samples are paired and the difference between the means of each pair is recorded, then according to what we have already learned, the average of all these differences is equal to ($\mu_1 - \mu_2$) and their standard error σ_d is equal to

$$\sqrt{\frac{\sigma^2}{n_1} + \frac{\sigma^2}{n_2}}.$$

But since the standard deviations in the two populations are not known, we may use the standard deviation of the sample as an approximation for the standard deviation of the population. Using this approximation, we find that the *estimated* standard error of the difference between sample means, denoted by $\hat{\sigma}_d$, is

$$\sqrt{\frac{s_1^2}{n_1} + \frac{s_2^2}{n_2}},$$

whereas the true standard error of the difference between two means is

$$\sqrt{\frac{\sigma_1^2}{n_1} + \frac{\sigma_2^2}{n_2}}.$$

Now, if the null hypothesis is true, that is, $\mu_1 = \mu_2$, then the distribution of the difference between the sample means is described by Fig. 8–7. Figure 8–7 essentially indicates that if the null hypothesis is true, although the difference between the means of any paired sample may not be zero, the average of all such differences is equal to zero. Furthermore, these differences are approximately normally distributed and their estimated standard error is .812 year.

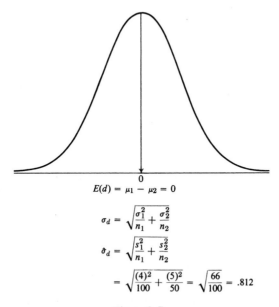

$$E(d) = \mu_1 - \mu_2 = 0$$

$$\sigma_d = \sqrt{\frac{\sigma_1^2}{n_1} + \frac{\sigma_2^2}{n_2}}$$

$$\hat{\sigma}_d = \sqrt{\frac{s_1^2}{n_1} + \frac{s_2^2}{n_2}}$$

$$= \sqrt{\frac{(4)^2}{100} + \frac{(5)^2}{50}} = \sqrt{\frac{66}{100}} = .812$$

Figure 8–7

To evaluate the validity of the null hypothesis in light of the two samples actually selected ($\bar{x}_1 = 23$, $\bar{x}_2 = 21$, and $d = 2$), let us calculate the z-value for an observed difference of 2 years or more, assuming that the null hypothesis is true (see Fig. 8-8). We have

$$z = \frac{d - E(d)}{\hat{\sigma}_d}$$

$$= \frac{2 - 0}{.812} = 2.46.$$

Since the value of z is 2.46, which is greater than 1.96, the null hypothesis is rejected and the difference between the two sample means is deemed significant at the .05 level. It is not significant at the .01 level, since the actual z-value is less than 2.58. (In testing the significance of the difference between two sample means, we always formulate the alternative hypothesis as $\mu_1 \neq \mu_2$. Consequently, the test is two-tailed.)

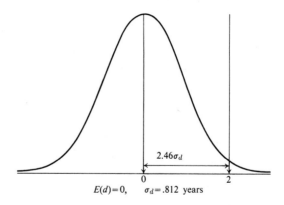

2.46σ_d

0   2

$E(d) = 0$, $\sigma_d = .812$ years

Figure 8–8

Example 1

To compare the average life of two brands of 9-volt batteries, a sample of 100 batteries from each brand is tested. The sample selected from the first brand shows an average life of 47 hours and a standard deviation of 4 hours. A mean life of 48 hours and a standard deviation of 3 hours are recorded for the sample from the second brand. Is the observed difference between the means of the two samples significant at the .05 level?

There are two hypotheses,

$$H_0: \mu_1 = \mu_2,$$
$$H_1: \mu_1 \neq \mu_2,$$

and if the null hypothesis is true, then $E(d) = 0$, and $\hat{\sigma}_d = .5$ hour, as shown in our calculations below and in Fig. 8–9.

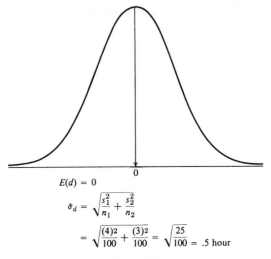

$$E(d) = 0$$

$$\hat{\sigma}_d = \sqrt{\frac{s_1^2}{n_1} + \frac{s_2^2}{n_2}}$$

$$= \sqrt{\frac{(4)^2}{100} + \frac{(3)^2}{100}} = \sqrt{\frac{25}{100}} = .5 \text{ hour}$$

Figure 8–9

The null hypothesis can be evaluated by computing the z-value for the observed difference between the means of the two samples ($\bar{x}_1 = 47$, $\bar{x}_2 = 48$, $d = -1$). (See Fig. 8–10.) We have

$$z = \frac{d - E(d)}{\hat{\sigma}_d}$$

$$= \frac{(-1) - 0}{.5} = -2.0.$$

Since the absolute z-value is greater than 1.96, the null hypothesis is rejected, and the difference between the means of the two samples is regarded as significant at the .05 level.

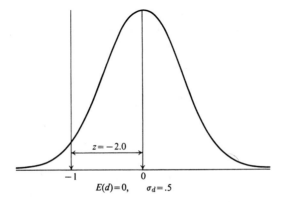

$$z = -2.0$$

$$-1 \qquad\qquad 0$$
$$E(d) = 0, \qquad \sigma_d = .5$$

Figure 8–10

Comparing the Means of Two Small-Size Samples

In testing the difference between two sample means, we have tacitly assumed that both samples are large-size samples. We have, furthermore, utilized a technique that essentially consists of computing a value of z, where

$$z = \frac{d}{\hat{\sigma}_d} = \frac{(\bar{x}_1 - \bar{x}_2)}{\sqrt{(s_1^2/n_1) + (s_2^2/n_2)}}.$$

When the above absolute value of z exceeds 1.96, the difference between the two sample means is regarded as significant at the .05 level. Similarly, the difference between the two sample means is regarded as significant at the .01 level when the absolute value of z exceeds 2.58.

The above technique is slightly modified when the means to be compared belong to small-size samples. Instead of the normal distribution, a t-distribution with $(n_1 + n_2 - 2)$ degrees of freedom is used.* Consequently, we must compute a t-value, where

$$t = \frac{d}{\hat{\sigma}_d} = \frac{(\bar{x}_1 - \bar{x}_2)}{\sqrt{\dfrac{(n_1 - 1)s_1^2 + (n_2 - 1)s_2^2}{(n_1 + n_2 - 2)}} \sqrt{\dfrac{1}{n_1} + \dfrac{1}{n_2}}}.$$

Using $(n_1 + n_2 - 2)$ degrees of freedom, we compare the above value of t with a critical value of $t_{.025}$ or $t_{.005}$, depending on whether the .05 or .01 level of significance is used. The difference between the means of the two samples is regarded as significant when the *observed* absolute value of t exceeds $t_{.025}$. Similarly, when the observed absolute value of t exceeds $t_{.005}$, the difference between the two sample means is regarded as significant at the .01 level. We now clarify this procedure with an illustrative problem.

Example 2

A reading test is given to an elementary school class that consists of 12 Anglo-American children and 10 Mexican-American children. The results of the test are

Anglo-American	Mexican-American
$\bar{x}_1 = 74$	$\bar{x}_2 = 70$
$s_1 = 8$	$s_2 = 10$

Is the difference between the means of the two groups significant at the .05 level?

---

*   The use of the t-distribution requires the following assumptions:
1. The two samples selected are independent random samples.
2. The parent populations are normally distributed with equal variances.

Solution:

$$H_0: \mu_1 = \mu_2,$$
$$H_1: \mu_1 \neq \mu_2,$$
$$\text{Level of significance} = .05.$$

To test the null hypothesis, we compute the observed value of t as

$$t = \frac{(\bar{x}_1 - \bar{x}_2)}{\sqrt{\dfrac{(n_1 - 1)s_1^2 + (n_2 - 1)s_2^2}{(n_1 + n_2 - 2)}}\sqrt{\dfrac{1}{n_1} + \dfrac{1}{n_2}}}$$

$$= \frac{(74 - 70)}{\sqrt{\dfrac{(12 - 1)(8)^2 + (10 - 1)(10)^2}{(12 + 10 - 2)}}\sqrt{\dfrac{1}{12} + \dfrac{1}{10}}}$$

$$= \frac{4}{\sqrt{\dfrac{1604}{20}}\sqrt{\dfrac{22}{120}}}$$

$$= \frac{4}{\sqrt{14.703}}$$

$$= 1.043.$$

With 20 degrees of freedom (df $= 12 + 10 - 2$), the critical value of $t_{.025}$ is 2.086. Since the actual value of t is 1.043 and the critical value is 2.086, the difference between the means of the two groups of children is not regarded as significant at the .05 level.

EXERCISES, SET 1

1. The mean height of a soldier in the U.S. Army is 69 inches with a standard deviation of 2 inches. The mean height of a soldier in the French army is 67 inches with a standard deviation of 1 inch. If a random sample of 80 soldiers is selected from each army, what is the probability that the mean height

   a) in the United States sample is greater than that of the French sample by 2.4 inches or more?

   b) in the French sample is smaller than that of the United States sample by 1.8 inches or less?

2. Given the following data on the lives of two different brands of nine-volt batteries:

Brand A, hours	Brand B, hours
$\mu_1 = 51$	$\mu_2 = 50$
$\sigma_1 = 8$	$\sigma_2 = 6$

If a random sample of 100 batteries is selected from each brand, what is the probability that the mean life of sample B is

a) *greater* than that of A by .6 hour or more?

b) *smaller* than that of A by .6 or more?

3. To assess the impact of windowless schools on the psychological development of school children, an anxiety test was given to a class of 40 children in a windowless school. The same test was also given to a similar class of 30 students in a school with windows. The results of the test are shown below.

Windowless schools	Schools with windows
$\bar{x}_1 = 117$	$\bar{x}_2 = 112$
$s_1 = 10$	$s_2 = 12$
$n_1 = 40$	$n_2 = 30$

If you are willing to reject a true hypothesis no more than 5 times in 100, can you conclude that the impact of the two types of school on the anxiety of the children is not the same?

4. The systolic blood pressure of a group of 64 persons from a South Pacific island was compared with that of a similar group of 64 persons living in New York City. The data below summarize the findings.

	South Pacific island	New York
$\sum x$	9088	9472
$\sum (x - \bar{x})^2$	9072	16128
n	64	64

Is the difference between the means of the two sample groups significant at the .05 level?

5. To compare the vitamin-A content of two different brands of vitamin capsules, a sample of 6 capsules was selected from each brand. The results are given below.

	First brand, USP units	Second brand, USP units
Mean	5000	4800
Standard deviation	400	300

If you are willing to reject a true hypothesis no more than once in 100, can you conclude that the vitamin-A contents of the two brands of capsules are not the same?

6. Given: The nicotine content of several cigarettes taken from two different brands.

$$\text{Brand } x: 17, 20, 20, 23$$
$$\text{Brand } y: 18, 20, 21, 22, 24$$

Using the .05 level of significance, can you conclude that the nicotine content of the two brands of cigarettes is not the same?

COMPARING SEVERAL SAMPLE MEANS—ANALYSIS OF VARIANCE

A sample of five batteries was selected from each of three different brands. The lifetime in hours was recorded for each individual battery. The results are shown below.

Brand A, hours	Brand B, hours	Brand C, hours
40	50	60
30	50	60
40	50	60
60	60	70
30	40	50
$\bar{x}_1 = 40$	$\bar{x}_2 = 50$	$\bar{x}_3 = 60$

Are these brands different with respect to their average life?

When several samples reveal different means, an investigator has two hypotheses to explore. First, there is the null hypothesis that these samples are from populations with equal means, that is, $\mu_1 = \mu_2 = \mu_3$. If this is the case, then the observed differences among *sample means* are due to sampling fluctuations and hence are not significant.

The alternative hypothesis to be explored is that each of the three samples is obtained from a given population, and that means of the three populations are *not all* equal. According to this hypothesis, the differences between the various *sample means* are significant.

In comparing the several sample means, the investigator tests the validity of the null hypothesis, utilizing the information contained in the samples. On the basis of such information, he either rejects the null hypothesis or reserves judgment. If the null hypothesis is rejected, the differences between the various sample means are regarded as significant. On the other hand, if the investigator is compelled to reserve judgment, the differences are not regarded as significant.

We must know clearly two basic assumptions underlying our method of analysis before we can proceed. First, our method of analysis assumes that each of our three samples is obtained from a normally distributed population. Second, it is assumed that the three populations have the same standard deviation, that is, $\sigma_1 = \sigma_2 = \sigma_3$. If these two assumptions can be justified, then the null hypothesis which states that $\mu_1 = \mu_2 = \mu_3$ can be extended to mean that the three samples are from one and the same population.

The general procedure used in our test is to assume first that the null hypothesis is true, i.e. to assume that all the samples belong to the same population. We then make two estimates of that population variance, σ^2. These two estimates are called the *variance within samples* and the *variance between samples*. If the disparity between the two estimates is "considerable," the null hypothesis is rejected and the differences between the various sample means are regarded as significant.

Assuming that our three samples do belong to the same population, how can we estimate the variance of this population, σ^2? A logical and a reasonable approach is to compute the variance of each sample and to use the average of the variances of the three samples as an estimate of σ^2. This estimate, called the *variance within samples*, is then contrasted with a second estimate of σ^2 obtained by means of an alternative procedure to be discussed later in this section. The computations used in determining the variance within samples are carried out below.

Estimating the Population Variance—Variance Within Samples

Recalling the fact that the variance is the square of the standard deviation, we compute the variance of each of the three samples, using the formula

$$s^2 = \frac{\sum (x - \bar{x})^2}{n - 1}.$$

Table 8–2

Sample 1		Sample 2		Sample 3	
x	$(x - \bar{x})^2$	x	$(x - \bar{x})^2$	x	$(x - \bar{x})^2$
40	0	50	0	60	0
30	100	50	0	60	0
40	0	50	0	60	0
60	400	60	100	70	100
30	100	40	100	50	100
$\bar{x}_1 = 40$	600	$\bar{x}_2 = 50$	200	$\bar{x}_3 = 60$	200
$n = 5$		$n = 5$		$n = 5$	
$s_1^2 = \dfrac{\Sigma (x - \bar{x}_1)^2}{n - 1}$		$s_2^2 = \dfrac{\Sigma (x - \bar{x}_2)^2}{n - 1}$		$s_3^2 = \dfrac{\Sigma (x - \bar{x}_3)^2}{n - 1}$	
$s_1^2 = \frac{600}{4}$		$s_2^2 = \frac{200}{4}$		$s_3^2 = \frac{200}{4}$	
$s_1^2 = 150$		$s_2^2 = 50$		$s_3^2 = 50$	

The population variance can be estimated as the average of the three sample variances:

$$\hat{\sigma}^2 = \frac{s_1^2 + s_2^2 + s_3^2}{3}$$

$$= \frac{150 + 50 + 50}{3}$$

$$= \frac{250}{3} = 83.33.$$

We obtained this estimate of the population variance utilizing the three sample *variances*. Alternatively, we can estimate the population variance, utilizing the *means* of the three samples. The estimate obtained in this way is called *variance between samples*. As mentioned earlier, the two estimates for the population variance obtained by these two methods are compared, and the decision of whether or not to reject the null hypothesis depends on the magnitude of the disparity between the two estimates.

Estimating Population Variance—Variance Between Samples

Three samples have been selected. Their arithmetic means are 40, 50, and 60 hours, respectively. If these samples do belong to the same population, what is our best estimate of the mean of that population? The mean of the population can best be estimated by averaging the three sample means.

Denoted by $\hat{\mu}$, the estimated population mean is therefore

$$\hat{\mu} = \frac{\bar{x}_1 + \bar{x}_2 + \bar{x}_3}{3}$$

$$= \frac{40 + 50 + 60}{3}$$

$$= \tfrac{150}{3} = 50 \text{ hours.}$$

In general, the estimated population mean is

$$\hat{\mu} = \frac{\bar{x}_1 + \bar{x}_2 + \cdots + \bar{x}_k}{k},$$

where there are k sample means to be compared.

Now we have three sample means: $\bar{x}_1 = 40$, $\bar{x}_2 = 50$, $\bar{x}_3 = 60$, which, according to the null hypothesis, belong to a population whose mean has been estimated as $\hat{\mu} = 50$. Using these three sample means, let us calculate the square of their standard deviation or the square of their standard error.

$$\hat{\sigma}_{\bar{x}}^2 = \frac{(\bar{x}_1 - \hat{\mu})^2 + (\bar{x}_2 - \hat{\mu})^2 + (\bar{x}_3 - \hat{\mu})^2}{\text{No. of samples} - 1}$$

$$= \frac{(40 - 50)^2 + (50 - 50)^2 + (60 - 50)^2}{3 - 1}$$

$$= \frac{100 + 0 + 100}{2} = 100.$$

In general,

$$\hat{\sigma}_{\bar{x}}^2 = \frac{(\bar{x}_1 - \hat{\mu})^2 + (\bar{x}_2 - \hat{\mu})^2 + \cdots + (\bar{x}_k - \hat{\mu})^2}{k - 1},$$

where k is the number of sample means to be compared and $\hat{\sigma}_{\bar{x}}^2$ is the square of their standard error.

But we have already learned from the theory dealing with the distribution of the sample means (Chapter 6) that when all possible random samples are selected from a given population, then the standard deviation or the standard error of these sample means is equal to the standard deviation in the *population* divided by the square root of the sample size. Symbolically,

$$\sigma_{\bar{x}} = \sigma / \sqrt{n}.$$

Squaring both sides, we obtain

$$\sigma_{\bar{x}}^2 = \sigma^2/n,$$

and solving for σ^2, we have

$$\sigma^2 = n\sigma_{\bar{x}}^2.$$

Thus we conclude that the variance in the population from which the three samples originate can be estimated as the product of the sample size times the standard error of the mean squared.

We have already determined that $\hat{\sigma}_{\bar{x}}^2$ is 100. With a sample size of 5, the variance in the population is estimated as follows:

$$\hat{\sigma}^2 = n\hat{\sigma}_{\bar{x}}^2$$
$$= (5)\cdot(100) = 500.$$

Thus utilizing the sample means, we estimate the variance in the population as 500.

Now we are ready to compare the two estimates obtained for the variance in the population. Utilizing the *variances* in the sample, we have obtained an estimate of 83.33 hours (variance within samples). An estimate of 500, on the other hand, is obtained when the sample means are utilized (variance between samples). The disparity between the two estimates can be measured by what is called the *F*-ratio, where

$$F = \frac{\text{variance between samples}}{\text{variance within samples}}$$

$$= \frac{500}{83.33} = 6.0.$$

If the null hypothesis is true, i.e. if there is no significant difference between the three brands, then the two estimates for the population variance must be reasonably close and the *F*-ratio is expected to be unity.

If the three brands are different, however, the variance *between* samples would be larger than the variance *within* samples. This is due to the fact that while the variance *within* samples is a good estimate of the population variance, the variance *between* samples consists of the population variance plus an additional variance reflecting the differences between the three brands of batteries. Thus, when the three brands differ with respect to their average life, the *F*-ratio is expected to be greater than unity. Furthermore, the greater is the difference between the various brands, the greater the *F*-ratio.

In order to determine whether or not the differences between brands are significant at the .05 level, the *actual F-ratio* is compared against what is called a *critical* $F_{.05}$-ratio (Table B). From Table B, we see that each critical

$F_{.05}$-ratio is determined by a combination of horizontal and vertical degrees of freedom. The *vertical* degrees of freedom are equal to $(k - 1)$, where k is the number of samples to be compared. The *horizontal* degrees of freedom are equal to $k(n - 1)$, where n is the size of each sample.

In order to determine the critical $F_{.05}$-ratio in our problem, the vertical and horizontal degrees of freedom must be determined first:

$$\text{Vertical degrees of freedom} = k - 1$$
$$= 3 - 1$$
$$= 2;$$

$$\text{Horizontal degrees of freedom} = k(n - 1)$$
$$= 3(5 - 1)$$
$$= 12.$$

For a combination of 2 vertical and 12 horizontal degrees of freedom, Table B lists a critical $F_{.05}$-ratio of 3.89, which means that with this particular combination of degrees of freedom, there is only a .05 probability of obtaining an F-ratio of 3.89 or more as a result of random sampling fluctuations. Furthermore, the probability of obtaining an F-ratio that is greater than 3.89 is less than .05 (see Fig. 8–11).

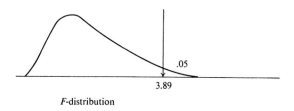

F-distribution

Figure 8–11

Thus, whenever our actual F-ratio exceeds the critical $F_{.05}$-ratio, the null hypothesis is rejected and the differences between the several sample means are regarded as significant at the .05 level. However, the differences between the various sample means are not significant when the actual F-ratio is less than the critical $F_{.05}$-ratio.

Similarly, the differences between several sample means are regarded as significant at the .01 level when the actual F-ratio is greater than the $F_{.01}$-ratio in Table C.

The actual F-ratio in our problem is 6.0, while the critical $F_{.05}$- and $F_{.01}$-ratios are 3.89 and 6.93, respectively. Since our actual F-ratio is greater than $F_{.05}$ but less than $F_{.01}$, the differences between the means of the three samples are significant at the .05 level but not at the .01 level. In other words, if we conclude that the three brands of batteries have different average lives, there is less than a .05 (but more than a .01) chance that our conclusion is wrong.

Before concluding this chapter, we wish to summarize the various steps used in comparing several means, as well as the solution to our problem.

Step 1. Compute the variance within samples.

$$\text{Variance within samples} = \frac{S_1^2 + S_2^2 + \cdots + S_k^2}{k},$$

where

$$S^2 = \frac{\Sigma (x - \bar{x})^2}{n - 1},$$

$k = $ no. of samples to be compared,
$n = $ size of each sample.

Step 2. Compute the variance between samples.

Variance between samples $= n\hat{\sigma}_{\bar{x}}^2$

$$= n\left[\frac{(\bar{x}_1 - \hat{\mu})^2 + (\bar{x}_2 - \hat{\mu})^2 + \cdots + (\bar{x}_k - \hat{\mu})^2}{k - 1}\right],$$

where

$$\hat{\mu} = \frac{\bar{x}_1 + \bar{x}_2 + \cdots + \bar{x}_k}{k}.$$

Step 3. Determine the actual F-ratio.

$$F = \frac{\text{variance between samples}}{\text{variance within samples}}.$$

Step 4. Determine the degrees of freedom.

vertical degrees of freedom $= k - 1$,
horizontal degrees of freedom $= k(n - 1)$.

Step 5. Compare the actual F-ratio with the critical F-ratio.

If $F > F_{.05}$, the difference is significant at the .05 level.
If $F > F_{.01}$, the difference is significant at the .01 level.

Summary of the Solution

Data and preliminary computations

Brand A		Brand B		Brand C	
x	$(x - \bar{x}_1)^2$	x	$(x - \bar{x}_2)^2$	x	$(x - \bar{x}_3)^2$
40	0	50	0	60	0
30	100	50	0	60	0
40	0	50	0	60	0
60	400	60	100	70	100
30	100	40	100	50	100
$\bar{x}_1 = 40$	600	$\bar{x}_2 = 50$	200	$\bar{x}_2 = 60$	200
$n = 5$		$n = 5$		$n = 5$	

$$s_1^2 = \frac{\sum (x - \bar{x}_1)^2}{n - 1} \qquad s_2^2 = \frac{\sum (x - \bar{x}_2)^2}{n - 1} \qquad s_3^2 = \frac{\sum (x - \bar{x}_3)^2}{n - 1}$$

$$= \tfrac{600}{4} = 150 \qquad\qquad = \tfrac{200}{4} = 50 \qquad\qquad = \tfrac{200}{4} = 50$$

$$\hat{\mu} = \frac{\bar{x}_1 + \bar{x}_2 + \bar{x}_3}{3}$$

$$= \frac{40 + 50 + 60}{3} = 50.$$

Step 1.

$$\text{Variance within samples} = \frac{s_1^2 + s_2^2 + s_3^2}{3}$$

$$= \frac{150 + 50 + 50}{3}$$

$$= \frac{250}{3} = 83.33.$$

Step 2.

$$\text{Variance between samples} = n\left[\frac{(\bar{x}_1 - \hat{\mu})^2 + (\bar{x}_2 - \hat{\mu})^2 + (\bar{x}_3 - \hat{\mu})^2}{3 - 1}\right]$$

$$= 5\left[\frac{(40 - 50)^2 + (50 - 50)^2 + (60 - 50)^2}{2}\right]$$

$$= 5\left[\frac{100 + 0 + 100}{2}\right] = 500.$$

Step 3.

$$F = \frac{\text{variance between samples}}{\text{variance within samples}}$$

$$= \frac{500}{83.33} = 6.0.$$

Step 4.

$$\text{Vertical degrees of freedom} = (k - 1)$$
$$= (3 - 1)$$
$$= 2,$$

$$\text{Horizontal degrees of freedom} = k(n - 1)$$
$$= 3(5 - 1)$$
$$= 12.$$

Step 5.

$$F_{.05} = 3.89 \text{ (Table B)},$$
$$F_{.01} = 6.93 \text{ (Table C)}.$$

Conclusion

Since the actual F-ratio is 6.0, which is greater than $F_{.05}$ but less than $F_{.01}$, the differences between the three sample means are significant at the .05 level but not significant at the .01 level.

EXERCISES, SET 2

1. To compare the effectiveness of three distinct methods of teaching arithmetic, 15 homogeneous fourth-grade students were divided into three equal subgroups, and each subgroup was taught arithmetic using one of these three methods. After one year, the pupils were given the same examination. The examination scores were as follows:

First method	Second method	Third method
77	73	76
78	76	77
80	76	82
82	77	83
83	78	87

If the subgroups were taught by teachers who were very much alike in all pertinent attributes, is there evidence that the effectiveness of the three teaching methods is not the same? Use a .05 level of significance.

2. Three cars from each of the four different brands, *F*, *C*, *D*, and *R* were driven in the same manner with one gallon of regular gasoline. The distance traveled by each car, in miles, is shown below.

F	C	D	R
18	21	23	20
20	22	26	19
22	26	26	21

If you are willing to reject a true hypothesis no more than 5 times in 100, do you conclude that the average mileage of the four brands of cars is not the same?

3. Each of three fertilizer mixtures, *X*, *Y*, and *Z* was applied to six test plots planted to a certain variety of wheat. The wheat yields, in bushels per acre. for the 18 test plots are given below.

Mixture X	Mixture Y	Mixture Z
50	55	70
55	57	68
57	59	66
56	60	71
59	57	67
53	54	66

What is your conclusion regarding the effectiveness of the three fertilizer mixtures? Use a .01 level of significance.

4. The means and the standard deviations of social maturity scores received by individuals belonging to four occupational groups are as follows:

	Dentists	Bankers	Machinists	Research Scientists
n	6	6	6	6
$\bar{x}$	57	59	43	61
s	10	10	10	10

Test the hypothesis that social maturity of the four occupational groups is the same, against the alternative hypothesis that their social maturity is not the same. Use a level of significance of .05.

REVIEW EXERCISES II

1. A large shipment of ball bearings has a mean diameter of 2.00 inches with a standard deviation of 0.02 inches ($\mu = 2.00$ and $\sigma = 0.02$).

a) Assuming that the diameters are normally distributed, establish a range where the probability is 0.95 that the diameter of a *single* ball bearing is expected to fall.

b) Establish a range where the probability is 0.95 that the mean diameter of a sample of 400 ball bearings is expected to fall.

c) What is the probability that the mean diameter of a random sample of 100 ball bearings will be greater than 2.003 inches?

2. The lives of 100-watt electric bulbs are normally distributed. The mean is 800 hours and the standard deviation is 50 hours ($\mu = 800$, $\sigma = 50$).

a) Establish an interval such that the probability is 0.95 that the life of a *single* bulb is within such an interval.

b) Establish an interval such that the probability is 0.99 that the average life of a random sample of 100 bulbs is within such an interval.

c) What is the probability that any single bulb will last over 820 hours?

d) What is the probability that the mean life of a simple random sample of 100 bulbs will be longer than 820 hours?

3. A manufacturer of instant coffee packages his product in 12-oz. jars. In order to control his filling process a sample of 36 jars is selected every hour. If the average weight in the sample, $\bar{x}$, is between 12.04 oz. and 12.08 oz., continue the process; otherwise stop and readjust the process.

a) What is the probability of stopping a process operating with a mean of 12.06 oz. per jar and a standard deviation of 0.3 oz. ($\mu = 12.06$, $\sigma = 0.3$)?

b) What is the probability of continuing a process operating with a mean of 12.08 oz. per jar and a standard deviation of 0.3 oz. ($\mu = 12.08$, $\sigma = 0.3$)?

4. The average age of all students attending UCLA is 22 years with a standard deviation of 4 years ($\mu = 22$, $\sigma = 4$).

a) What is the probability that the mean age in a random sample of 64 students will be between 21.5 and 23.0 years?

b) What sample size is required in order to estimate the average age of a UCLA student if it is desired that there would be a 0.99 confidence that the error in our estimate does not exceed 1 year?

5. In a very large shipment of batteries, the useful life of a battery is normally distributed with a mean of 400 hours. It is also known that 90 percent of the batteries have a useful life which ranges between 318 and 482 hours. If a random sample of 100 batteries is selected from this shipment, what is the probability that the sample mean is greater than 420 hours?

6. A random sample of 64 key punch operators employed by commercial banks in Los Angeles revealed the following information:

$$x = \text{hourly wage rate of a key punch operator}$$
$$\sum x = \$192$$
$$\sum (x - \bar{x})^2 = \$63$$

Establish a 0.99 confidence interval estimate of the average hourly earnings of all key punch operators employed by commercial banks in Los Angeles.

7. A random sample of a new variety of apples showed a mean weight of 10 ounces and a standard deviation of 2 ounces. Establish a 0.95 confidence interval estimate of the average weight of the variety of apples,

a) assuming that the sample consisted of 64 apples,

b) assuming that the sample consisted of 16 apples.

8. A random sample of 5 fuses of a certain brand is tested in order to determine the average breaking point. The breaking points measured in amperes are as follows:

$$18, \quad 22, \quad 20, \quad 14, \quad 26$$

With what degree of confidence can we assert that the average breaking point of this brand of fuse is between 15.736 and 24.264?

9. Mr. Smith, a farmer, desires to determine the total weight of a lot of 10,000 oranges. Since Mr. Smith has only a small scale, he selected a random sample of 16 oranges. The sample showed a mean of 7 oz. and a standard deviation of 1 oz. Establish a 0.95 confidence interval estimate of the *total* weight of the entire lot.

10. The ABC Shoe Company operates a chain of retail shoe stores in the state of Illinois. It is the policy of the ABC Company that no branch is established in any given community unless the company is 0.99 sure that aggregate expenditures on shoes in that community is at least $1 million. The company is now considering the establishment of a branch in Springdale, a community of 20,000 families. In order to estimate aggregate expenditures on shoes in Springdale, a random sample of 49 families was selected. The sample revealed an average annual expenditure of $60 per family with a standard deviation of $21. Based on this sample information, should the ABC Company establish a branch in Springdale?

11. The Food and Drug Administration suspects that a newly developed birth control drug has the undesirable effect of raising body temperature. The drug is administered to 64 healthy females and the following information is obtained:

$$x = \text{body temperature measured in degrees Fahrenheit}$$
$$\sum x = 6323.2$$
$$\sum (x - \bar{x})^2 = 10.08$$

Using the 0.05 level of significance, is the Food and Drug Administration correct in its belief concerning the side effect of the new drug? The average body temperature of a healthy female is 98.6 degrees Fahrenheit.

12. The Ka-Chu Nut Company markets its mixed nuts in 16 oz. jars. The jars are claimed to contain, on the average, 50 percent cashews and 50 percent peanuts. Suspecting that the mixture contains more peanuts than cashews, Mrs. Nader examined the contents of 5 jars. The proportion of cashews in these jars are:

$$0.51, \quad 0.47, \quad 0.45, \quad 0.43, \quad 0.39$$

If Mrs. Nader is willing to reject a true claim no more than once in 100, would she reject the Ka-Chu Nut Company claim?

13. A random sample of 100 college professors in the United States shows the following:

$$x = \text{age of a college professor}$$
$$\sum x = 4000 \text{ years}$$
$$\sum (x - \bar{x})^2 = 9900 \text{ years}$$

a) Establish a 0.95 confidence-interval estimate of the average age of an American college professor.

b) With what degree of confidence can we assert that the average age of an American college professor is somewhere between 38.36 and 41.64 years?

c) An official of the American Association of University Professors (AAUP) claims that the average age of a college professor is 43 years. What would you conclude about the validity of this claim, using a level of significance of 0.01?

14. The 1980 census shows that average annual family income in San Francisco is $14,200. A random sample of 100 families from the city showed an average annual income of $14,050 with a standard deviation of $2000 in 1981. Using a level of significance of 0.05, test the hypothesis that mean family income has declined in 1981.

15. An arithmetic test is administered to a random sample of 100 eighth graders in the Los Angeles school district. The sample showed an average of 72 with a standard deviation of 8. When the same test is administered to a random sample of 100 eighth graders in the Pasadena School District, the sample showed an average of 75 with a standard deviation of 6. Using a level of significance of 0.01, can you conclude that the arithmetic performance of eighth graders in the two school districts is *not* the same?

16. In random samples of 100 blondes and 100 brunettes, the following information is revealed:

Blondes
$x = $ number of dates per month
$\bar{x} = 7.5$ dates
$s = 4$ dates

Brunettes
$x = $ number of dates per month
$\bar{x} = 6$ dates
$s = 3$ dates

Using a level of significance of 0.01, test the hypothesis that "blondes have more fun."

17. In order to compare the average nicotine content of four brands of cigarettes, 6 cigarettes of each brand were tested and the following information obtained:

	Brand			
	A	B	C	D
Mean (in milligrams)	28	27	31	30
Standard deviation (in milligrams)	2	2	2	2

Are the four brands of cigarettes significantly different with respect to their average nicotine content? Use a level of significance of 0.05.

18. A random sample of 400 families in Los Angeles shows an average weekly expenditure of $80 on food with a standard deviation of $20.

   a) Establish a .95 confidence interval estimate of average weekly family expenditures on food in Los Angeles.

   b) What can we assert with .98 confidence about the possible size of our error if we estimate the mean expenditure to be $80?

   c) It is claimed that average family expenditures on food in Los Angeles is $85 per week. Use the sample information to test the hypothesis that it is less than $85.

19. A random sample of 400 families is selected from a community of 50,000 families. The sample shows an average annual expenditure of $94 per family on books and magazines, with a standard deviation of $20. Establish a .95 confidence interval estimate of *aggregate* annual expenditure on books and magazines for the entire community.

9
Estimating
the Population
Proportion
from a
Sample

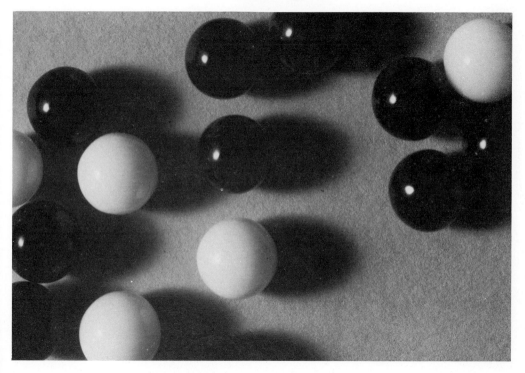

Estimating the Population Proportion from a Sample

In the preceding three chapters we treated exclusively problems concerning the mean of a population. In this chapter and the succeeding one we shall examine problems dealing with the *population proportion.*

The relationship between the sample proportion, p, and the population proportion, π, is described in the first part of this chapter. In the remaining sections we shall concern ourselves with the problem of estimating the population proportion from a simple random sample.

THE SAMPLING DISTRIBUTION OF THE PROPORTION

The relationship between the population proportion and the proportions of the various samples that can be selected from such a population can best be illustrated by describing a sampling operation from a known population. Let the known population consist of five marbles: one of the marbles is white and the remaining four marbles are black. Denoting a white marble by W and a black marble by B, we display our known population below.

Population of five marbles

Marble	Color
1	*W*
2	*B*
3	*B*
4	*B*
5	*B*

The proportion of white marbles, denoted by π, in the above population is .20.

Now, let us select *all* possible samples of four marbles from this population, and compute the proportion of white marbles, p, for each sample.

Possible samples ($n = 4$)	Sample proportions (p)
1, 2, 3, 4 (W, B, B, B)	.25
1, 2, 3, 5 (W, B, B, B)	.25
1, 2, 4, 5 (W, B, B, B)	.25
1, 3, 4, 5 (W, B, B, B)	.25
2, 3, 4, 5 (B, B, B, B)	.00

Thus, while the true population proportion π of white marbles is .20, the sample proportion p is .25 in four of the above samples, and it is zero in the last sample.

To examine these five possible sample proportions in greater detail, let us first calculate their arithmetic mean, or their average. The average of all sample proportions, denoted by $E(p)$, is

$$E(p) = \frac{p_1 + p_2 + p_3 + p_4 + p_5}{5}$$

$$= \frac{.25 + .25 + .25 + .25 + .00}{5}$$

$$= .20 = \pi.$$

We can thus say that although individual sample proportions may either overstate or understate the true population proportion, their average is always equal to the population proportion.

Let us now calculate the standard deviation of these five possible sample proportions. Their standard deviation, called the *standard error of the proportion* (denoted by σ_p) is computed below.

p	$(p - \pi)$	$(p - \pi)^2$
.25	.05	.0025
.25	.05	.0025
.25	.05	.0025
.25	.05	.0025
.00	−.20	.0400

$$\Sigma (p - \pi)^2 = .0500$$

$$\sigma_p = \sqrt{\frac{\Sigma (p - \pi)^2}{\text{no. of samples}}}$$

$$= \sqrt{\frac{.0500}{5}} - \sqrt{.01} = .10.$$

The standard error of the proportion indicates the "average" disparity between the various p's and π. The first four sample proportions are different from the population mean by .05 and the fifth sample proportion by .20. However, on the average, each sample proportion is different from the population proportion by .10.

A small value of σ_p indicates two facts: (1) the various values of p are close to each other; (2) the average difference between these p's and π is small. Consequently any one p is a good estimate of π.

Although the standard error of the proportion measures the "average" difference of all possible sample proportions from the population proportion, it is not necessary to actually select all such samples in order to determine its value. Fortunately, there is an alternative way of determining the standard error of the proportion. It has been found that

$$\sigma_p = \sqrt{\frac{\pi(1 - \pi)}{n}} \sqrt{\frac{N - n}{N - 1}},$$

where π is the population proportion, N is the population size, and n is the sample size.

For our population of five marbles, where $\pi = .20$, the standard error of the proportions for all possible samples of size 4 is

$$\sigma_p = \sqrt{\frac{\pi(1 - \pi)}{n}} \sqrt{\frac{N - n}{N - 1}}$$

$$= \sqrt{\frac{(.20)(.80)}{4}} \sqrt{\frac{5 - 4}{5 - 1}}$$

$$= \sqrt{\frac{.16}{4}} \sqrt{\frac{1}{4}} = \sqrt{.01} = .10.$$

The above value is exactly identical to the one previously obtained.

Thus we can say that if all possible samples of size n are selected from a given population, then

$$E(p) = \pi, \quad \text{and} \quad \sigma_p = \sqrt{\frac{\pi(1 - \pi)}{n}} \sqrt{\frac{N - n}{N - 1}}.$$

Furthermore, the *finite population correction*, $\sqrt{(N - n)/(N - 1)}$, will approach unity whenever we are sampling from an infinite population or whenever the sample size is less than 10% of the population. Under either of these conditions, therefore, the standard error of the proportion becomes

$$\sigma_p = \sqrt{\frac{\pi(1 - \pi)}{n}}.$$

Concluding our examination of the properties of the various sample proportions, we may add one final but important property: The sample proportions are approximately normally distributed whenever the sample size is fairly large.

In summary, we may therefore state this general theorem: *If all possible simple random samples of size n are selected from a population having a proportion of* π, *then the resulting sample proportions are approximately normally distributed, have an* <u>*average*</u> *equal to* π and a standard deviation or a standard error equal to $\sqrt{\pi(1 - \pi)/n}$. (See Fig. 9–1.)

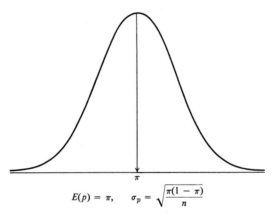

$$E(p) = \pi, \qquad \sigma_p = \sqrt{\frac{\pi(1 - \pi)}{n}}$$

Figure 9–1

To clarify the implications of the above theorem, let us assume that all possible random samples of 100 students are selected from a college where the actual proportion of senior students is 20%. According to our theorem, the various sample proportions would be approximately normally distributed, have an average of 20%, and a standard error of 4%. (See computations in Fig. 9–2.)

Since the various sample proportions are approximately normally distributed, 68% of all such proportions will be in the range

$$\pi \pm 1.0\sigma_p,$$
$$.20 \pm 1.0(.04),$$

or between 16% and 24% (Fig. 9–3). In the same manner, 95% of all such sample proportions will be in the range

$$\pi + 1.96\sigma_p,$$
$$.20 \pm 1.96(.04),$$
$$.20 \pm .08,$$

or approximately between 12% and 28% (Fig. 9–4).

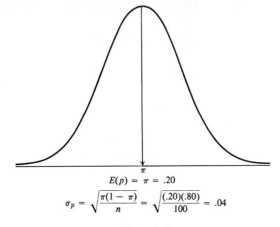

$$E(p) = \pi = .20$$

$$\sigma_p = \sqrt{\frac{\pi(1-\pi)}{n}} = \sqrt{\frac{(.20)(.80)}{100}} = .04$$

Figure 9–2

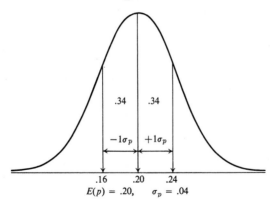

$$E(p) = .20, \qquad \sigma_p = .04$$

Figure 9–3

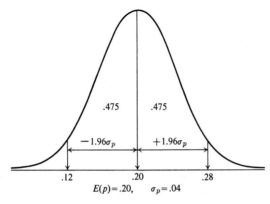

$$E(p) = .20, \qquad \sigma_p = .04$$

Figure 9–4

We now present a few examples illustrating the distribution of sample proportions.

Example 1

It is known that 64% of all registered voters in a certain U.S. Congressional District are Democrats. What is the probability that a simple random sample of 100 registered voters from this district will reveal a proportion of Democrats of 60% or less?

As we have already learned, if all possible samples of size 100 registered voters are selected from this district, then the various sample proportions are approximately normally distributed, have an average of 64% and a standard error of 4.8% (Fig. 9–5).

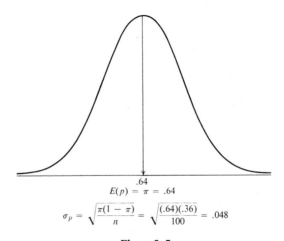

$$.64$$
$$E(p) = \pi = .64$$

$$\sigma_p = \sqrt{\frac{\pi(1-\pi)}{n}} = \sqrt{\frac{(.64)(.36)}{100}} = .048$$

Figure 9–5

The probability that the proportion of Democrats in a sample of 100 registered voters will be 60% or less is calculated below (see also Fig. 9–6).

$$z = \frac{p - \pi}{\sigma_p}$$

$$= \frac{.60 - .64}{.048} = \frac{-.04}{.048} = -.83.$$

Therefore

$$P(p \le .60) = .50 - .2967 = .2033.$$

Let us suppose now that a simple random sample of 400 registered voters is to be selected from this district. What is the probability that the sample proportion is 60% or less?

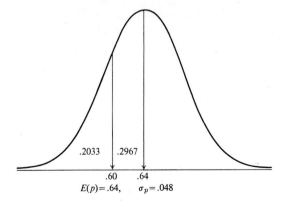

.2033 | .2967

.60   .64

$E(p) = .64,$ $\sigma_p = .048$

Figure 9–6

If many samples of size 400 registered voters are selected, then the various sample proportions are approximately normally distributed, have an average of 64%, and a standard error of 2.4% (see Fig. 9–7).

The probability that the proportion of Democrats in a sample of 400 registered voters is 60% or less is calculated below (see also Fig. 9–8).

$$z = \frac{p - \pi}{\sigma_p}$$

$$= \frac{.60 - .64}{.024} = \frac{-.04}{.024} = -1.67.$$

Therefore

$$P(p \le .60) = .50 - .4525 = .0475.$$

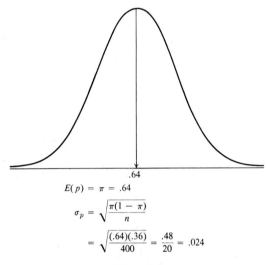

.64

$E(p) = \pi = .64$

$$\sigma_p = \sqrt{\frac{\pi(1 - \pi)}{n}}$$

$$= \sqrt{\frac{(.64)(.36)}{400}} = \frac{.48}{20} = .024$$

Figure 9–7

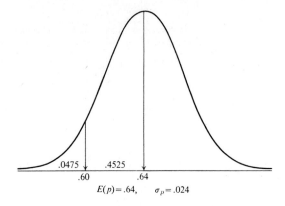

$$E(p) = .64, \qquad \sigma_p = .024$$

Figure 9–8

Thus, while the probability that a sample of 100 voters will show a proportion of Democrats of 60% or less is .2033, this probability is reduced to .0475 when the sample size increases to 400 voters.

Example 2

Forty percent of all graduate students on this campus are married. If 100 graduate students are selected at random, what is the probability that the proportion of married students in this particular sample will be between 32% and 47%?

When several samples of 100 graduate students are selected, the distribution of the various sample proportions can be described by Fig. 9–9.

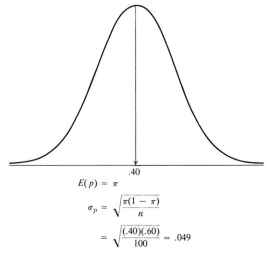

$$E(p) = \pi$$

$$\sigma_p = \sqrt{\frac{\pi(1 - \pi)}{n}}$$

$$= \sqrt{\frac{(.40)(.60)}{100}} = .049$$

Figure 9–9

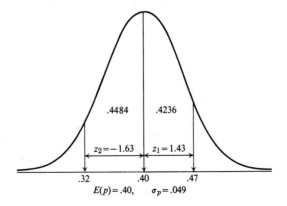

Figure 9–10

The probability that the proportion of married students in a sample of size 100 will be between 32% and 47% is determined as follows (see Fig. 9–10):

Left side

$$z_2 = \frac{p - \pi}{\sigma_p}$$

$$= \frac{.32 - .40}{.049} = \frac{-.08}{.049} = -1.63$$

Right side

$$z_1 = \frac{p - \pi}{\sigma_p}$$

$$= \frac{.47 - .40}{.049} = \frac{.07}{.049} = 1.43$$

Area under the normal curve = .4484   Area under the normal curve = .4236

Hence the probability that the sample proportion will be between 32% and 47% is the sum of these two areas, that is,

$$P(.32 \le p \le .47) = .4484 + .4236 = .8720.$$

Example 3

The receiving department of a large television manufacturer uses the following rule in deciding whether to accept or reject a shipment of 100,000 small parts shipped every week by a supplier: Select a sample of 400 parts from each lot received. If 3% or more of the selected parts are defectives, reject the entire lot; if the proportion of defectives is less than 3% accept the lot. What is the probability of rejecting a lot that *actually* contains 2% defectives?

Although it is known that this lot contains 2% defectives, $\pi = .02$, it is possible that a sample of 400 parts will reveal a proportion of 3% or more defectives, and consequently the entire lot is rejected. To determine the

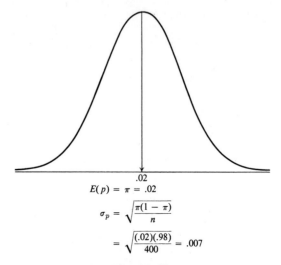

$$E(p) = \pi = .02$$

$$\sigma_p = \sqrt{\frac{\pi(1 - \pi)}{n}}$$

$$= \sqrt{\frac{(.02)(.98)}{400}} = .007$$

Figure 9–11

probability of rejecting this lot, let us first describe the distribution of the various sample proportions that could be selected from such a lot (Fig. 9–11).

The probability of obtaining a sample proportion of 3% or more and consequently rejecting the lot is calculated below (see also Fig. 9–12).

$$z = \frac{p - \pi}{\sigma_p}$$

$$= \frac{.03 - .02}{.007} = 1.43.$$

Hence

$$P(p \geq .03) = .50 - .4236 = .0764.$$

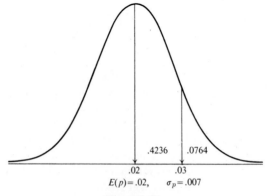

$$E(p) = .02, \quad \sigma_p = .007$$

Figure 9–12

EXERCISES, SET 1

1. Our population consists of the five faculty members in the Department of Psychology at Bradford College. The marital status of each member is given below.

Faculty member	Marital status
A	Married
B	Single
C	Married
D	Single
E	Single

a) Determine the proportion of married members in the population.
b) Select all possible samples of two members from this population and compute the proportion of married members in each sample.
c) Compute the mean $E(p)$ and the standard deviation σ_p of the 10 sample proportions computed in part (b).
d) Show that

$$\sigma_p = \sqrt{\frac{\pi(1 - \pi)}{n}} \sqrt{\frac{N - n}{N - 1}}.$$

2. Given the following information concerning a population of six persons:

Person	Education
A	College graduate
B	High school graduate
C	College graduate
D	High school graduate
E	College graduate
F	High school graduate

a) Determine the proportion of college graduates in the population.
b) Select all possible samples of three persons from this population and compute the proportion of college graduates in each sample.
c) What is the probability that the sample proportion ($n = 3$) will be over 40%? [*Hint:* How many of the 20 sample proportions are over .40?]
d) What is the probability that the sample proportion ($n = 3$) will differ from the population proportion by 20% or less?

3. Ten percent of all the people in a certain community have Type B blood. If a random sample of 900 persons from this community is selected, what is the probability that the proportion of people with Type B blood in the sample is

a) more than 12%,   b) less than 10.5%,   c) between 11% and 13%?

4. To decide whether or not to endorse gubernatorial candidate X, the Society of Independent Voters adopts the following procedure: A random sample of 400 registered voters is to be selected. The Society will endorse candidate X only when 50% or more of the sample voters are in his favor. What is the probability that the Society

   a) will come out for candidate X when he is actually endorsed by only 45% of all voters?

   b) will fail to come out for candidate X when he is actually endorsed by 55% of all voters?

5. The following rule is used in controlling the operation of a process that produces certain parts: Select a sample of 36 parts from the process. If the percentage of defective parts in the sample is $p^*\%$ or more, stop the process; otherwise continue the operation. Determine the value of p^* such that there is a .95 chance that a process producing, on the average, 10% defectives will be stopped.

6. To decide whether or not to endorse gubernatorial candidate X, the Society of Independent Voters adopts the following procedure: A random sample of n registered voters is to be selected. The Society will endorse the candidate only if $p^*\%$ of the sample voters are in his favor. Determine the values of n and p^* such that there is only a 0.05 risk of *endorsing* the candidate when 45% of all voters are in his favor and only a 0.01 risk of *not* endorsing the candidate when 50% of all voters come out for him.

7. Fifty percent of all families in Los Angeles own at least two cars. A random sample of 400 families is selected from Los Angeles.

   a) What is the probability that the sample proportion of families owning at least two cars is less than 45 percent?

   b) The probability is .9544 that sample proportion of these families owning at least 2 cars is between p_1 and p_2. Find p_1 and p_2 using symmetrical limits around π.

8. A manufacturer receives a shipment of 100,000 small parts each week. He adapted the following rule in deciding on whether to accept or to reject each shipment: Select a random sample of 100 parts from the shipment. If the proportion of defective parts in the sample is .06 or more, reject the entire shipment; if the proportion of defective parts is less than .06, accept the shipment.

   a) What is the probability of accepting a lot that *actually* contains 10 percent defective parts?

   b) What is the probability of rejecting a lot that *actually* contains 2 percent defective parts?

   c) What is the probability of accepting a lot that *actually* contains 6 percent defective parts?

9. Assume that 60% of all California voters favor a certain property tax relief initiative on the June primary ballot. If a random sample of 150 voters is selected at random, what is the probability that the proportion of those favoring the initiative is between 65% and 68%?

10. In deciding on whether or not to introduce a new welfare legislation program to Congress, the administration adopted the following sampling rule: Select a random sample of 150 families at random. If 10% or more of the families selected earn below poverty level, introduce the new welfare legislation program to Congress; otherwise, do not introduce the program.

   a) What is the probability of not introducing the new legislation to Congress assuming that 15% of all families in the U.S. earn below poverty level?

   b) What is the probability of introducing the program assuming only 8% of all families earn below poverty level?

ESTIMATING THE POPULATION PROPORTION

We have already stated that when repeated samples of size n are selected from a population having a proportion of π, then the various sample proportions, or p's, are approximately normally distributed. Furthermore, these sample proportions have an average which is equal to π and a standard error which is equal to $\sqrt{\pi(1 - \pi)/n}$ (Fig. 9–13).

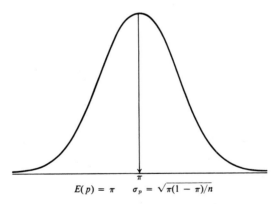

$$E(p) = \pi \qquad \sigma_p = \sqrt{\pi(1 - \pi)/n}$$

Figure 9–13

Since these sample proportions are approximately normally distributed, we expect .95 of all such proportions to fall within the range defined by

$$\pi \pm 1.96\sigma_p \qquad \text{or} \qquad \pi \pm 1.96\sqrt{\pi(1 - \pi)/n}$$

(see Fig. 9–14). We can also conclude that if one sample of size n is to be selected, there is a .95 chance that the proportion in this sample will fall within the interval $\pi \pm 1.96\sigma_p$. The probability that it will fall outside this range is .05.

Now, let us assume that we *actually* select a simple random sample of size n and calculate the proportion p. The interval $p \pm 1.96\sigma_p$ *may* or *may not* contain π.

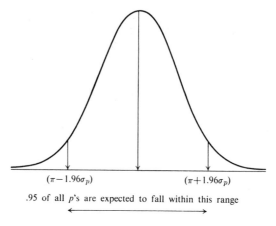

$(\pi-1.96\sigma_p)$ $(\pi+1.96\sigma_p)$

.95 of all p's are expected to fall within this range

Figure 9–14

If the proportion of the sample selected happens to fall within any part of the shaded area of the normal curve (selected from *within* the interval $\pi \pm 1.96\sigma_p$) in Fig. 9–15, the interval $p \pm 1.96\sigma_p$ will certainly contain π.

If, on the other hand, p happens to fall within the unshaded area of the normal curve (selected from *outside* the interval $\pi \pm 1.96\sigma_p$) in Fig. 9–16, the interval $p \pm 1.96\sigma_p$ will fail to contain π.

We may therefore conclude that if all possible samples of size n are selected and the interval $p \pm 1.96\sigma_p$ is established for each sample, then .95

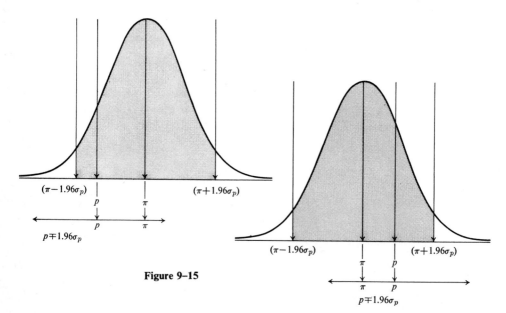

$(\pi-1.96\sigma_p)$ $(\pi+1.96\sigma_p)$

p π

p π

$p \mp 1.96\sigma_p$

Figure 9–15

$(\pi-1.96\sigma_p)$ $(\pi+1.96\sigma_p)$

π p

π p

$p \mp 1.96\sigma_p$

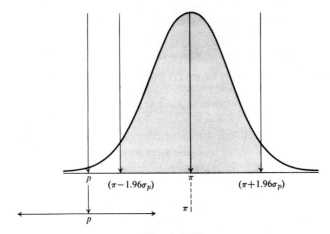

Figure 9–16

of all such intervals are expected to contain π. The interval $p \pm 1.96\sigma_p$ or $p \pm 1.96\sqrt{\pi(1 - \pi)/n}$ is called a 95% confidence interval estimate of π.

To come up with the above interval estimate of π, we must know the value of π. This, however, is not known. Hence we are forced to use the sample proportion p as an approximation of π. The approximate .95 confidence interval estimate of π thus becomes $p \pm 1.96\sqrt{p(1 - p)/n}$. In the same manner, the range $p \pm 2.58\sqrt{p(1 - p)/n}$ is called an approximate 99% confidence interval estimate of π.

Example 4

In assessing the desirability of windowless schools, officials asked 144 elementary school children whether or not they like windows in their classrooms. Thirty percent of the children preferred windows. Establish a 0.95 confidence-interval estimate of the proportion of elementary school children who like windows in their classrooms.

The .95 confidence interval estimate of π is

$$p \pm 1.96\sqrt{\frac{p(1 - p)}{n}},$$

$$.30 \pm 1.96\sqrt{\frac{(.30)(.70)}{144}} = .30 \pm 1.96\frac{.458}{12}$$

$$= .30 \pm 1.96(.038) = .30 \pm .074,$$

or between 22.6% and 37.4%.

Example 5

In a public opinion poll, 320 out of 400 persons interviewed supported the administration's policy on disarmament.

a) Establish a 95% confidence interval estimate of the proportion of persons supporting the government's stand on disarmament.

b) What can we conclude with .99 confidence about the maximum error in our estimate if the proportion of those supporting the administration policy is estimated as 80%?

c) With what degree of confidence can we assert that the proportion of persons supporting the administration policy is somewhere between 77% and 83%?

Solution (a): The sample proportion is determined as

$$p = \tfrac{320}{400} = .80.$$

The .95 confidence interval estimate of π is therefore

$$p \pm 1.96 \sqrt{\frac{p(1-p)}{n}},$$

$$.80 \pm 1.96 \sqrt{\frac{(.80)(.20)}{400}} = .80 \pm 1.96(.02) = .80 \pm .0392,$$

or between 76.08% and 83.92%.

Solution (b): A .99 confidence interval estimate of π is

$$p \pm 2.58 \sqrt{\frac{p(1-p)}{n}},$$

$$.80 \pm 2.58 \sqrt{\frac{(.80)(.20)}{400}} = .80 \pm 2.58(.02) = .80 \pm .0516.$$

Since $2.58\sqrt{(.80)(.20)/400}$, or .0516, has been added to or subtracted from the sample proportion .80 in order to establish a .99 confidence interval estimate of π, we can assert with a probability of .99 that our maximum error is .0516 if π is estimated as 80%. In general, the maximum expected error in a point estimate of π is determined by the expression $z\sqrt{p(1-p)/n}$.

Solution (c): If π is estimated as being between .77 and .83, then .03 must have been added to or subtracted from the sample proportion .80 in order to establish this interval estimate of π. In other words, our interval estimate

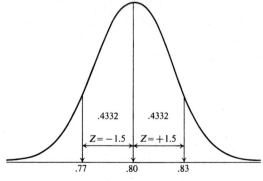

Figure 9–17

was obtained as follows:

$$p \pm z \sqrt{\frac{p(1-p)}{n}},$$

$$.80 \pm .03.$$

Therefore,

$$z \sqrt{\frac{p(1-p)}{n}} = .03,$$

$$z \sqrt{\frac{(.80)(.20)}{400}} = .03,$$

$$z(.02) = .03,$$

$$z = \frac{.03}{.02} = 1.5.$$

Using the area under the normal curve in Table A, we conclude that the range .77 to .83 is the .8664 confidence interval estimate of π (see Fig. 9–17).

Example 6

In a community of 10,000 families, a random sample of 400 families showed that 20 percent of the families have an annual family income exceeding $30,000. Establish a 0.95 confidence-interval estimate of the number of families earning over $30,000 in that community.

Solution: This problem is solved in two steps. In the first step we will establish a 0.95 confidence-interval estimate of the *proportion* of families in the community with income exceeding $30,000 (a 0.95 confidence interval estimate of π). In the second step, we will multiply the range established for π by the number of families in the community, to obtain a range for the number of families with income exceeding $30,000 in the entire community.

Step 1: The 0.95 confidence-interval estimate of π is

$$p \pm 1.96\left(\sqrt{\frac{p(1-p)}{n}}\right) = 0.20 \pm 1.96\left(\sqrt{\frac{(0.20)(0.80)}{400}}\right)$$

$$= 0.20 \pm 1.96(0.02) = 0.20 \pm 0.0392$$

or between 16.08% and 23.92%.

Step 2. Since the community consists of 10,000 families, the 0.95 confidence interval estimate of the number of families with income exceeding $30,000 is obtained as

$$10,000 \ (0.1608 \leq \pi \leq 0.2392),$$

or between 1608 and 2392 families.

It may be useful now to summarize the solution to our problem. A 0.95 confidence-interval estimate of the number of families earning over $30,000 in the entire community can be determined as

$$N\left[p \pm z\left(\sqrt{\frac{p(1-p)}{n}}\right)\right] = 10,000\left[0.20 \pm 1.96\sqrt{\frac{(0.20)(0.80)}{400}}\right],$$

or between 1608 and 2392 families.

Determining the Sample Size

Whenever we are asked to estimate a population proportion from a simple random sample, we must first decide on how large our sample should be. Generally speaking, the sample size is determined by the degree of accuracy required in estimating the population proportion. The higher the required degree of accuracy in our estimate, the larger the size of the sample we must select.

There are two aspects to the degree of accuracy required in estimating the population proportion from a sample:

1. the magnitude of the maximum allowable error, and
2. the degree of confidence that the error in the estimate does not exceed the maximum allowable error.

If, for example, we are asked to estimate, within 2%, the proportion of the television viewers tuned in to "Peyton Place" on Monday night, the maximum allowable error is .02. However, regardless of the sample size, there is always a chance that the error in our estimate will exceed .02. But the larger the sample size, the smaller the risk that the allowable error will be exceeded, or the greater the confidence that our estimate will be within the allowable error. If it is also specified that we must have .95 confidence that the error in our estimate will not exceed .02, the two aspects of accuracy in this case are as stated below:

1. the maximum allowable error, denoted by e, is .02.
2. the degree of confidence that the maximum allowable error is not exceeded is .95.

Having specified the accuracy requirements of our estimate, we must determine the sample size that satisfies these two requirements. Specifically, the sample size n is manipulated until $1.96\sigma_p = .02$ (see Fig. 9–18).

In general, the size of the sample is manipulated until

$$z\sigma_p = e,$$

where z is determined by the degree of confidence. Since

$$\sigma_p = \sqrt{\pi(1 - \pi)/n},$$

it follows that

$$z\sqrt{\pi(1 - \pi)/n} = e.$$

Solving for n, we have

$$n = z^2\pi(1 - \pi)/e^2.$$

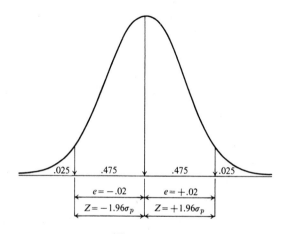

Figure 9–18

Thus the sample size is determined by z, e, and π, where e is equal to the maximum allowable error, z is determined by the degree of confidence, and π is the population proportion which we want to estimate.

At this point the reader may think that we have reached an impasse: to determine the size of the sample needed to estimate the population proportion π, the value of π itself must be known. This is a problem! However, it may be overcome if the value of π can be estimated from surveys taken previously. Thus, if a previous estimate of π exists, we can use it to obtain the sample size from the formula

$$n = \frac{z^2 \pi (1 - \pi)}{e^2}.$$

But what if no previous surveys were taken and the value of π is unknown? Examining the expression $\pi(1 - \pi)$ in our formula, we can show that such an expression attains its maximum value when π is equal to 50%.

π	$\pi(1 - \pi)$
.10	.10 × .90 = .09
.20	.20 × .80 = .16
.30	.30 × .70 = .21
.40	.40 × .60 = .24
.50	.50 × .50 = .25 (maximum)
.60	.60 × .40 = .24
.70	.70 × .30 = .21
.80	.80 × .20 = .16
.90	.90 × .10 = .09

Therefore we may conclude that, other things being equal, a larger sample size is required when π equals 50% than when π assumes any other value. Hence, in the absence of any previous knowledge about the value of π, we can safely set π equal to 50%, and the sample size thus obtained will be as large or larger than required, regardless of the actual value of π.

Example 7

An advertising agency desires to estimate the proportion of the Los Angeles television audience viewing " Layton Place," a weekly television program.

a) Determine the sample size needed to ensure .95 confidence that the error in our estimate will not exceed .02. A survey made last season showed that 20% of the families in Los Angeles viewed "Layton Place."

b) What would be the sample size if no previous survey had been taken?

Solution (a):

$$n = \frac{z^2\pi(1 - \pi)}{e^2}$$

$$= \frac{(1.96)^2(.20)(.80)}{(.02)^2} = \frac{(3.8416)(.16)}{.0004}$$

$$= 1536.64 = 1537 \text{ families.}$$

Solution (b): In the absence of any previous knowledge concerning the value of π, we assume that $\pi = 50\%$, and obtain the following sample size:

$$n = \frac{z^2\pi(1 - \pi)}{e^2}$$

$$= \frac{(1.96)^2(.50)(.50)}{(.02)^2} = \frac{(3.8416)(.25)}{.0004} = 2401 \text{ families.}$$

The Distribution of Sample Proportions and the Binomial Distribution

Note that the distribution of sample proportions is merely a modified version of the binomial distribution. To examine the relationship between these two distributions, let us first summarize what we know about them.

	Binomial distribution	Distribution of sample proportions
Mean	$n\pi$	π
Standard deviation	$\sqrt{n\pi(1 - \pi)}$	$\sqrt{\dfrac{\pi(1 - \pi)}{n}}$

The mean and the standard deviation of the distribution of sample proportions are obtained by dividing the mean and the standard deviation of the binomial distribution by n.

$$\text{Mean} = \frac{n\pi}{n} = \pi.$$

$$\text{Standard deviation} = \frac{\sqrt{n\pi(1 - \pi)}}{n} = \sqrt{\frac{\pi(1 - \pi)}{n}}.$$

Thus the distribution of sample proportions is the binomial distribution expressed in percentage form.

The relationship between the two distributions can, perhaps, be more easily understood if we consider the following problem.

Example 8

A machine produces, on the average, 20% defective parts. If a random sample of 64 parts is selected, what is the probability that the proportion of defective parts in the sample is 25% (16 parts) or more?

We will solve this problem, using two alternative methods: the binomial distribution, and the distribution of sample proportions. The results of the two methods can then be compared.

Solution 1 (the binomial distribution): First, we must determine the mean and the standard deviation of the binomial distribution with $\pi = .20$ and $n = 64$. The mean of the distribution is

$$\mu = n\pi$$
$$= 64(.20) = 12.8,$$

and the standard deviation is

$$\sigma = \sqrt{n\pi(1 - \pi)}$$
$$= \sqrt{64(.20)(.80)} = 3.2.$$

Using the normal curve approximation to the binomial distribution, we can now determine the probability of obtaining 16 defective parts or more as follows (Fig. 9–19):

$$z = \frac{x - \mu}{\sigma}$$

$$= \frac{16 - 12.8}{3.2} = \frac{3.2}{3.2} = 1.0.$$

Therefore

$$P(x \geq 16) = .5000 - .3413 = .1587.$$

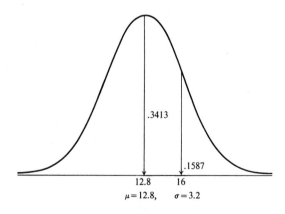

.3413

.1587

12.8   16

$\mu = 12.8,$ $\sigma = 3.2$

Figure 9–19

Solution 2 (distribution of sample proportions): First, we must determine the mean and the standard deviation of the distribution of sample proportions. The mean of the distribution is

$$E(p) = \pi$$
$$= .20,$$

and the standard deviation of the distribution is

$$\sigma_p = \sqrt{\frac{\pi(1 - \pi)}{n}}$$

$$= \sqrt{\frac{(.20)(.80)}{64}} = .05.$$

The probability that the sample proportion is 25% or more can now be determined as follows (Fig. 9–20):

$$z = \frac{p - \pi}{\sigma_p}$$

$$= \frac{.25 - .20}{.05} = 1.0.$$

Therefore

$$P(p \geq .25) = .5000 - .3413 = .1587.$$

In comparing the results of the two solutions to our problem, we can only conclude that the two distributions are essentially the same.

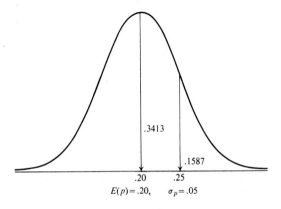

.3413

.1587

.20   .25

$E(p) = .20,$ $\sigma_p = .05$

Figure 9–20

EXERCISES, SET 2

1. In a random sample of 400 families living in Northridge, California, 80 families have more than one television set. Establish a .98 confidence interval estimate of the proportion of Northridge families with more than one television set.

2. A random sample of 100 male students from the University of Illinois shows that 10 students belong to a social fraternity.

   a) Establish a .90 confidence interval estimate of male students who belong to a social fraternity at the University of Illinois.

   b) What can we assert with probability .95 about the maximum possible error if the proportion of male students who belong to a social fraternity at the University of Illinois is estimated as 10%?

3. Two hundred and fifty-six patients suffering from a certain disease were treated with a newly developed drug. The drug was effective in curing 128 of these cases. With what degree of confidence can we assert that the effectiveness of this drug is between 45% and 55%?

4. A government agency wishes to determine the proportion of families earning less than $5000 a year in a given community. Previous studies have shown that this proportion is 20%.

   a) What sample size is required to ensure .95 confidence that the error in estimating such a proportion will not exceed .05?

   b) How will the required sample size change if the maximum allowable error is reduced to .01?

5. The Dean of Students is considering a new policy for student housing. Before making a final decision, he wants to select a random sample of students in order to estimate the proportion of students in favor of the new policy. What is the sample size needed to ensure that the risk of exceeding an error of .10 or more is only .05?

6. A random sample of 200 voters in a community of 5000 registered voters showed 110 voters in favor of candidate X. Establish a 0.99 confidence-interval estimate of the number of voters favoring candidate X in the entire community.

7. In a university with a total enrollment of 20,000 students, a random sample of 100 students produced 64 students in favor of a new policy for student housing. Establish a 0.95 confidence-interval estimate of the number of students favoring the new housing policy in the entire university.

8. A random sample of 400 students attending this college shows 80 married students.

   a) Establish a .95 confidence interval estimate of the proportion of married students attending this college.

   b) With what degree of confidence can we assert that the proportion of married students attending this college is between .16 and .24?

9. Bank A maintains 20,000 customers' saving accounts. A random sample of 400 accounts shows that 144 accounts are in excess of $10,000. Establish a .99 confidence interval estimate of the *number* of saving accounts held by Bank A that exceed $10,000.

10
Testing
Hypotheses
Concerning
Proportions

Testing
Hypotheses
Concerning
Proportions

This chapter, which deals with the general question of testing hypotheses concerning proportions, is organized into four parts. In the first section, we test hypotheses concerning a single population proportion. The theorectical tools needed for testing the significance of the difference between two sample proportions are treated in the second part of the chapter, while their application is the subject matter of the third part. Finally, the Chi-square test, which is designed to compare several proportions, is discussed in the last section.

Testing Hypotheses Concerning a Single Proportion

The principal objective of some statistical investigations is to test a hypothesis concerning a population proportion. We may want to evaluate, for example, a politician's claim that he commands 55% of all votes in a certain U.S. Congressional district. Or we may desire to test the hypothesis that a given drug is 90% effective in the cure of a certain disease.

How can we test a politician's claim that he commands 55% of all votes in a given district? A random sample of registered voters can be selected and interviewed. And based on the result of this sample, the politician's claim is either rejected or not rejected.

Stated briefly and rigorously, the general procedure used in testing a hypothesis concerning a population proportion consists of the following steps: The investigator must first formulate his null hypothesis and an appropriate alternative hypothesis. Then, on the basis of his sample information, he will either reject the null hypothesis or reserve judgment. As in any other testing problem, either a .05 or a .01 level of significance is used as the decision criterion.

To be specific, let us illustrate the above procedure with three examples: a one-sided test is deemed appropriate in the first two examples, and a two-sided test in the third one.

Example 1

Mr. Dixon, a Republican, claims that he has the support of 55% of all voters in the 23rd U.S. Congressional district. What will the party Central Committee conclude if, out of a random sample of 500 registered voters, only 245 expressed their preference for Dixon? Use a level of significance of .01.

The null hypothesis (Dixon's claim), which states that $\pi = .55$, is tested against the appropriate alternative hypothesis that $\pi < .55$.

If the null hypothesis is true, that is, $\pi = .55$, and if many samples of 500 voters are selected from this district, then the various sample proportions are distributed as shown in Fig. 10–1.

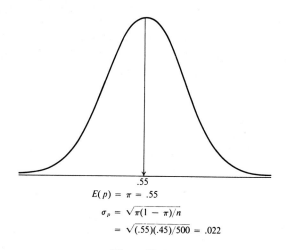

$$E(p) = \pi = .55$$

$$\sigma_p = \sqrt{\pi(1 - \pi)/n}$$

$$= \sqrt{(.55)(.45)/500} = .022$$

Figure 10–1

The sample shows that $\frac{245}{500}$ or 49% of the voters favor Mr. Dixon. The probability of obtaining a sample proportion p of 49% or less when the null hypothesis is true is only .0032 (see Fig. 10–2 and the computations below).

$$z = \frac{p - \pi}{\sigma_p} = \frac{.49 - .55}{.022} = \frac{-.06}{.022} = -2.73.$$

Therefore

$$P(p \le .49) = .50 - .4968 = .0032.$$

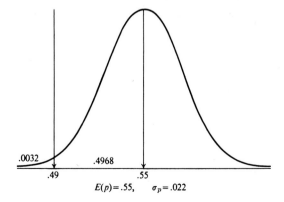

.0032

.4968

.49   .55

$E(p) = .55, \qquad \sigma_p = .022$

Figure 10–2

Since the probability of obtaining a sample proportion of 49% or less is .0032, which is less than .01, the null hypothesis is rejected, using .01 level of significance.

It may be useful to point out again that the decision whether or not to reject the null hypothesis becomes obvious once the z-value is determined. The null hypothesis is rejected whenever the *absolute* value of z exceeds a critical level. These critical levels are:

Level of significance	One-sided test	Two-sided test
.05	1.64	1.96
.01	2.33	2.58

Once again, using a .01 level of significance, we reject the null hypothesis in our problem, since the absolute value of z is 2.73, which is greater than 2.33.

Example 2

The Department of Health, Education, and Welfare reports that only 10% of all persons over 65 years old are covered by adequate private health insurance. What would the American Medical Association (AMA) conclude about the Department's claim if, out of a random sample of 900 elderly persons, 99 possessed adequate private health insurance? Use a level of significance of .05.

The two hypotheses can be formulated as

$$H_0: \pi = .10, \qquad H_1: \pi > .10.$$

If the null hypothesis is true and many samples of size 900 are selected, then the various sample proportions are distributed as shown in Fig. 10–3.

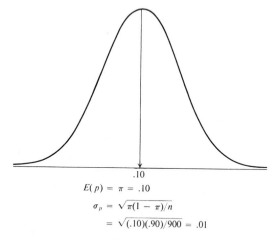

$$E(p) = \pi = .10$$

$$\sigma_p = \sqrt{\pi(1 - \pi)/n}$$

$$= \sqrt{(.10)(.90)/900} = .01$$

Figure 10–3

The sample selected by the AMA shows a proportion of $\frac{99}{900}$ or 11%. To test the validity of the Department's claim, let us calculate the deviation of an 11% sample proportion from the mean of the distribution which is 10% (the deviation is measured in z-units). We obtain (Fig. 10–4).

$$z = \frac{p - \pi}{\sigma_p}$$

$$= \frac{.11 - .10}{.01} = 1.0.$$

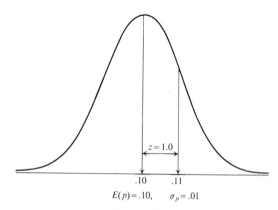

$$E(p) = .10, \qquad \sigma_p = .01$$

Figure 10–4

Since z is 1.0, which is less than 1.64, the null hypothesis cannot be rejected using the .05 level of significance. In other words, the AMA does not have enough evidence to reject the claim made by the Department of Health, Education, and Welfare.

Example 3

The sponsor of a weekly television show would like the studio audience to consist of an equal number of men and women. Out of 400 persons attending the show on a given night, 220 are men. Using a level of significance of .01, can the sponsor conclude that the desired sex composition of the audience is not properly maintained?

Let $\pi = 50\%$ stand for the desired proportion of males in the audience. The null hypothesis which states that $\pi = .50$ is then tested against the alternative hypothesis that $\pi \neq .50$. The test is two-sided, since the null hypothesis is rejected whenever the sample proportion is significantly above or below 50%.

If the null hypothesis is true, then the distribution of the various sample proportions is as shown in Fig. 10–5.

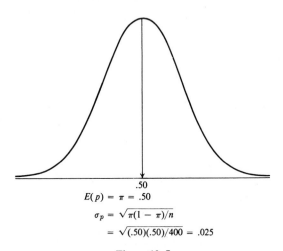

$$.50$$
$$E(p) = \pi = .50$$
$$\sigma_p = \sqrt{\pi(1 - \pi)/n}$$
$$= \sqrt{(.50)(.50)/400} = .025$$

Figure 10–5

The sponsor found that the proportion of men in a sample of 400 persons was $\frac{220}{400}$ or 55%. To test the validity of the null hypothesis, let us calculate the deviation of a sample proportion of 55% from the mean of the distribution which is 50% (the deviation is measured in z-units). We have (Fig. 10–6).

$$z = \frac{p - \pi}{\sigma_p} = \frac{.55 - .50}{.025} = 2.0.$$

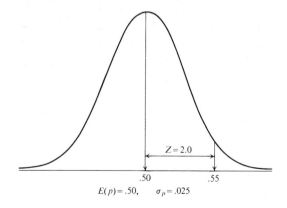

Figure 10–6

Since z is 2.0, which is less than 2.58, the null hypothesis cannot be rejected using the .01 level of significance. (The test is two-tailed.)

EXERCISES, SET 1

1. A television manufacturer claims that, on the average, 90% of his color television sets do not require any repair during the first two years of operation. The Consumer Protection Union selects a random sample of 100 sets and finds that 15 sets require some repair within the first two years of operation. If the Consumer Protection Union is willing to reject a true claim no more than 5 times in 100, will the Union reject the manufacturer's claim?

2. The Johnson Pharmaceutical Laboratory has developed a certain drug for the treatment of high blood pressure. The Laboratory asserts that the drug effectively lowers the blood pressure in 80% of all high blood pressure cases. If 175 out of 225 patients treated with this drug report a substantial decline in their blood pressure, do you conclude that the Laboratory has exaggerated the effectiveness of the drug? Use a level of significance of .01.

3. The Institute of Family Relations reports that 50% of all couples living in Los Angeles County reach a divorce court within the first year of their marriage. What do you conclude about the validity of this report if, out of a random sample of 400 marriages, only 193 reach the divorce court within the first year of marriage? Use a level of significance of .01.

4. At a press conference, the Press Secretary of the President states that 90% of all American voters support the President's "High Society Program." An independent poll of 625 American voters shows that 550 voters are in favor of the President's program. If you are willing to reject a true hy-

pothesis no more than once in 100, do you conclude that the popularity of the High Society Program is exaggerated by the Press Secretary?

5. When a special coin was tossed 64 times, heads came up 28 times. Using a level of significance of .05, do you conclude that the coin is biased?

6. A discount house is considering the purchase of a large stock of records from a supplier who claims that, on the average, only 2% of the records are imperfect. In examining 400 of these records, the discount house finds that 15 are imperfect. Will the discount house reject the supplier's claim if it is willing to reject a true claim no more than once in 10?

7. It is claimed that 20% of all families in Los Angeles are Mexican-Americans.
   a) What do you conclude about the validity of this claim if a random sample of 400 families shows 90 Mexican-American families?
   b) What do you conclude about the validity of this claim if a random sample of 400 families shows 120 Mexican-American families?

8. Twenty percent of all color televisions sold in California in 1979 are of Brand X. A random sample of 100 color television sets sold in California in 1980 shows that only 17 sets are of Brand X. Using a level of significance of 0.01, can you conclude that the Brand X color television share of the California market has declined between 1979 and 1980?

9. The student newspaper at the state college just published the results of a recent student poll concerning a proposal that all student dorms become coed. In this poll, 100 students were sampled and 80 students stated that they were in favor of the proposal.
   a) Construct a .80 confidence interval estimate for the proportion of the student population in favor of the proposal.

   The College president quickly ruled against the proposal claiming that the sample result did not prove beyond a reasonable doubt that the majority of the students favor it.

   b) If a level of .05 level of significance is used, is the president's conclusion valid?

10. Kelly Log owns a large breakfast-food company. One of the cereals, Trixx, is quite popular. Trixx comes in a 16 ounce package. Suppose government, regulations require that at least 80% of the boxes must contain 16 ounces or more of cereal if the packages are to be labeled with the 16-ounce label. A recent analysis, conducted by government inspectors, revealed that of 400 boxes of Trixx, 300 boxes contained 16 ounces or more of cereal. Using a .05 level of significance, can the government prove a violation?

The Sampling Distribution of the Difference Between Two Sample Proportions

Let π_1 stand for the proportion of senior students in San Francisco State College, and let π_2 stand for the proportion of senior students at Los Angeles

State College. Let us further assume that these two proportions are 24% and 20% respectively.

If we select many random samples of 200 students from San Francisco State College and record the proportion of senior students in each sample, the various sample proportions may show the following results:

San Francisco sample proportions (p_1)
.24
.22
.26
.21
.25
.23
$\vdots$
.25

Similarly, if an equal number of samples of 400 students are selected from Los Angeles State College, the proportions of senior students in the various samples might be as shown in the table.

Los Angeles sample proportions (p_2)
.20
.19
.22
.23
.19
.17
$\vdots$
.22

In comparing the results of these two sets, let us randomly pair these samples and compute the difference d between the proportions of each pair.

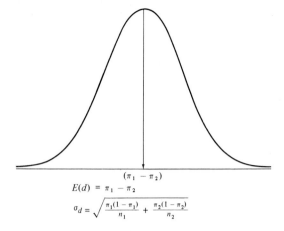

$$E(d) = \pi_1 - \pi_2$$

$$\sigma_d = \sqrt{\frac{\pi_1(1 - \pi_1)}{n_1} + \frac{\pi_2(1 - \pi_2)}{n_2}}$$

Figure 10-7

p_1	p_2	$(p_1 - p_2) = d$
.24	.20	.04
.22	.19	.03
.26	.22	.04
.21	.23	− .02
.25	.19	.06
.23	.17	.06
⋮	⋮	⋮
.25	.22	.03

Remembering that the p_1's are the proportions of samples drawn from a population with $\pi_1 = .24$, and that the p_2's are the proportions of samples drawn from a population with $\pi_2 = .20$, let us examine the properties of the differences recorded in the last column above.

First, although the difference between the proportions of the first pair of samples is .04, of the second pair. 03, and of the last pair .03, we expect these differences to have an average of .04 (that is, .24 − .20). Hence the first property of the differences between two sample proportions is

$$E(d) = \pi_1 - \pi_2,$$

where $E(d)$ is the average of all differences, π_1 is the proportion in the population from which the first set of samples is selected (San Francisco State), and

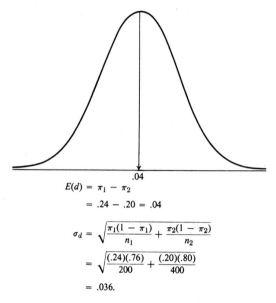

$$E(d) = \pi_1 - \pi_2$$
$$= .24 - .20 = .04$$

$$\sigma_d = \sqrt{\frac{\pi_1(1 - \pi_1)}{n_1} + \frac{\pi_2(1 - \pi_2)}{n_2}}$$

$$= \sqrt{\frac{(.24)(.76)}{200} + \frac{(.20)(.80)}{400}}$$

$$= .036.$$

Figure 10–8

π_2 is the proportion in the population from which the second set of samples is selected (Los Angeles State).

Statisticians have studied these differences in great detail and have found that they are approximately normally distributed and have a standard deviation or a standard error equal to

$$\sqrt{\frac{\pi_1(1 - \pi_1)}{n_1} + \frac{\pi_2(1 - \pi_2)}{n_2}}.$$

Denoting the standard error of these differences by σ_d, we summarize their properties in Fig. 10–7.

For our samples from San Francisco State and Los Angeles State, these properties imply that the average of all differences between the proportions of paired samples is .04 and their standard error is .036 (Fig. 10–8).

Having determined the mean and the standard error of the differences between the proportions of the paired samples, we can draw some interesting conclusions. Since these differences are approximately normally distributed, we expect 68% of the differences to fall within the range

$$E(d) \pm 1.0\sigma_d,$$
$$.04 \pm .036,$$

or between .004 and .076 (Fig. 10–9).

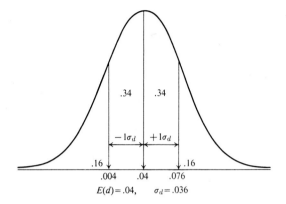

Figure 10–9

In the same manner, we expect 95% of all these differences to fall within the range

$$E(d) \pm 1.96\sigma_d,$$
$$.04 \pm 1.96(.036),$$
$$.04 \pm .07,$$

or between $-.03$ and $.11$ (Fig. 10–10).

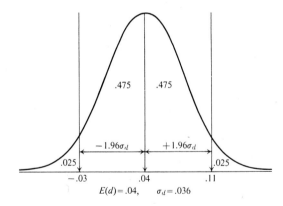

Figure 10–10

Let us now summarize the implications of the preceding discussion. We assume that the proportions of senior students in San Francisco State and Los Angeles State are 24% and 20%, respectively; then, if a random sample of 200 students is selected from San Francisco State and a sample of 400 students from Los Angeles State, we expect the proportion of seniors in the San Francisco State sample to be, on the average, .04 higher than the proportion of seniors in the Los Angeles State College sample. However, the difference between the two sample proportions may very well be smaller or greater than .04. It is even possible, for example, that the proportion of seniors in the San Francisco sample will be .11 (or more) *higher* than that in the Los Angeles sample, and it is also possible that the San Francisco sample proportion will be .03 (or more) *lower* than that of Los Angeles. But the probability of either of these extreme results is only .025 (see Fig. 10–10).

Test of Hypotheses Concerning the Difference Between Two Sample Proportions

If the proportion of senior students in a random sample of 200 students from UCLA is 18% and the same proportion in a random sample of 400 students from USC is 15%, can we safely conclude that the proportions of senior students in the two schools are not the same?

When two sample proportions p_1 and p_2 differ, we have two hypotheses to explore. First, there is the null hypothesis that the two populations from which the two samples originate have the same proportion ($\pi_1 = \pi_2$). If this is the case, then the observed difference between the two sample proportions is *not significant* and is attributed to chance or random sampling fluctuations. The alternative hypothesis to be explored is that the two samples are drawn from two different populations, each having a distinct proportion ($\pi_1 \neq \pi_2$). If this alternative hypothesis is true, the observed difference between the two sample proportions is deemed significant.

Stated briefly, therefore, the two hypotheses in our example are

$$\text{null hypothesis: } \pi_1 = \pi_2,$$
$$\text{alternative hypothesis: } \pi_1 \neq \pi_2,$$

where π_1 is the proportion of seniors in the first population (UCLA), and π_2 is the proportion of seniors in the second population (USC).

Now let us assume that the null hypothesis is true, that is, π_1 and π_2 are the same, or each is equal to π. If this is the case, then the sampling distribution of the differences between two sample means can be described by Fig. 10–11.

The above result can be interpreted in the following manner: Let us assume that the proportions of seniors at UCLA and USC are the same; and that many samples of 200 students are selected from the first school and equally many samples of 400 students are selected from the second school. When these two sets of samples are paired and the difference between the proportions of seniors in each pair is recorded, then the average of all these differences is equal to zero and their standard error is

$$\sqrt{\pi(1 - \pi)\left(\frac{1}{n_1} + \frac{1}{n_2}\right)} \ .$$

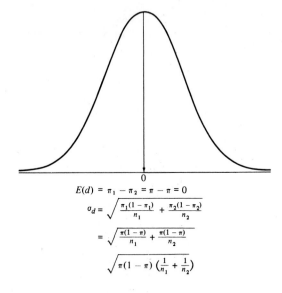

$$E(d) = \pi_1 - \pi_2 = \pi - \pi = 0$$

$$\sigma_d = \sqrt{\frac{\pi_1(1 - \pi_1)}{n_1} + \frac{\pi_2(1 - \pi_2)}{n_2}}$$

$$= \sqrt{\frac{\pi(1 - \pi)}{n_1} + \frac{\pi(1 - \pi)}{n_2}}$$

$$\sqrt{\pi(1 - \pi)\left(\frac{1}{n_1} + \frac{1}{n_2}\right)}$$

Figure 10–11

If the null hypothesis is true, the proportion of seniors in both schools is the same. This proportion, denoted by π, is certainly unknown. But it can be approximated by

$$\hat{\pi} = \frac{n_1 p_1 + n_2 p_2}{n_1 + n_2}$$

$$= \frac{200(.18) + 400(.15)}{200 + 400}$$

$$= \frac{96}{600} = 16\%,$$

and the sampling distribution of the difference between the two sample proportions is as described by Fig. 10–12.

The validity of the null hypothesis is evaluated in terms of the results of the two samples actually selected. The difference between these two sample proportions is .03(.18 − .15), which deviates from the mean of the distribution by .94 unit of z (see Fig. 10–13 and computations below).

$$z = \frac{d - E(d)}{\hat{\sigma}_d}$$

$$= \frac{.03 - 0}{.032} = .94$$

Since the test is two-sided, and $z < 1.96$, the difference between the two sample proportions is not significant at the .05 level. In other words, on the

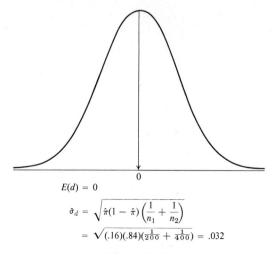

$$E(d) = 0$$

$$\hat{\sigma}_d = \sqrt{\hat{\pi}(1 - \hat{\pi})\left(\frac{1}{n_1} + \frac{1}{n_2}\right)}$$

$$= \sqrt{(.16)(.84)(\tfrac{1}{200} + \tfrac{1}{400})} = .032$$

Figure 10–12

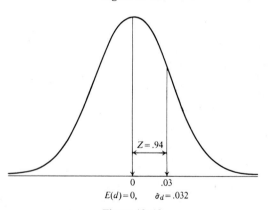

$Z = .94$

$$0 \qquad .03$$

$$E(d) = 0, \qquad \hat{\sigma}_d = .032$$

Figure 10–13

basis of the evidence provided by the two samples, one cannot reject the hypothesis that the proportion of seniors in the two schools is the same.

Summarizing the procedure used in testing the significance of the difference between two sample proportions, one must first calculate the value of z:

$$z = \frac{d}{\hat{\sigma}_d}$$

$$= \frac{p_1 - p_2}{\sqrt{\hat{\pi}(1 - \hat{\pi})\left(\frac{1}{n_1} + \frac{1}{n_2}\right)}},$$

where

$$\hat{\pi} = \frac{n_1 p_1 + n_2 p_2}{n_1 + n_2}.$$

If $z > 1.96$, the difference is significant at the .05 level. If $z > 2.58$, the difference is significant at the .01 level.

Example 4

In a public opinion poll of 400 men and 600 women, 70% of the men and 75% of the women expressed approval of the "Voting Right Act." Using a level of significance of .05, can we conclude that the observed difference between these two proportions is significant?

Solution: Given $p_1 = .70$, $p_2 = .75$, $n_1 = 400$, $n_2 = 600$. Therefore

$$\hat{\pi} = \frac{n_1 p_1 + n_2 p_2}{n_1 + n_2}$$

$$= \frac{400(.70) + 600(.75)}{400 + 600} = .73,$$

and

$$z = \frac{p_1 - p_2}{\sqrt{\hat{\pi}(1 - \hat{\pi})\left(\dfrac{1}{n_1} + \dfrac{1}{n_2}\right)}}$$

$$= \frac{.70 - .75}{\sqrt{(.73)(.27)(\frac{1}{400} + \frac{1}{600})}}$$

$$= \frac{-.05}{.028} = -1.72.$$

Since the *absolute* value of z is less than 1.96, the difference between the two sample proportions is not significant at the .05 level.

EXERCISES, SET 2

1. It is known that the percentage of families with an annual income of $10,000 or more in Los Angeles and San Francisco is 25% and 20%, respectively. If a random sample of 100 families is selected from each of these two cities and the proportions of families earning $10,000 or more in the two samples are compared, what is the probability that the Los Angeles sample proportion will be

   a) greater than that of San Francisco's by 3% or more?
   b) smaller than that of San Francisco's by 3% or more?

2. A researcher selected random samples of 120 psychologists and 80 psychiatrists to investigate their views on whether schizophrenia is a biochemical

abnormality or a maladjustment to others dating from childhood. The data below are the result of this investigation.

	Psychologists	Psychiatrists
Biochemical abnormality	60	50
Childhood maladjustment	60	30
Total	120	80

If you are willing to reject a true hypothesis no more than once in 100, will you reject the hypothesis that the views of the psychologists and psychiatrists on the cause of schizophrenia are the same?

3. On August 20 a poll of 200 registered voters showed 120 voters in favor of candidate X. A month later, a poll of 300 registered voters showed only 150 voters in his favor. Does this indicate that the popularity of candidate X has changed during the time between the two polls? Use a level of significance of .05.

4. A random sample of 400 housewives selected by a market research organization showed that 20% preferred brand X coffee to all other competing brands. After completion of an intensive radio and television advertising campaign, a second sample of 600 housewives was selected to assess the effectiveness of the campaign. The post-advertising sample showed 22% of the housewives in favor of brand X. If you are willing to reject a true hypothesis no more than once in 10, will you reject the hypothesis that the advertising program was ineffective?

5. A random sample of 1000 state employees revealed 400 men and 600 women. The sample, furthermore, showed that 70% of the men preferred a new retirement plan. Of the women, 80% preferred the new retirement plan. Using the .05 level of significance, test the hypothesis that the new retirement plan is equally preferred by men and women.

Comparison of Several Proportions—Chi Square Analysis

Random samples of 160, 240, and 200 persons were selected from Los Angeles, Chicago, and New York, respectively. The persons selected were asked "What type of television program do you like best: Drama, Western, Documentary, or Comedy?" The responses are summarized below.

Type of program	Los Angeles		Chicago		New York	
	No. of persons	%	No. of persons	%	No. of persons	%
Drama	60	37.50	100	41.67	80	40.00
Western	30	18.75	30	12.50	30	15.00
Documentary	30	18.75	40	16.67	50	25.00
Comedy	40	25.00	70	29.16	40	20.00
Total	160	100	240	100	200	100

Examining the above results, we note that Drama was preferred by 37.50% of persons interviewed in Los Angeles, 41.67% of those interviewed in Chicago and 40% of those interviewed in New York. The question, therefore, arises whether the observed difference between these three proportions is due to a real difference in television preferences among the residents of the three cities; or to random sampling fluctuations. Similarly, one wonders whether or not the observed difference between the sample proportions of those who preferred Westerns (Los Angeles, 18.75%; Chicago, 12.50%; and New York, 15.00%) is significant.

Although there are many sample proportions to be compared, the central issue in this problem can be stated as follows: From the gathered sample information, can we safely conclude that the residents of these three metropolitan areas differ in their preference for the various types of television programs? Specifically, the investigator has two hypotheses to explore. First, there is the null hypothesis that preferences for the various types of television programs in the three cities are essentially the same and that any observed differences are due to random sampling fluctuations and hence are not significant. The alternative hypothesis is that there exists a real difference in preferences. On the basis of the results of the three samples selected, the investigator may either reject the null hypothesis or not reject it.

Before testing the validity of the null hypothesis, let us organize our sample data in a manner that will facilitate analysis.

Type of program	Number of persons		
	Los Angeles	Chicago	New York
Drama	60	100	80
Western	30	30	30
Documentary	30	40	50
Comedy	40	70	40
Total	160	240	200

We see that the data are organized in a table that has four rows (Drama, Western, Documentary, and Comedy) and three columns (Los Angeles, Chicago, and New York). This table is said to be a 4-by-3 array, since it contains four rows and three columns. Furthermore, it contains 12 cells. Note that the number in each cell indicates the *number of people* (not percent of people) residing in a certain city and preferring a certain television program. The number 60, for example, which is located in the first cell indicates that 60 persons out of the 160 interviewed in Los Angeles prefer Drama.

Let us proceed as in any other testing problem, that is, let us assume that the null hypothesis is true, i.e. there is no difference in preferences among the three citities. If this is the case, then one expects the proportion of those who prefer drama to be the same in all three cities. We estimate this proportion as follows:

$$\frac{60 + 100 + 80}{160 + 240 + 200} = \frac{240}{600}$$

$$= 40\%;$$

that is, out of *all* persons interviewed, the proportion of those who prefer Drama is 40%. We next estimate the proportions of those who prefer Westerns, Documentaries, and Comedies:

$$\frac{30 + 30 + 30}{160 + 240 + 200} = \frac{90}{600}$$

$$= 15\%;$$

that is, out of *all* persons interviewed, the proportion of those who prefer Westerns is 15%.

$$\frac{30 + 40 + 50}{160 + 240 + 200} = \frac{120}{600}$$

$$= 20\%;$$

that is, out of *all* persons interviewed, the proportion of those who prefer Documentaries is 20%.

$$\frac{40 + 70 + 40}{160 + 240 + 200} = \frac{150}{600}$$

$$= 25\%;$$

that is, out of *all* persons interviewed, the proportion of those who prefer Comedy is 25%.

These calculations are summarized below.

Actual number of responses

	Los Angeles	Chicago	New York	Total	Percent
Drama	60	100	80	240	$\frac{240}{600} = 40\%$
Western	30	30	30	90	$\frac{90}{600} = 15\%$
Documentary	30	40	50	120	$\frac{120}{600} = 20\%$
Comedy	40	70	40	150	$\frac{150}{600} = 25\%$
Total	160	240	200	600	

If the null hypothesis is true, then in any sample (whether from Chicago, Los Angeles, or New York), we expect 40% to prefer Dramas, 15%, Westerns, 20%, Documentaries, and 25%, Comedies. Applying these percentages to the three samples actually selected, we expect the following responses in each cell:

Expected number of responses

	Los Angeles	Chicago	New York
Drama	$160 \times .40 = 64$	$240 \times .40 = 96$	$200 \times .40 = 80$
Western	$160 \times .15 = 24$	$240 \times .15 = 36$	$200 \times .15 = 30$
Documentary	$160 \times .20 = 32$	$240 \times .20 = 48$	$200 \times .20 = 40$
Comedy	$160 \times .25 = 40$	$240 \times .25 = 60$	$200 \times .25 = 50$
Total	160	240	200

To find the expected number of responses in the first cell, i.e. the Drama fans in Los Angeles, we multiply the number of persons interviewed in Los Angeles by the expected proportion of those who prefer Drama. Thus the expected number of responses in the first cell is

$$160 \times .40 = 64 \text{ persons.}$$

This result differs from the actual number of responses in that cell, which is only 60.

Now, if the null hypothesis is true, then the actual and expected numbers of responses in each cell should be reasonably close. The greater the disparity between these two sets of responses, the more doubtful the null hypothesis.

To facilitate the analysis of the data, let us list both the actual and the expected response distributions in one table. The expected responses are enclosed in parentheses.

Actual and expected number of responses

	Los Angeles	Chicago	New York
Drama	60 (64)	100 (96)	80 (80)
Western	30 (24)	30 (36)	30 (30)
Documentary	30 (32)	40 (48)	50 (40)
Comedy	40 (40)	70 (60)	40 (50)

We can measure the disparity between the actual and expected distributions of responses by computing the term

$$\frac{(\text{actual} - \text{expected})^2}{\text{expected}}$$

for each cell and adding these 12 terms. The sum, called Chi square and denoted by x^2, is computed as follows:

$$x^2 = \frac{(60 - 64)^2}{64} + \frac{(100 - 96)^2}{96} + \frac{(80 - 80)^2}{80} + \frac{(30 - 24)^2}{24} + \frac{(30 - 36)^2}{36}$$

$$+ \frac{(30 - 30)^2}{30} + \frac{(30 - 32)^2}{32} + \frac{(40 - 48)^2}{48} + \frac{(50 - 40)^2}{40}$$

$$+ \frac{(40 - 40)^2}{40} + \frac{(70 - 60)^2}{60} + \frac{(40 - 50)^2}{50}$$

$$= .25 + .16 + 0 + 1.38 + 1.00 + 0 + .12 + 1.33 + 2.50 + 0 + 1.66$$

$$+ 2.00$$

$$= 10.40.$$

If the preferences for television programs are the same in all three cities, the actual and expected responses must be reasonably close and the value of x^2 must be very small. Furthermore, the greater the disparity between the two sets of responses, the larger the value of x^2 and the more doubtful the hypothesis that the preferences are the same. But how large a value should x^2 reach before we can safely conclude that the preferences are not the same? This question can be answered after we have examined the theoretical sampling distribution of x^2 in Table D.

Table D shows the various values of x^2 that can be obtained as a result of pure sampling fluctuations. The probability given at the top of each column indicates the chance of obtaining a larger value than the one specified in the table. In addition, each row in the table is associated with the number of degrees of freedom (to be explained later).

Thus, with six degrees of freedom, for example, there is a .05 chance that a $x^2 > 12.592$ could occur as a result of pure random sampling fluctuations (Fig. 10–14). With the same number of degrees of freedom, the probability is only .01 that $x^2 > 16.812$ as a result of sampling fluctuations (Fig. 10–15).

We have already mentioned that each row in Table D is associated with a certain number of degrees of freedom. What are these degrees of freedom? The number of degrees of freedom in any given problem is determined as follows:

$$\text{Degrees of freedom} = (r - 1)(c - 1),$$

where *r* is the number of rows in the problem, and *c* is the number of columns in the problem.

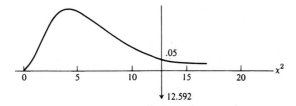

Fig. 10–14 Chi square distribution (6 degrees of freedom).

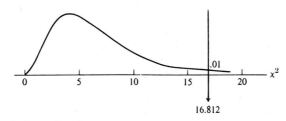

Fig. 10–15 The chance is only 0.01 that x^2 is larger than 16.812.

In our problem, there are four rows (drama, western, documentary, and comedy) and three columns (Los Angeles, Chicago, and New York). Therefore, the number of degrees of freedom is

$$\begin{aligned} \text{df} &= (r - 1)(c - 1) \\ &= (4 - 1)(3 - 1) \\ &= 6. \end{aligned}$$

The reader may recall that we have obtained an *actual* x^2-value of 10.40. With six degrees of freedom, Table D shows that there is a .10 chance that $x^2 > 10.645$ could occur as a result of sampling fluctuations. The probability is greater than .10, therefore, that we will obtain $x^2 > 10.40$ although the preference for the various television programs is the same in the three cities.

Generally speaking, the null hypothesis is rejected only when the chance of obtaining an observed value of x^2 is less than .05 or .01. These are the two levels of significance most frequently used.

Thus, using either the .01 or .05 level of significance, we cannot reject our null hypothesis that preferences are the same.

Example 5

In a public opinion poll, 1000 Americans were asked, "How do you rate the job Mr. X is doing as Secretary of State: good, fair, or poor?" The responses classified according to the educational level of the respondents are shown below.

Actual distribution of responses by educational level

	8th grade or less	High school	College	Total
Good	82	427	191	700
Fair	10	110	60	180
Poor	8	63	49	120
Total	100	600	300	1000

Using a level of significance of .05, can we conclude that the rating is independent of the educational level of the respondents?

If the Secretary of State rating is the same regardless of the educational level of the respondent (the null hypothesis), then we expect

$$\frac{700}{1000} = 70\% \text{ to respond ``good,''}$$
$$\frac{180}{1000} = 18\% \text{ to respond ``fair,''}$$
$$\frac{120}{1000} = 12\% \text{ to respond ``poor.''}$$

Applying these three ratios to the number of people selected from each of the three educational levels, we obtain the following expected or theoretical distribution of responses:

Expected distribution of responses

	8th grade or less	High School	College
Good	$100 \times .70 = 70$	$600 \times .70 = 420$	$300 \times .70 = 210$
Fair	$100 \times .18 = 18$	$600 \times .18 = 108$	$300 \times .18 = 54$
Poor	$100 \times .12 = 12$	$600 \times .12 = 72$	$300 \times .12 = 36$
Total	100	600	300

Before computing the observed value of χ^2, let us combine the actual and the expected response distributions in one table. The expected number of responses is placed in parentheses.

Actual and expected distribution of responses

	8th grade or less	High School	College
Good	82 (70)	427 (420)	191 (210)
Fair	10 (18)	110 (108)	60 (54)
Poor	8 (12)	63 (72)	49 (36)

$$\chi^2 = \frac{(82-70)^2}{70} + \frac{(427-420)^2}{420} + \frac{(191-210)^2}{210} + \frac{(10-18)^2}{18}$$

$$+ \frac{(110-108)^2}{108} + \frac{(60-54)^2}{54} + \frac{(8-12)^2}{12} + \frac{(63-72)^2}{72}$$

$$+ \frac{(49-36)^2}{36}$$

$$= 2.507 + .116 + 1.719 + 3.555 + .037 + .666 + 1.333$$
$$+ 1.125 + 4.694$$

$$= 15.752.$$

Before we can determine the probability of obtaining $\chi^2 > 15.752$ as a result of sampling fluctuations, we must determine the number of degrees of freedom in our problem:

$$\text{degrees of freedom} = (r-1)(c-1)$$
$$= (3-1)(3-1)$$
$$= 4.$$

With four degrees of freedom, Table D shows that there is only a .01 chance that $\chi^2 > 13.277$ could occur as a result of sampling fluctuations. Thus the probability is less than .01 that $\chi^2 > 15.752$ will occur if rating and educational level are independent. Using the .05 level of significance (or even the .01 level), one must, therefore, conclude that the Secretary of State rating is *not* independent of the educational level of the respondent.

EXERCISES, SET 3

1. In a public opinion poll, 500 American adults were asked, "Do you approve of the escalation of the war in Vietnam? Yes, no, or no opinion." The responses, cross-classified by party affiliation, are shown below.

	Yes	No	No opinion
Democrats	95	110	45
Republicans	80	75	45
Independents	25	15	10

Test the hypothesis that the attitude of American adults toward the escalation of the war in Vietnam is independent of their political party affiliations. Use a level of significance of .05.

2. Three drugs, Cloroquine, Pyrimethamine, and Sulformethoxine, were used in the treatment of 240 cases of Falciparum malaria. The number of patients cured with each drug, as well as the relapse rates, are shown below.

	Cloroquine	Pyrimethamine	Sulformethoxine
Number of cured cases	60	80	100
Relapse rate	40%	30%	12%

Using a level of significance of .01, are the data evidence that the relapse rates of the three drugs are not the same?

3. The following is a *percentage* distribution, by home ownership and income level, of a random sample of 400 families in Los Angeles:

	Annual income		
	Less than $5000	$5000–$10,000	More than $10,000
Home owners	5%	35%	10%
Renters	15%	25%	10%

Test the hypothesis that in Los Angeles home ownership is independent of the family income level. Use a level of significance of .01.

4. The Committee to Reelect mailed letters soliciting political contributions to a random sample of 1000 Republicans. Of these 1000 letters, 400 are originally typed personal letters, and 600 are mimeographed form letters. The committee received the following responses from the two kinds of letters:

	Typed letters	Mimeographed forms
Contribution under $100	200	200
Contribution of $100 or more	100	200
No response	100	200
	400	600

Test the hypothesis that political contributions are not influenced by the type of printing used in the solicitating letter. Use a level of significance of .01.

REVIEW EXERCISES III

1. It is known that 64 percent of all registered voters in the 27th Congressional District are Democrats. What is the probability that in a random sample of 400 voters from this district there would be 240 or more Democrats?

2. It is known that 10 percent of all families in a certain city are Mexican-American. If one hundred families are selected at random from this city, what is the probability that the percentage of Mexican-Americans would be

   a) greater than 13 percent?
   b) between 13 and 16 percent?

3. Assume that 50 percent of all adults in California favor admitting the newly formed Republic of Bangal to the United Nations.

   a) What is the probability that in a random sample of 300 adult Californians, 55 percent or more of the adults interviewed would favor admitting the Republic of Bangal to the United Nations?
   b) The probability is ninety percent that the proportion of adult Californians who favor admitting the new state to the United Nations in a random sample of 100 is somewhere between p_1 and p_2. Find p_1 and p_2 using symmetrical limits around the mean.

4. Twenty percent of all students attending this college are nonresident students.

   a) What is the probability that a random sample of 500 students would reveal 22 percent or more nonresident students?
   b) The probability is 0.50 that the percentage of nonresident students in a random sample of 1,600 is between p_1 and p_2. Find p_1 and p_2 using symmetrical limits around the mean of the sampling distribution of the proportion.
   c) Construct an interval where there is 0.80 chance that the proportion of non-residents in a random sample of 100 students is expected to fall.

5. A very large shipment of small parts contains 10 percent defectives. If a random sample of 225 parts is selected from this shipment,

   a) what is the probability that the sample proportion would be between 0.07 and 0.13?
   b) Establish an interval where there is 0.95 confidence that the sample proportion would fall within that interval.

6. A random sample of 200 housewives showed that 72 preferred Brand X coffee to all other competing brands. Establish a 0.95 confidence-interval estimate of the proportion of housewives favoring Brand X coffee.

7. A random sample of 400 families in Los Angeles revealed that 200 families have two cars or more.

   a) Establish a 0.90 confidence-interval estimate of the proportion of families owning two cars or more in Los Angeles.
   b) With what degree of confidence can we assert that the proportion of families with 2 cars or more in Los Angeles is somewhere between 45 and 55 percent?

8. A random sample of 400 files of students who visited the Student Health Center at this College last year shows that 80 students have an illness of a psychosomatic nature.

   a) With what degree of confidence can we assert that 16 to 24 percent of *all* students who visited the Health Center last year have a psychosomatic illness?

b) Assume that 2,000 students have visited the Health Center last year. How many of these students have a psychosomatic illness? What degree of confidence can you attach to your estimate?

9. It is claimed that chloroquine is 80 percent effective in the cure of malaria. 800 malaria cases were treated with chloroquine and only 600 were cured. Would you conclude that the effectiveness of chloroquine in the cure of malaria is exaggerated?

10. Mr. Smith, a life insurance salesman, reports to the regional sales manager that 20 percent of the prospects assigned to him would purchase a life insurance policy.

a) What would the manager conclude about the salesman's claim if a random sample of 100 prospects assigned to Mr. Smith shows that 18 percent of these prospects purchased a life insurance policy?

b) Would the manager's conclusion be different if only 7 percent of the sampled prospects purchased a life insurance policy? Justify your answer.

11. Machine #36 produces product X. Several years of experience have shown that machine #36 produces, on the average, 10 percent defective units. Recently, the quality-control engineer suspected that the quality of product X has deteriorated. Hence, he selected a random sample of 100 units. The sample revealed 14 defective units. Using a level of significance of 0.05, can the quality control engineer conclude that the quality of product X has deteriorated?

12. A random sample of 1,000 college professors in the United States shows 200 females.

a) Establish 0.95 confidence-interval estimate of female college professors in the United States.

b) An official of the AAUP claims 24 percent of all college professors in the United States are females. Test the validity of this claim, using a level of significance of 0.01.

13. A random sample of 500 students from this College shows 100 freshmen students.

a) Establish a 0.95 confidence-interval estimate of the proportion of freshmen students in this college.

b) The Dean of Students claims that the proportion of freshmen is only 15 percent. Test the validity of this claim using 0.01 level of significance.

c) Estimate the number of freshmen assuming that the college has a total enrollment of 20,000 students. Justify your answer.

14. A sample survey is conducted in a midwestern city of 10,000 families in order to determine the proportion of families who own a color television set. A random sample of 200 families shows 160 families own color television.

a) Establish a 0.95 confidence-interval estimate of the proportion of families who own color television in the city.

b) The president of the local Chamber of Commerce claims that at least 8,500 families in the city own color television. Test this claim using 0.01 level of significance.

15. A random sample of 400 new automobiles sold in the United States in 1980 showed 96 foreign-made cars.

    a) Establish a 0.95 confidence-interval estimate of the proportion of new foreign-made automobiles sold in the U.S. in 1980.

    b) Assuming that 20 percent of *all* new automobiles sold in the U.S. in 1979 are foreign-made, test the hypothesis that the proportion of foreign-made cars has increased between 1979 and 1980.   Use a level of significance of 0.05.

16. A random sample of 500 college professors in the United States revealed 400 men and 100 women.   The sample, furthermore, showed that of the men, 240 hold the doctorate degree.   Of the women professors, 80 have their doctorate. Using the 0.05 level of significance can we conclude that the percentage of those holding the doctorate degree is different between men and women?

17. A sample survey of 100 female and 100 male students from this college showed 55 females and 45 males in favor of coed housing.   If you are willing to reject a true hypothesis no more than once in 100, would you reject the hypothesis that there is no real difference of opinion between male and female students in regard to coed housing?

18. Random samples of 40, 40, and 120 parts are taken from the weekly productions of machines A, B, and C, respectively.   The number of *defective* parts found in the samples are: 15, 5, and 30 for machines A, B, and C, respectively.   Using the 0.05 level of significance can you conclude that the percentage of defective parts turned out by the three machines is essentially the same?

11
Linear
Regression and
Correlation

Linear Regression and Correlation

Since this chapter deals with simple linear regression and correlation, we shall begin by briefly discussing the concept of *simple linear equations*. Thus the first section of this chapter deals with different aspects of simple linear equations, including their graphical representation. The remaining sections of the chapter are devoted to the main topics: regression and correlation.

Simple Linear Equations

Let us assume that two variables, X and Y, are related in the following fashion: $Y = 3 + 1.5X$. This relationship enables us to compute the value of Y for any given value of X. If X is known to be 2, for example, $Y = 6$. Similarly, $Y = 9$ when X is known to be 4. The data below show various values of Y, each of which is associated with a given value of X.

Predicted value of Y	Known value of X	Change in Y
4.5	1	–
6.0	2	1.5
7.5	3	1.5
9.0	4	1.5
10.5	5	1.5

The last column of the above data indicates, furthermore, that the value of Y increases by 1.5 units as the value of X is increased by one unit.

An equation like $Y = 3 + 1.5X$ not only enables us to determine the value of Y associated with any given value of X, but it also describes the effect of a change in variable X on the values to be assumed by variable Y. Specifically, the coefficient of the variable X indicates the change in the value of Y caused by one unit change in X. Thus, as mentioned above, in the equation $Y = 3 + 1.5X$, the value of Y increases by 1.5 units as the value of X is increased by 1 unit. Similarly, the value of Y in the equation $Y = 6 + 0.2X$ increases by .2 unit as the value of X is increased by 1 unit. In both of these two equations, variable Y increases as variable X increases, and the two variables are said to be *directly* related. On the other hand, in the equation $Y = 100 - 2X$, the value of Y decreases by 2 units as X is increased by 1 unit, and X and Y are said to be *inversely* related. The sign of the coefficient of X in the equation indicates whether the two variables are directly or inversely related. A positive coefficient indicates a direct relationship, while a negative coefficient indicates an inverse one.

The equation $Y = 3 + 1.5X$ is a *first-degree* equation, while the equations $Y = 3 + 2X + 4X^2$ and $Y = 3 + 7X + 5X^3$ are called *second-degree* and *third-degree* equations, respectively. The degree of an equation refers to the highest power of X in the equation. All these three equations, however, are called *simple equations* because they express a relationship between only two variables, Y and X. An equation such as $Y = 3X + 8Z$, which expresses a relationship between *more than two* variables, is called a *multiple* equation. In this chapter our discussion is limited to simple first-degree relationships.

Let us see how a simple first-degree equation such as $Y = 3 + 1.5X$ can be represented graphically. To represent any equation by a graph, one must first prepare a table that shows several values for variable X and the corresponding values of Y.

X	1	2	3	4	5
Y	4.5	6.0	7.5	9.0	10.5

Next we measure off the value of X on the horizontal axis and the value of Y on the vertical axis. Then each pair of values of X and Y such as (1, 4.5), (2, 6), and (3, 7.5) represent a point in the graph. The five paired values above are plotted in Fig. 11–1. Examining the diagram, we can also readily observe that the five points lie on a straight line. This line is the graphical representation of the equation $Y = 3 + 1.5X$.

The equation $Y = a + bX$ is the general form of a simple first-degree equation (in the equation $Y = 3 + 1.5X$, we have $a = 3$ and $b = 1.5$). Since the graphic representation of such an equation is a straight line, a simple first-degree equation is also called a simple linear equation.

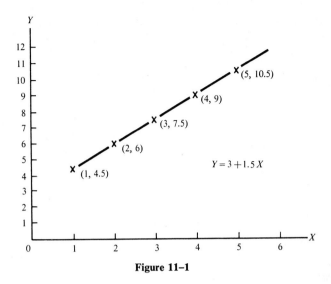

Figure 11–1

Simple Linear Regression

The principal objective in simple regression analysis is to establish a quantitative relationship (in the form of an equation) between two related variables. Once this relationship is established, one can predict the value of one of the variables if the value of the other variable is known. The variable to be predicted is called the dependent variable and is denoted by Y. The known variable is called the independent variable and is denoted by X.

In order to establish the quantitative relationship between Y and X, one must obtain certain sample data. These data consist of a number of *paired observations* on Y and X, where each of these pairs belongs to one particular elementary unit in the sample.

To be specific, let us assume that we desire to determine the simple linear relationship between the height and the age of a papaya tree, based on a sample of five trees. If the height and age of each tree are recorded, our data will then consist of five paired observations, with each pair referring to the height and age of one tree. These five paired observations are shown below.

Y, feet	9	5	7	14	10
X, years	3	1	2	5	4

Thus the first pair of observations ($Y = 9$, $X = 3$) indicates that the first tree is 9 feet high and 3 years old. In the same manner, the pair ($Y = 5$,

$X = 1$) refers to the second tree whose height is 5 feet and whose age is 1 year. Finally, the pair ($Y = 10$, $X = 4$) refers to the height and age of the fifth tree.

Once the necessary data are obtained, the relationship between the two variables can be determined either graphically, or mathematically by the *least-squares method*. These two methods are discussed below.

The Graphic Method

In the graphic method, each pair of observations is represented by a point on a chart. A point is obtained by plotting the independent variable, X, along the horizontal axis and the dependent variable, Y, along the vertical axis. The pair ($Y = 9$, $X = 3$), for example, is represented by point (1) in Fig. 11–2; point (1) is located at the intersection of 3 on the horizontal axis and 9 on the vertical axis. The remaining four pairs of observations are plotted in a similar fashion. A plot of all observations is called a *scatter diagram*.

Once all observations are plotted, we draw a straight line that best represents the relationship between the two variables. Such a line (Fig. 11–2) is called the *estimated regression line*. It enables us to predict the value of Y for any given value of X. If X is known to be 7, for example, the value of Y is expected to be approximately 17.

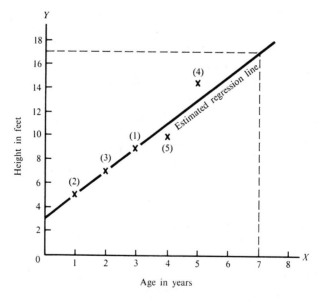

Figure 11–2

The graphic method for determining the relationship between two variables is certainly simple and straightforward. Its main drawback, however, is the subjective nature of the procedure. Different people will most likely draw different regression lines. The mathematical or least-squares method, on the other hand, will always yield the same equation, an equation that best describes the linear relationship between two variables.

The Least-Squares Method

We have already learned that a linear relationship between two variables is represented by a straight line whose general equation is $Y = a + bX$. The least-squares method is the mathematical procedure used to determine the numerical value of the constants a and b in this equation. The general equation $Y = a + bX$ is called the *regression equation* and the constants a and b are called the *regression coefficients*.

Like the graphic method, the least-squares method utilizes a set of paired observations on Y and X. Returning to our example of determining the simple linear relationship between the height and age of a papaya tree, we repeat the set of five paired observations:

Y, feet	9	5	7	14	10
X, years	3	1	2	5	4

A few preliminary computations facilitate the procedure used in determining the numerical values of the constants a and b. These calculations are carried out, explained, and summarized below.

Preliminary Computations

(1) Y	(2) X	(3) Y^2	(4) X^2	(5) XY
9	3	81	9	27
5	1	25	1	5
7	2	49	4	14
14	5	196	25	70
10	4	100	16	40
$\sum Y = 45$	$\sum X = 15$	$\sum Y^2 = 451$	$\sum X^2 = 55$	$\sum XY = 156$

Explanation

Column (1) contains the various values of *Y*, together with their sum.

Column (2) contains the various values of *X*, together with their sum.

Column (3) lists the squares of the individual values of *Y* in column (1), together with their sum. .Thus, for example, the 81 in column (3) is obtained by squaring the 9 in column (1).

Column (4) lists the squares of the individual values of *X* in column (2), together with their sum. Thus, for example, the 9 in column (4) is obtained by squaring the 3 in column (2).

Column (5) lists the cross products of the corresponding *Y*- and *X*-values in columns (1) and (2), together with their sums. Thus, for example, the 27 in column (5) is the cross product of the 9 and 3 from columns (1) and (2).

Summary of computations

$$\sum Y = 45 \qquad \sum Y^2 = 351$$
$$\sum X = 15 \qquad \sum X^2 = 55$$
$$\sum XY = 156$$

Once the sum of each variable, the sum of each of their squares, and the sum of their cross products are computed, we go to the next step which is to substitute these values in what is called the two *normal equations.*
The two normal equations are

$$\sum Y = na + b \sum X,$$
$$\sum XY = a \sum X + b \sum X^2,$$

where *n* refers to the number of paired observations used in the regression (we have five paired observations here, that is, $n = 5$) and *a* and *b* are the regression coefficients which we desire to determine.
Substituting the values $\sum Y = 45$, $n = 5$, $\sum X = 15$, $\sum XY = 156$, $\sum X^2 = 55$ in the two normal equations, we obtain the following two simultaneous equations:

$$45 = 5a + 15b, \tag{1}$$

$$156 = 15a + 55b. \tag{2}$$

Solving these two equations, we find that the regression coefficients *b* and *a* are 2.1 and 2.7, respectively, and hence the regression equation is $Y = 2.7 + 2.1X$. This equation enables us to estimate the value of *Y* for any given value of *X*. If *X* is known to be 10 years, for example, the estimated value of *Y* is 23.7 feet. The regression equation, furthermore, indicates that the value of Y increases by 2.1 feet as the value of X increases by 1 year. In other words, each additional year will add 2.1 feet to the height of a papaya tree.

Before concluding this discussion, we must make a clear distinction between the *observed* value of Y and the *estimated* value of Y. While the observed value of Y refers to the *actual* height of a given tree, the estimated value of Y is the estimated height of a tree, using the regression equation as a means of estimation. The observed and the estimated height of a tree may or may not be the same. This point becomes clear if one refers to any of the trees in our sample. For example, the second tree which is 1 year old has an observed, or actual, height of 5 feet. However, using the regression equation $Y = 2.7 + 2.1X$, we obtain its estimated height as 4.8 feet (the estimated height is obtained by substituting $X = 1$ in the regression equation). Thus, while the observed value of Y is 5 feet, the estimated value is 4.8 feet; the two values are not the same. To distinguish between the observed and the estimated height of a given tree, we use Y to denote the observed height and Y' to denote the estimated height. Thus the regression equation, which always refers to the *estimated* value of the dependent variable, must be written $Y' = 2.7 + 2.1X$ (see Fig. 11–3).

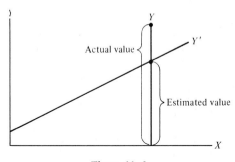

Figure 11–3

Standard Error of the Estimate

The *standard error of the estimate*, denoted by S_{YX}, measures the "average" disparity between the actual and estimated values of the variable Y. Hence, to determine the standard error of the estimate, we must begin by computing the estimated height of each tree in our sample, that is, we must substitute the age of each tree in the regression equation $Y' = 2.7 + 2.1X$. The observed height, Y, the age, X, and the estimated height, Y', of the five trees are shown in the first, second, and third columns below.

(1) Y	(2) X	(3) Y'	(4) $(Y - Y')$	(5) $(Y - Y')^2$
9	3	9.0	.0	0
5	1	4.8	+.2	.04
7	2	6.9	+.1	.01
14	5	13.2	+.8	.64
10	4	11.1	−1.1	1.21
				$\sum (Y - Y')^2 = 1.90$

Column (4) above shows the difference between the observed and estimated height of each tree. These differences are then squared and entered in column (5). The total of column (5) is $\sum (Y - Y')^2$. The standard error of the estimate can now be computed as

$$S_{YX} = \sqrt{\frac{\sum (Y - Y')^2}{n - 2}}$$

$$= \sqrt{\frac{1.90}{5 - 2}} = \sqrt{\frac{1.90}{3}} = \sqrt{.633} = .80 \text{ feet.}$$

Since the standard error of the estimate measures the "average" disparity between the actual and estimated heights of a tree in the sample, it can be regarded as an indication of how well the regression equation describes the relationship between the two variables. If S_{YX} is small, the actual and estimated values of Y must be reasonably close, and the regression equation is a good description of the relationship. When $S_{YX} = 0$, for example, the actual and estimated values of Y must have been identical, and the regression equation is a perfect description of the relationship, which, in this case, is called an *exact relationship*. A large standard error of the estimate, on the other hand, means a large disparity between the actual and estimated values of Y; in this case, the regression equation is regarded as a poor description of the relationship between the two variables.

The standard error of the estimate is measured in the same units as Y; in the case of our papaya tree, both are measured in feet. If we had measured the height of each tree in inches, the standard error of the estimate would also be in inches. Furthermore, its numerical value would be 12 times as large. In other words, the standard error is affected by changes in the units of Y. It is for this reason that it is not regarded as a good measure of the degree of a relationship. What we need instead is a measure that is abstract and unaffected by any change in the units in which Y is measured. Such a measure is called the *correlation coefficient*.

Coefficient of Correlation

The regression equation $Y' = 2.7 + 2.1X$ was derived from a sample of five trees whose actual or observed heights are $Y_1 = 9$ feet, $Y_2 = 5$ feet, $Y_3 = 7$ feet, $Y_4 = 14$ feet, and $Y_5 = 10$ feet. The average height of a tree in this sample can be computed as

$$\overline{Y} = \frac{Y_1 + Y_2 + Y_3 + Y_4 + Y_5}{5}$$

$$= \frac{9 + 5 + 7 + 14 + 10}{5} = \frac{45}{5} = 9 \text{ feet.}$$

We measure the variation in the heights of these five trees by computing the sum of the squared deviations, i.e., differences between the actual height of the individual tree and the mean height, $\sum (Y - \overline{Y})^2$:

Y	$\overline{Y}$	$(Y - \overline{Y})$	$(Y - \overline{Y})^2$
9	9	0	0
5	9	-4	16
7	9	-2	4
14	9	$+5$	25
10	9	$+1$	1
			$\sum (Y - \overline{Y})^2 = 46$

The sum, $\sum (Y - \overline{Y})^2$, which measures variations in the heights of the five trees is called *total variation* (Fig. 11–4).

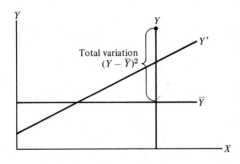

Figure 11–4

We can now ask an interesting question: Can we account for or explain this variation in the heights of these trees? There are certainly many factors responsible for it, such as, for example, the variation in the soil in which these trees are planted, the variation in climate, and, above all, the variation in the ages of the individual trees. The next question to be asked is: Since we are interested in the relationship between height and age, why not separate the effect of variation in age from the effect of all other factors? This is precisely what we will attempt to do. The total variation in heights will be subdivided into two components. The first component is the variation in height which is associated with, or could be explained by, variation in age. This component is called *explained variation*. The remaining component of total variation is attributed to all other factors that influence height such as soil and climate. The latter component is called the *unexplained variation*, since it is not explained by variation in age, but by variations in all other factors.

We measure the explained variation by computing the sum of the squared deviations, as the difference between the *estimated* height of each individual tree and the mean height in the sample, i.e., $\sum (Y' - \bar{Y})^2$:

Y'	$\bar{Y}$	$(Y' - \bar{Y})$	$(Y' - \bar{Y})^2$
9.0	9	0	.00
4.8	9	-4.2	17.64
6.9	9	-2.1	4.41
13.2	9	$+4.2$	17.64
11.1	9	$+2.1$	4.41
			$\sum (Y' - \bar{Y})^2 = 44.10$

Explained variation is shown in Fig. 11–5.

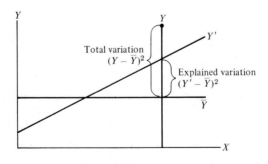

Figure 11–5

To measure the unexplained variation, we compute the sum of the squared deviations, as differences between the *actual* and estimated height of each tree, i.e., $\sum (Y - Y')^2$:

Y	Y'	$(Y - Y')$	$(Y - Y')^2$
9	9.0	.0	.00
5	4.8	+.2	.04
7	6.9	+.1	.01
14	13.2	+.8	.64
10	11.1	−1.1	1.21
			$\sum (Y - Y')^2 = 1.90$

Unexplained variation is shown in Fig. 11–6.

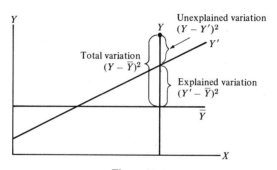

Figure 11–6

Let us now summarize the computations required to obtain the total variation, explained variation, and unexplained variation.

Y	$\bar{Y}$	Y'	$(Y - \bar{Y})$	$(Y - \bar{Y})^2$	$(Y' - \bar{Y})$	$(Y' - \bar{Y})^2$	$(Y - Y')$	$(Y - Y')^2$
9	9	9.0	0	0	0	.00	.0	.00
5	9	4.8	−4	16	−4.2	17.64	+.2	.04
7	9	6.9	−2	4	−2.1	4.41	+.1	.01
14	9	13.2	+5	25	+4.2	17.64	+.8	.64
10	9	11.1	+1	1	+2.1	4.41	−1.1	1.21
				$\sum (Y - \bar{Y})^2 = 46$ Total variation		$\sum (Y' - \bar{Y})^2 = 44.10$ Explained variation		$\sum (Y - Y')^2 = 1.90$ Unexplained variation

The reader may have already observed that

Total variation = Explained variation + Unexplained variation,

46 = 44.10 + 1.90.

This is, by no means, a coincidence. The total variation in height has indeed been subdivided into two components: the explained variation, which is associated with age, and the unexplained variation, which is associated with other factors.

The ratio of explained variation to total variation is called the *coefficient of determination*. Denoted by r^2, the coefficient of determination is

$$r^2 = \frac{\text{explained variation}}{\text{total variation}}$$

$$= \frac{\sum (Y' - \bar{Y})^2}{\sum (Y - \bar{Y})^2}$$

$$= \frac{44.10}{46.0} = .96.$$

It expresses the proportion of the variation in Y which is associated with, or related to, the variation in X. In our particular example, we may, therefore, conclude that 96% of the variation in the heights of the five papaya trees is related to, or explained by, the variation in the ages of these trees.

Now let us consider the possible numerical values that can be assumed by the coefficient of determination. If there is an exact relationship between X and Y, then, as we have already learned, the actual and estimated values of Y will be identical for every observation in the sample. In this case, the unexplained variation $\sum (Y - Y')^2 = 0$, and the explained variation will be equal to the total variation. Consequently, $r^2 = 1$, which is the highest value the coefficient of determination can have. It indicates a perfect correlation between the two variables, since all the variation in Y is associated with, or explained by the variation in X. On the other hand, if X and Y are unrelated, the explained variation is zero, and hence $r^2 = 0$.

Thus we can conclude that the value assumed by the coefficient of determination is between 0 and 1. While $r^2 = 1$ indicates the highest degree of correlation, $r^2 = 0$ indicates no correlation at all. The stronger the relationship between X and Y, the higher the value of r^2.

The *coefficient of correlation* r is the square root of the coefficient of determination r^2. Its numerical value, therefore, is also between 0 and 1. But while r^2 is always positive, r can be either positive or negative. We find that r is positive when X and Y are directly related, and is negative when X and Y are inversely related. Stated differently, r always assumes the same sign as the regression coefficient b in the regression equation $Y' = a + bX$.

For our problem of the papaya tree, $r^2 = .96$. Hence $r = \sqrt{.96}$, or .98. Furthermore, r is positive because X and Y are directly related as indicated by the positive sign of the regression coefficient b in the regression equation $Y' = 2.7 + 2.1X$.

Before concluding this discussion, we wish to issue a few words of caution. Both r^2 and r measure the strength of a linear relationship between two variables. Both measures assume the value 1 when the correlation is perfect and the value 0 when there is absolutely no correlation. Except in these two extreme cases, however, r^2 is the more meaningful of the two measures, because it is r^2 and not r that measures the percentage of the variation in Y explained by the variations in X.

Correlation and Regression. Short-cut Method

We can calculate the three measures, the regression coefficients, the standard error of the estimate, and the correlation coefficient, using an alternative method, called the *short-cut* method. It consists of the following steps:

Step 1. Construct five columns to record the *actual* values of X and Y, their squares, and their cross products. (Use upper-case letters for X and Y.) Each column is then totaled.

Y	X	Y^2	X^2	XY
9	3	81	9	27
5	1	25	1	5
7	2	49	4	14
14	5	196	25	70
10	4	100	16	40
$\sum Y = 45$	$\sum X = 15$	$\sum Y^2 = 451$	$\sum X^2 = 55$	$\sum XY = 156$

Step 2. Compute the mean of Y and the mean of X.

$$\overline{Y} = \frac{\sum Y}{n}$$

$$= \tfrac{45}{5} = 9 \text{ feet};$$

$$\overline{X} = \frac{\sum X}{n}$$

$$= \tfrac{15}{5} = 3 \text{ years.}$$

Step 3. Compute $\sum y^2$, $\sum x^2$, and $\sum xy$. (Note that we use lower-case letters for these terms which are completely different from the terms denoted by upper-case letters that we computed previously.) Use the following three equations:*

$$\sum y^2 = \sum Y^2 - n(\bar{Y})^2$$
$$= 451 - 5(9)^2 = 46; \tag{1}$$

$$\sum x^2 = \sum X^2 - n(\bar{X})^2$$
$$= 55 - 5(3)^2 = 10; \tag{2}$$

$$\sum xy = \sum XY - n\bar{X}\bar{Y}$$
$$= 156 - 5(9)(3) = 21. \tag{3}$$

Step 4. Compute the regression coefficient b:

$$b = \frac{\sum xy}{\sum x^2} = \tfrac{21}{10} = 2.1.$$

Step 5. Compute the regression coefficient a:

$$a = \bar{Y} - b\bar{X}$$
$$= 9 - (2.1)(3) = 9 - 6.3 = 2.7.$$

Hence the regression equation is

$$Y' = 2.7 + 2.1X.$$

Step 6. Compute the standard error of the estimate:

$$S_{YX} = \sqrt{\frac{\sum y^2 - b \sum xy}{n - 2}}$$

$$= \sqrt{\frac{46 - (2.1)(21)}{5 - 2}} = \sqrt{\frac{46 - 44.1}{3}}$$

$$= \sqrt{\frac{1.90}{3}} = .80.$$

---

*   The terms $\sum x^2$, $\sum y^2$, and $\sum xy$ are obtained as follows:

$$\sum x^2 = \sum (X - \bar{X})^2, \qquad \sum y^2 = \sum (Y - \bar{Y})^2, \qquad \sum xy = \sum (X - \bar{X})(Y - \bar{Y}).$$

However, we can readily compute them, using the three equations cited under Step 3.

Step 7. Compute the correlation coefficient:

$$r = \frac{\sum xy}{\sqrt{\sum x^2 \sum y^2}}$$

$$= \frac{21}{\sqrt{(10)(46)}} = \frac{21}{\sqrt{460}} = \frac{21}{21.16} = .98.$$

The Significance of Correlation

Since the correlation coefficient r is computed from sample data, an interesting question always arises: How is the sample correlation coefficient related to the *true population* correlation coefficient? If the correlation coefficient is .60, for example, what can we say about the magnitude of the population correlation coefficient?

The relationship between the sample correlation coefficient and the population correlation coefficient is very complex and its examination in depth is beyond the scope of this book. However, we shall discuss some of its aspects in a somewhat simplified form.

Let us assume that the *population* correlation coefficient is equal to zero, i.e., there is no correlation between two variables in the population. Now, if a random sample of n paired observations is drawn from this population, will the sample correlation coefficient be equal to zero? Of course not. Stated differently, the mere fact that the absolute value of a sample correlation coefficient is greater than zero is, by no means, conclusive evidence that a correlation exists in the population.

When the absolute value of the sample correlation coefficient is greater than zero, we have essentially two hypotheses to explore. First, there is the null hypothesis that there is no correlation between the two variables in the population. Consequently, the correlation coefficient of *any* sample drawn from this population is not significant, regardless of the magnitude of that sample correlation coefficient. The second hypothesis to be explored is that a correlation does exist in the population. If this is the case, then the correlation coefficient of any sample drawn from this population is indeed significant. But if the *population* correlation coefficient is never known, how can we decide whether or not a particular sample correlation coefficient is significant? Table G enables us to infer, with specified risk, whether or not a particular correlation coefficient is significant.

The first column in Table G lists various numbers of degrees of freedom. In a simple correlation problem, the number of degrees of freedom is equal to the sample size minus 2, that is $(n - 2)$. Thus in a correlation problem with 19 paired observations, the number of degrees of freedom is 17.

The entries in the column labeled $r^*_{.05}$ are critical levels of r, each associated with a given number of degrees of freedom. Each of these critical levels can be interpreted as follows: If the null hypothesis is true, i.e. there is no correlation in the population, then the probability is only .05 that a sample correlation coefficient will be as large as or larger than this critical level. Therefore, when the sample correlation coefficient exceeds this critical level, the null hypothesis is rejected and the sample correlation coefficient is regarded to be significant at the .05 level (of course, there is still a less than 5% risk that we are rejecting a true hypothesis). Thus, in a simple correlation problem of 19 paired observations (df = 17), the sample correlation coefficient is regarded as significant at the .05 level whenever its absolute value exceeds .456. Likewise, in a correlation problem of 30 paired observations (df = 28), the correlation coefficient is regarded as significant at the .05 level whenever its absolute value exceeds the critical level of .361.

The entries in the last column, labeled $r^*_{.01}$, can be interpreted in a similar fashion. These entries, however, are the critical 1% levels of r. Consequently, a sample correlation coefficient is regarded as significant at the .01 level only when its absolute value exceeds the corresponding critical level in this column.

Before concluding this discussion, let us reexamine the degree of the relationship between the height and the age of our papaya trees. Based on a sample of five trees, the correlation coefficient was .98. With three degrees of freedom this correlation coefficient is significant at the .01 level (r is greater than the critical 1% level of .959).

EXERCISES

1. A medical experiment was conducted in order to determine the effect of the drug ephedrine on the heart rate. A patient was given various daily dosages of the drug for six days. The data below summarize the results of the experiment.

Total daily dosage of ephedrine, grains (X)	Number of heart beats/minute (Y)
3	70
2	60
1	50
3	80
5	100
4	90

a) Plot a scatter diagram for the data and draw the regression line (free hand). Use the regression line to estimate the patient's heart rate for a total daily dosage of 3.5 grains of ephedrine.

b) Determine the estimated regression equation by means of the two normal equations. Use the regression equation to estimate the patient's heart rate for a total daily dosage of 3.5 grains of ephedrine.

2. Based on the data in Exercise 1, the estimated regression equation is

$$Y' = 36 + 13X.$$

a) What is the average increase in the number of heart beats associated with an *additional* dosage of one grain of ephedrine?

b) Estimate the heart rate of the patient when the total daily dosage is 50 grains.

c) A physician asserts that a total daily dosage of 50 grains of ephedrine reduces the heart rate to zero. In the light of this statement, discuss the validity of the estimate made in part (b) of this question.

3. Given the data and the results of Exercise 1 and without the use of the short-cut method,

a) Compute the total variation, explained variation, and unexplained variation. How are the three variations related?

b) Compute the standard error of the estimate and explain its shortcomings as an indicator of the degree of relationship.

c) Compute the coefficient of determination and correlation coefficient. Which one of the two coefficients is a more meaningful measure of the degree of relationship and why?

4. A random sample of 5 families shows the following information concerning annual family income and annual expenditure on durable goods (refrigerators, washing machines, stereos, etc.):

Family	Annual income (thousands of dollars)	Expenditures on durable goods (hundreds of dollars)
Addison	5	1
Baum	8	2
Cleary	7	1
Dunn	10	2
Evans	15	4

a) Determine the estimated regression equation $Y' = a + bX$

b) Estimate the annual expenditure on durable goods of a family earning $12,000 per year.

5. An economist selected a random sample of five firms from a very large industry

in order to study the relationship between total cost and volume of production. The following information was obtained:

Firm	Number of units produced	Total cost (in 1000's of dollars)
A	2	9
B	4	12
C	6	15
D	5	14
E	3	10

a) Determine the estimated linear regression equation:

$$Y' = a + bX$$

where

Y = total cost of production in 1000's of dollars,
X = number of units produced.

b) What is the estimated total cost of producing 10 units?
c) Determine the fixed and variable costs.

6. A random sample of 82 supermarkets was selected to measure the relationship between annual sales (Y) and annual advertising expenditures (X). The sample revealed the following information:

$$\text{Explained variation} = \Sigma (Y' - \bar{Y})^2 = \$49 \text{ (million)}$$
$$\text{Unexplained variation} = \Sigma (Y - Y')^2 = \$51 \text{ (million)}$$

a) Compute the correlation coefficient.

b) Test the significance of r using the .01 level of significance.

7. A random sample of 37 male students from the university is selected to study the relationship between the weight and the height of a male college student. The heights of the students selected in the sample range from 66 to 72 inches. The results of the regression and correlation in the study are

$$Y' = -110 + 3.8X;$$
$$(r = .51, S_{YX} = 4.2);$$

where

Y = weight in pounds,
X = height in inches.

a) What is the meaning of the regression coefficient b, in terms of the data of this problem?

b) Mr. Smith's height is 70 inches, his weight is 140 lbs. Is Mr. Smith's weight normal? Explain.

c) How do you justify that the value of regression coefficient a is negative (i.e., $a = -110$)?

d) Test the significance of the correlation coefficient using the .01 level of significance.

8. Data for United States disposable personal income and personal consumption expenditures from 1950 through 1977 are given below.

United States disposable personal income and personal consumption expenditures, 1950–1977

Year	Personal consumption expenditures, billions Y	Disposable income, billions X
1950	191	207
1951	206	227
1952	217	238
1953	230	253
1954	237	257
1955	254	275
1956	267	293
1957	281	309
1958	290	319
1959	311	337
1960	325	350
1961	335	364
1962	355	385
1963	375	405
1964	401	438
1965	433	473
1966	466	512
1967	492	546
1968	536	591
1969	580	634
1970	618	692
1971	667	746
1972	729	803
1973	805	903
1974	877	979
1975	980	1084
1976	1094	1186
1977	1217	1322

a) Determine the linear regression equation with personal consumption expenditures as the independent variable. Use the short-cut method.

b) If the U.S. disposable personal income is estimated as 1550 billion dollars in 1980, what is the amount of personal consumption expenditures in 1980?

c) Determine the degree of this relationship and test its significance.

12
Inferences
in Regression

Inferences
in Regression

Introduction

We have introduced the subject of regression and correlation in the previous chapter. Specifically, we have learned how to determine the linear regression equation $Y' = a + bX$ from sample data. This equation is used to estimate the average value of Y for any given value of X. In addition, we have also learned how to compute the standard error of the estimate S_{YX} and the correlation coefficient r. The meanings of these two measures were also explained.

In this chapter, we will extend our treatment of regression. In the first place, we will discuss the reliability of the regression coefficient b. Second, we will investigate the reliability of estimating the average value of Y for any given value of X.

Based on a sample of n paired observations on the variables X and Y, we have used the least-squares method to obtain the coefficients a and b in the linear regression equation $Y' = a + bX$. Now, since the values of these regression coefficients are computed from sample data, we expect these values to vary from sample to sample. In other words, the values of regression coefficients a and b depend on the sample chosen; and an alternative sample may result in different values for these coefficients.

Throughout this text, an attempt has been made to distinguish between the population and the sample. This distinction must be brought into focus once more.

To begin, let us assume that *in the population*, the average relationship between Y and X is of the form

$$Y = A + BX.$$

343

This relationship is a population relationship and both A and B are population parameters. The population relationship, $Y = A + BX$, is estimated from a sample of n paired observations by means of the least-squares method. The estimated relationship is

$$Y' = a + bX,$$

where the regression coefficients a and b are sample estimates of the parameters A and B, respectively. In fact, a is called a point estimate of A and b is the point estimate of B.

In the linear relationship $Y = A + BX$, the regression coefficient B is of prime importance. The coefficient B describes the effect of a change in variable X on the values to be taken by the variable Y. Specifically, the coefficient B measures the average change in the value of Y caused by one unit change in the value of X. For example, in the relationship $Y = 3 + 7X$, the value of Y increases by 7 units as the value of X is increased by one additional unit. It is for this reason, therefore, that the next part of our discussion will be devoted to the problem of estimating the value of B and to tests of hypothesis concerning the value of B.

Estimating the Regression Coefficient B

Once again, the population relationship $Y = A + BX$ is estimated from a sample of n paired observations on the variables X and Y. Using the least-squares method, the estimated relationship is $Y' = a + bX$. The value of b varies from sample to sample; and to establish a confidence interval estimate of B, one must take this variability into account.

As an initial step in establishing a confidence interval estimate of B, it is necessary first to define the statistic t as

$$t = \frac{b - B}{\hat{\sigma}_b}, \tag{1}$$

where $\hat{\sigma}_b$ is an estimate of the standard error of b, a measure of the variations in the value of b from sample to sample; and the statistic t has a t-distribution with $n - 2$ degrees of freedom. For illustrative purposes, let $n = 7$, and consider the problem of finding a 95% confidence interval estimate of B. From Table E, with $df = 7 - 2 = 5$, we obtain $t_{.025} = 2.571$. Now, since

$$t = \frac{b - B}{\hat{\sigma}_b}$$

has a *t*-distribution with 5 degrees of freedom, the probability is .95 that *t* will satisfy the inequalities

$$-2.571 < t < 2.571.$$

From (1) above, this is equivalent to

$$-2.571 < \frac{b - B}{\hat{\sigma}_b} < 2.571.$$

Solving these inequalities for *B* we arrive at

$$b - 2.571\hat{\sigma}_b < B < b + 2.571\hat{\sigma}_b.$$

Thus, $b \pm 2.571\hat{\sigma}_b$ is a 95% confidence interval estimate of *B*. In general, the confidence interval estimate of *B* is

$$b \pm t\hat{\sigma}_b.$$

As a final step, the estimated standard error, $\hat{\sigma}_b$, is given by

$$\hat{\sigma}_b = \frac{S_{YX}}{\sqrt{\sum x^2}},$$

where S_{YX} = standard error of the estimate and $\sum x^2$ = sum of squared derivations $\sum (X - \overline{X})^2$ (a term computed by the short-cut method as $\sum X^2 - n(\overline{X})^2$). Hence, the confidence interval estimate of *B* becomes

$$b \pm t \frac{S_{YX}}{\sqrt{\sum x^2}}.$$

We will illustrate the use of this formula with an example from the preceding chapter.

Example 1

The following data represents the heights and ages of a random sample of five papaya trees:

Y, feet	9	5	7	14	10
X, year	3	1	2	5	4

a) Find the regression coefficient *b* and the standard error of the estimate S_{YX}.

b) Establish a 95% confidence interval estimate of *B*.

Solution (a): We will use the short-cut method, which was explained in detail in the preceding chapter, to find the regression coefficient *b* and the standard error of the estimate S_{YX}.   Hence, our first step is to compute the terms $\Sigma X^2, \Sigma Y^2, \Sigma XY, \bar{X}, \bar{Y}$.   These computations are carried out below.

Y	X	Y^2	X^2	XY
9	3	81	9	27
5	1	25	1	5
7	2	49	4	14
14	5	196	25	70
10	4	100	16	40
$\Sigma Y = 45$	$\Sigma X = 15$	$\Sigma Y^2 = 451$	$\Sigma X^2 = 55$	$\Sigma XY = 156$

$$\bar{Y} = \frac{\Sigma Y}{n}$$

$$= \frac{45}{4} = 9 \text{ feet};$$

$$\bar{X} = \frac{\Sigma X}{n}$$

$$= \frac{15}{5} = 3 \text{ years.}$$

Our next step is to compute the terms Σx^2, Σy^2, and Σxy (use lower-case letters for these terms) as follows:

$$\Sigma y^2 = \Sigma Y^2 - n(\bar{Y})^2$$

$$= 451 - 5(9)^2 = 46; \tag{1}$$

$$\Sigma x^2 = \Sigma X^2 - n(\bar{X})^2$$

$$= 55 - 5(3)^2 = 10; \tag{2}$$

$$\Sigma xy = \Sigma XY - n\bar{X}\bar{Y}$$

$$= 156 - 5(9)(3) = 21. \tag{3}$$

We now compute the regression coefficient b:

$$b = \frac{\Sigma xy}{\Sigma x^2} = \frac{21}{10} = 2.1.$$

Next we compute the standard error of the estimate:

$$S_{YX} = \sqrt{\frac{\sum y^2 - b \sum xy}{n-2}}$$

$$= \sqrt{\frac{46 - (2.1)(21)}{5-2}} = \sqrt{\frac{46 - 44.1}{3}}$$

$$= \sqrt{\frac{1.90}{3}} = .80 \; .$$

Solution (b): In part (a) of our solution we obtained

$$b = 2.1, \qquad S_{YX} = .8, \qquad \sum x^2 = 10.$$

We will now establish a 95% confidence interval estimate of B as

$$b \pm t_{.025} \frac{S_{YX}}{\sqrt{\sum x^2}} \; .$$

With $(n-2) = (5-2) = 3$ degrees of freedom, the value of $t_{.025}$ from Table E is 3.182. Hence, the 95% confidence interval estimate of B becomes

$$b \pm 3.182 \frac{S_{YX}}{\sqrt{\sum x^2}} \; ;$$

$$2.1 \pm 3.182 \frac{.8}{\sqrt{10}}, \qquad 2.1 \pm .805,$$

or

$$1.295 \le B \le 2.905.$$

Hence, we may conclude that we are .95 confident that trees grow between 1.295 and 2.905 feet each year on the average.

Example 2

In the analysis of gross national product, a great deal of attention is paid to the nature of relationship between personal consumption expenditures and disposable personal income (disposable personal income refers to the income at the disposal of the consumers that the consumers may either spend or save; and personal consumption expenditures is the amount that the consumers spend on goods and services of all sorts). Various economic studies have suggested that these two variables are related in the following fashion:

$$Y = A + BX,$$

where

$$X = \text{disposable personal income,}$$

$$Y = \text{personal consumption expenditures.}$$

In this relationship the regression coefficient B is of a great importance to the economist. Called the *marginal propensity to consume*, the regression coefficient B represents the amount that individuals spend out of each additional dollar of income.   Economists believe that the magnitude of the

Table 1

United States disposable personal income and personal consumption expenditures, 1950–1974.

Year	Personal consumption expenditures, billions Y	Disposable income, billions X
1950	191	207
1951	206	227
1952	217	238
1953	230	253
1954	237	257
1955	254	275
1956	267	293
1957	281	309
1958	290	319
1959	311	337
1960	325	350
1961	335	364
1962	355	385
1963	375	405
1964	401	438
1965	433	473
1966	466	512
1967	492	546
1968	536	591
1969	580	634
1970	618	692
1971	667	746
1972	729	803
1973	805	903
1974	877	979

marginal propensity to consume, B, is a crucial factor in determining the effectiveness of governmental fiscal policies.*

In this example, we will establish a 95% confidence interval estimate of the regression coefficient B (marginal propensity to consume), based on a random sample of 25 paired observations on X and Y. These observations are obtained from the United States National Income accounts for the years 1950 through 1974 (see Table 1).

Since a 95% confidence interval for B is

$$b \pm t_{.025} \frac{S_{YX}}{\sqrt{\sum x^2}},$$

we must first calculate the regression coefficient, b, and the standard error of the estimate, S_{YX}. We will use the short-cut method to calculate these two measures. Hence, the initial step in our solution is to compute the following quantities (notice the use of upper case letters for X and Y):

$$\sum X = 11,536; \qquad\qquad \sum X^2 = 6,501,720;$$

$$\bar{X} = \frac{11,536}{25} = 461.44; \qquad \sum Y^2 = 5,321,776;$$

$$\sum Y = 10,478; \qquad\qquad \sum XY = 5,881,794.$$

$$\bar{Y} = \frac{10,478}{25} = 419.12;$$

Our next step is to compute the terms $\sum x^2$, $\sum y^2$, and $\sum xy$ (lower-case letters) as follows:

$$\sum x^2 = \sum X^2 - n(\bar{X})^2$$
$$= 6,501,720 - 25(461.44)^2$$
$$= 1,178,548;$$
$$\sum y^2 = \sum Y^2 - n(\bar{Y})^2$$
$$= 5,321,776 - 25(419.12)^2$$
$$= 930,237;$$
$$\sum xy = \sum XY - n(\bar{X})(\bar{Y})$$
$$= 5,881,794 - 25(461.44)(419.12)$$
$$= 1,046,826.$$

---

* Changes in government spending and changes in tax rates are the main instruments of governmental fiscal policies.

We are ready now to compute the regression coefficient b:

$$b = \frac{\sum xy}{\sum x^2} = \frac{1,046,826}{1,178,548} = .888.$$

And we also compute the standard error of estimate S_{YX}:

$$S_{YX} = \sqrt{\frac{\sum y^2 - b \sum xy}{n - 2}}$$

$$= \sqrt{\frac{930,237 - (.888)(1,046,826)}{25 - 2}} = 5.339.$$

Hence, so far, we have computed the following measures:

$$b = .888, \qquad S_{YX} = 5.339, \qquad \sum x^2 = 1,178,548.$$

Using these results, we will now establish a .95 confidence interval estimate of B as

$$b \pm t_{.025} \frac{S_{YX}}{\sqrt{\sum x^2}} .$$

With $(n - 2)$ or $(25 - 2) = 23$ degrees of freedom, the value of $t_{.025}$ from Table E is 2.069. Hence, the .95 confidence interval estimate of B becomes

$$b \pm 2.069 \frac{S_{YX}}{\sqrt{\sum x^2}} ;$$

$$.888 \pm 2.069 \frac{5.339}{\sqrt{1,178,548}} , \qquad .888 \pm .010,$$

or

$$.878 < B < .898$$

Test of Hypothesis Concerning the Value of B

In the population regression relationship

$$Y = A + BX,$$

the regression coefficient, B, indicates the average change in the value of Y for a unit change in X. The preceding section of this chapter dealt with the problem of estimating the value of the parameter B from a random sample of n paired observations on X and Y. In this section we will learn how to use this sample information to test a hypothesis concerning the value of B. For example, we may test the hypothesis that $B = 0$, i.e., there is no relationship between X and Y. We will also test the hypothesis that B is equal to some specific value, say B_0.

In general, in order to test a null hypothesis about the value of the population regression coefficient B, one must select a random sample from such a
population. Depending on the finding in the sample, one will either reject
the null hypothesis or not reject it (reserve judgment). The null hypothesis is
rejected if the sample findings are inconsistent with it. On the other hand,
the null hypothesis is not rejected if the sample findings do not contradict it.

The exact procedure to test the validity of the null hypothesis $H_0: B = B_0$
is to compute the measure

$$t = \frac{b - B_0}{S_{YX}/\sqrt{\sum x^2}},$$

where

$t = $ a t-distribution with $n - 2$ degrees of freedom,

$B_0 = $ the value of the regression coefficient B specified by the null
hypothesis,

$b = $ the sample regression coefficient,

$S_{YX} = $ the standard error of the estimate,

$\sum x^2 = $ the term $\sum (X - \overline{X})^2$ which can be computed by the short-cut
method as $\sum X^2 - n(\overline{X})^2$.

The above value of t (observed value) is then compared with a critical t value
obtained from Table E. The critical t value is determined by the level of
significance (e.g., .01 or .05), by the degrees of freedom $(n - 2)$, and by the
nature of the alternative hypothesis (one-sided vs. two-sided test). The null
hypothesis is rejected when the observed absolute value of t exceeds the
critical t value, otherwise we reserve judgment. We will clarify this procedure
with two illustrative problems.

Example 3

The following information related to the height in feet, Y, and the age in
years, X, of a random sample of five papaya trees:

$$Y' = 2.7 + 2.1X \text{ (sample regression equation)},$$

$$S_{YX} = .8 \text{ feet}, \sum x^2 = 10.$$

Use this information to test the null hypothesis $H_0: B = 0$, against the
alternative hypothesis $H_1: B \neq 0$. Use a level of significance of .01.

Solution:

$$H_0: B = 0,$$

$$H_1: B \neq 0,$$

Level of significance: $\alpha = .01$.

To test the null hypothesis that $B = 0$, we compute the observed value of t as

$$t = \frac{b - B}{S_{YX}/\sqrt{\sum x^2}}$$

$$= \frac{2.1 - 0}{.8 \div \sqrt{10}} = 8.30.$$

Since the alternative hypothesis is two-sided, and the level of significance is .01, the critical value of t is found under the column labeled $t_{.005}$ in Table E. With $(n - 2) = (5 - 2) = 3$ degrees of freedom, this critical value of $t_{.005}$ is 5.841. Now, since the observed value of t is 8.30, which is greater than the critical value of 5.841, the null hypothesis is rejected. In other words, we conclude that $B \neq 0$ and, therefore, that the height of a papaya tree is related to its age.

Example 4

Use the information provided in Example 3 above to test the null hypothesis that a papaya tree grows on the average 2 feet per year in height, against the alternative hypothesis that the average growth is greater than 2 feet per year. Use a level of significance of .05.

Solution:

$$H_0: B = 2 \text{ feet per year,}$$

$$H_1: B > 2 \text{ feet per year,}$$

Level of significance: $\alpha = .05$.

To test the null hypothesis that $B = 2$, we compute the observed value of t as

$$t = \frac{b - B_0}{S_{YX}/\sqrt{\sum x^2}}$$

$$= \frac{2.1 - 2.0}{.8/\sqrt{10}} = .40.$$

From Table E, we find that the critical value of t is 2.353 ($t_{.05}$ with 3 degrees of freedom in a one-sided test). Since the observed value of t is .40, which is less than the critical value of 2.353, one must reserve judgment. In other words, based on the information provided by the sample we do not have enough evidence to reject the null hypothesis that the average growth in height is 2 feet per year.

Estimating the Mean Value of Y

The regression equation $Y' = a + bX$ enables us to estimate the *average* value of Y for any given value of X. For example, the regression equation

$$Y' = 2.7 + 2.1X,$$

where

$$X = \text{the age of a papaya tree in years,}$$

$$Y = \text{the height of a tree in feet,}$$

enables us to estimate the average height of a papaya tree for any given age. According to this equation, for instance, the average height of a 7-year old tree is 17.4 feet.

Once more, the regression equation enables us to estimate the *average* value of Y for a given value of X. This estimate, furthermore, is *a point estimate*. A confidence interval estimate of the average value of Y for any given value of X, on the other hand, is established as follows:

$$Y' \pm tS_{YX}\sqrt{\frac{1}{n} + \frac{(X_g - \bar{X})^2}{\sum x^2}}$$

where

$\qquad Y' = $ a point estimate of the average value of Y,

$\qquad t = $ a t-distribution with $n - 2$ degrees of freedom,

$\quad S_{YX} = $ standard error of the estimate,

$\qquad X_g = $ the given value of X for which the average value of Y is to be predicted,

$\sum x^2 = $ the sum of the squared derivations $\sum (X - \bar{X})^2$.

We will illustrate the use of this formula with a familiar example.

Example 5

The following information relates to the height in feet, Y, and the age in years, X, of a random sample of five papaya trees:

$$Y' = 2.7 + 2.1X \text{ (regression equation)};$$

$$S_{YX} = .8 \text{ feet,} \qquad \bar{X} = 3 \text{ years,} \qquad \sum x^2 = 10.$$

Use this information to find

a) *a point estimate* of the average height of a 7-year-old papaya tree;

b) a 95% confidence interval estimate of the average height of a 7-year-old papaya tree;

c) a 95% confidence interval estimate of the average height of a 10-year-old papaya tree.

Solution (a): A point estimate of the average height of a 7-year-old papaya tree is obtained by substituting $X = 7$ in the regression equation $Y' = 2.7 + 2.1X$. Hence, this point estimate is

$$Y' = 2.7 + 2.1(7)$$

$$= 17.4 \text{ feet.}$$

Solution (b): A 95% confidence interval estimate of the average height of a 7-year-old papaya tree is established according to the formula

$$Y' \pm t_{.025}S_{YX}\sqrt{\frac{1}{n} + \frac{(X_g - \bar{X})^2}{\sum x^2}}.$$

Substituting the following values in this formula,

$Y' = 17.4$ (the point estimate of the average height of a 7-year-old tree),

$t_{.025} = 3.182$ (this is the value of $t_{.025}$ with $(5 - 2) = 3$ degrees of freedom),

$S_{YX} = .8$ (standard error of the estimate),

$n = 5$ (sample size),

$X_g = 7$ (the given value of X for which the average value of Y is to be predicted),

$\bar{X} = 3$ (mean of X),

$\sum x^2 = 10$ (sum of squared deviations),

would result in the range

$$Y' \pm t_{.025}S_{YX}\sqrt{\frac{1}{n} + \frac{(X_g - \bar{X})^2}{\sum x^2}};$$

$$17.4 \pm (3.182)(.8)\sqrt{\frac{1}{5} + \frac{(7 - 3)^2}{10}}, \qquad 17.4 \pm 3.4,$$

or between 14.0 feet and 20.8 feet. Hence, a 95% confidence interval of the average height of a 7-year-old tree is somewhere between 14.0 feet and 20.8 feet.

Solution (c): To establish a 95% confidence interval estimate of the average height of a 10-year-old tree, we must first determine the point estimate, Y', for a tree at that age. The point estimate is obtained by substituting $X = 10$ in the regression equation as follows:

$$Y' = 2.7 + 2.1X,$$
$$= 2.7 + 2.1 (10),$$
$$= 23.7 \text{ feet.}$$

Now, a .95 confidence interval estimate of the average height of a 10-year-old papaya tree can be established using the formula

$$Y' \pm t_{.025} S_{YX} \sqrt{\frac{1}{n} + \frac{(X_g - \bar{X})^2}{\sum x^2}}.$$

With $Y' = 23.7$ feet and $X_g = 10$ years, and with all other values the same as in part (b) of this example, our formula would result in the following range:

$$23.7 \pm (3.182)(.8) \sqrt{\frac{1}{5} + \frac{(10 - 3)^2}{10}}, \qquad 23.7 \pm 5.7,$$

or between 18.0 feet and 29.4 feet. Hence, a .95 confidence interval estimate of the average height of a 10-year-old papaya tree is somewhere between 18.0 feet and 29.4 feet.

EXERCISES

1. The following data gives the gross national product and the value of exports for a random sample of nine Western European Countries in a given year:

Country	GNP (in billions of dollars)	Exports (in billions of dollars)
Belgium	26	16
Ireland	5	2
Finland	13	3
Switzerland	30	7
Norway	15	3
United Kingdom	134	24
Germany	260	46
France	238	26
Spain	46	4

a) Determine the linear regression equations with GNP as the independent variable.

b) Establish a .95 confidence interval estimate of the population regression coefficient.

c) Test the null hypothesis that $B = .10$, against the alternative hypothesis that $B > .10$. Use a level of significance of .01.

2. A random sample of ten Western Countries shows the following information concerning their per capita gross national product and per capita consumption in a given year:

Country	GNP per capita (in thousands of dollars)	Consumption per capita (in thousands of dollars)
United States	5.2	3.3
Canada	4.3	2.4
Switzerland	4.2	2.5
Norway	3.5	1.9
West Germany	3.8	2.0
Belgium	3.3	2.0
United Kingdom	2.7	1.7
Italy	2.0	1.3
Greece	1.2	.8
France	3.5	2.0

a) Determine the linear regression equation with GNP as the independent variable.

b) Establish a .99 confidence interval estimate of the population regression coefficient.

c) Test the null hypothesis that $B = .60$, against the alternative hypothesis that $B < .60$. Use a level of significance of .05.

3. A random sample of 12 boutique shops in Atlanta, Georgia, shows the following information in regard to annual sales and annual expenditures on advertising:

Annual sales (thousands of dollars)	Advertising expenditures (thousands of dollars)
170	10
135	7
148	8
205	13
120	6
180	11
155	9
142	8
125	6
110	10
144	8
190	12

a) Use the method of least squares to determine the linear regression equation with advertising expenditures as the independent variable.

b) What is the average increase in annual sales associated with an additional advertising expenditure of $1000?

c) Use a level of significance of .05 to test the hypothesis that $B = 10$, against the alternative hypothesis that $B > 10$.

d) Establish a 90% confidence interval estimate of average annual sales when annual advertising expenditures is $15,000.

4. The general manager of a gambling casino in Las Vegas wishes to find an accurate method of forecasting the number of customers visiting his establishment each week. The manager noted that the number of customers in any given week seems to be linearly related to the number of conventions in Las Vegas at that time. Ten weeks, selected at random, reveal the following information:

Number of customers (in hundreds)	Number of conventions
27	3
46	7
27	4
31	4
26	3
48	8
52	9
42	6
41	6
60	10

a) Use the least-squares method to determine the linear relationship with the number of conventions as the independent variable.

b) Interpret the meaning of the regression coefficient b in terms of this problem.

c) Establish a .90 confidence interval estimate of the population regression coefficient.

d) Establish a 99% confidence interval estimate of the average number of customers visiting the gambling casino during a week for which eight conventions are scheduled.

5. The post office district of Los Angeles, California, maintains a large fleet of small mail delivery trucks. Records are kept for individual trucks to study the relationship between the age of a truck and its annual repair costs. The

following data are obtained from a random sample of 15 mail delivery truck records:

Truck	Age (in years)	Annual repair costs (in dollars)
1	5	525
2	1	80
3	4	421
4	4	356
5	6	684
6	3	222
7	4	490
8	2	144
9	4	473
10	3	384
11	5	507
12	2	203
13	3	332
14	2	206
15	2	219

a) Use the method of least squares to determine the linear regression equation with the age of the truck as the independent variable.

b) How do you justify that the value of a is negative?

c) What is the average increase in repair costs as a truck becomes one year older?

d) Test the null hypothesis that $B = \$100$, against the alternative hypothesis that $B > \$100$. Use a level of significance of .01.

e) What is the average repair cost of a 4-year-old truck? Establish a 95% confidence interval estimate of this cost.

6. Sixteen families are randomly selected in a given community. The sample shows the following information concerning annual family income and monthly housing expenditures:

Family	Annual income (thousands of dollars)	Housing expenditures
A	22	306
B	30	332
C	15	350
D	40	740
E	7	143
F	24	320
G	15	451
H	14	250
I	21	314
J	17	265
K	19	503
L	50	350
M	25	400
N	19	362
O	7	245
P	13	400

What is the average monthly housing expenditures of a family earning $15,000 a year? Establish a .99 confidence interval estimate of this average.

13
Nonparametric
Statistics

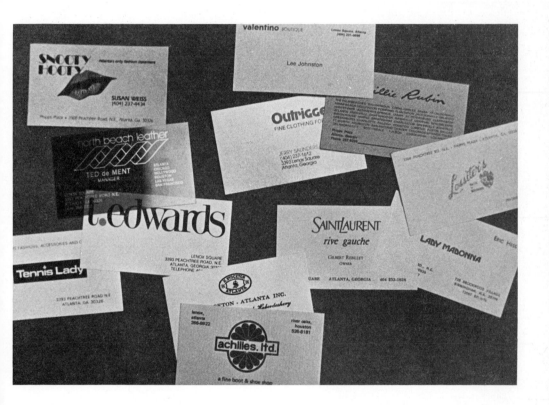

Nonparametric
Statistics

In the preceding chapters, we have developed various methods for testing hypotheses concerning population parameters. For example, we have used the normal distribution to test the hypothesis that a population parameter such as the arithmetic mean, μ, or the proportion, π, is equal to a specified value. We have also used the normal distribution or the t-distribution to test the hypothesis that two populations have equal means, that is, $\mu_1 = \mu_2$. Finally, we used the F-distribution to test the hypothesis that several populations have equal means, that is, $\mu_1 = \mu_2 = \mu_3 = \cdots \mu_k$. Tests in which the hypotheses deal with population parameters are called *parametric tests*. Hence, all the tests we have just described are parametric tests.

In contrast to parametric tests which deal with tests concerning population parameters, nonparametric tests deal with hypotheses concerning population *frequency* distributions. For example, a nonparametric test is used to test the hypothesis that two populations have the same frequency distributions.

Parametric tests often require certain assumptions about the population from which the sample is drawn. For example, the use of the t-distribution to test the significance of the difference between the means of two small samples requires that the two samples are independent samples selected from normally distributed populations with equal variances. Similarly, the use of the F-distribution to compare the means of several populations requires that the samples be drawn from normal populations with equal variances. When these stringent assumptions about the distributions of parent populations cannot be met, parametric tests are no longer applicable; and some alternative tests, called "distribution-free" tests, are employed instead. Although the term "nonparametric" originally referred only to tests dealing with hypotheses concerning population frequency distributions, distribution-free tests are also classified now as nonparametric tests.

Nonparametric tests have gained popularity in recent years for two main reasons. First, nonparametric tests require less restricting assumptions than comparable parametric tests, and often require very few arithmetic computations. Second, but more important, is the fact that nonparametric tests are found to be most suitable for analyzing data which consists of observations that can only be rated or ranked. For example, if a consumer panel ranks two products according to its personal preference, we will discover that a nonparametric test is the most appropriate means of comparing the ranks received by the two products.

Before concluding this discussion, we must clearly point out that nonparametric tests do not utilize all the information provided by the sample. Consequently, in situations where both parametric and nonparametric tests are applicable, nonparametric tests have the definite disadvantage of allowing a greater risk of accepting a false hypothesis (committing a type II error). Hence, nonparametric test should be reserved for situations where either the necessary conditions needed for a parametric test cannot be met, or where the nature of the data is not suitable for a parametric test, as in the case of ranked data.

The Sign Test

We have already mentioned that the use of the *t*-distribution to test the significance of the difference between the means of two small samples requires that the two samples be independent random samples selected from normally distributed populations with equal variances. If any of these assumptions is not met, the use of the *t*-test can no longer be justified; and an alternative nonparametric test called the *sign test* may be used in this case instead. A second use for the sign test is in the case of ranked data. We will illustrate each of these two cases in turn by an example.

Example 1

The pulse rates of 22 male students are measured before and after smoking. The results are shown in Table 13–1*. In addition, the rightmost column of Table 13–1 indicates whether the pulse rate has increased, decreased, or remained the same after smoking. An increase in pulse rate is indicated by a plus sign and a decrease is indicated by a minus sign. If we ignore the two cases where there are no changes in pulse rate, the sample size is reduced to 20 paired observations in which there are 15 plus signs and 5 minus signs.

---

* The *t*-test is not applicable here because the two samples are not independent.

Table 13–1

Effect of smoking on pulse rates

Student	Pulse rate before smoking	Pulse rate after smoking	Sign of change
1	72	74	+
2	70	72	+
3	68	69	+
4	67	68	+
5	73	72	−
6	71	72	+
7	72	72	No change
8	70	71	+
9	69	67	−
10	70	73	+
11	68	69	+
12	72	71	−
13	69	68	−
14	66	69	+
15	73	74	+
16	71	73	+
17	70	70	No change
18	72	74	+
19	70	68	−
20	69	71	+
21	72	74	+
22	73	74	+

Using the information presented in Table 13–1, let us now test the null hypothesis that smoking has no effect on the pulse rate, against the alternative hypothesis that the pulse rate increases after smoking. We will use a level of significance of 0.05.

If the null hypothesis is true (if smoking has no effect on the pulse rate), we would expect an equal number of plus signs and minus signs in the population. Stated in different terms, the null hypothesis states that the probability of getting a plus sign is $\frac{1}{2}$ and the probability of getting a minus sign is $\frac{1}{2}$ also.

Now, if the probability of getting a plus sign is $\frac{1}{2}$, then the number of plus signs in a random sample of size 20 is described by a binomial distribution with $\pi = \frac{1}{2}$ and $n = 20$. The mean and the standard deviation of this binomial distribution are

$$\mu = n\pi = 20(\tfrac{1}{2}) = 10,$$

and

$$\sigma = \sqrt{n\pi(1 - \pi)} = \sqrt{20(\tfrac{1}{2})(\tfrac{1}{2})} = \sqrt{5} = 2.24.$$

Thus, if the null hypothesis is true, the number of plus signs in a random sample of 20 is expected to be 10.   By contrast we have observed 15 plus signs in our sample.   In order to test whether such disparity between the expected number of plus signs and observed number of plus signs could be attributed to chance, we will use the normal-curve approximation to the binomial distribution to determine the probability of getting 15 plus signs or more.

Hence, using a normal distribution with $\mu = 10$ and $\sigma = 2.24$, we find that the probability of getting 15 plus signs or more is approximated by the area lying to the right of $x = 14.5$ (see shaded area in Fig. 13–1).

$$z = \frac{x - \mu}{\sigma} = \frac{14.5 - 10}{2.24} = \frac{4.5}{2.24} = 2.01;$$

$$A = 0.4778$$

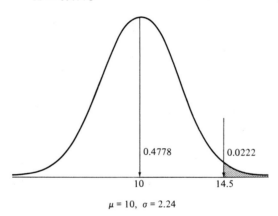

| 0.4778 | 0.0222 |

10   14.5

$\mu = 10, \ \sigma = 2.24$

Figure 13–1

Therefore,

$$P(X \geq 14.5) = 0.50 - 0.4778 = 0.0222.$$

Since the probability of getting 15 plus signs or more is less than 0.05, we reject the null hypothesis using the 0.05 level of significance.   (The reader may already recall that the decision whether to reject the null hypothesis or to reserve judgment becomes obvious once the z-value is calculated.   Using the 0.05 level of significance, when the test is one-sided, the null hypothesis is rejected when the absolute value of z exceeds 1.64.)

As a final remark, we must note that by using the normal curve, we can *approximate* the binomial probability of getting 15 plus signs or more, as 0.0222. Using the cumulative binomial probabilities in Table J instead, we find that the exact probability is 0.0207. Comparing the approximated probability with the exact probability (0.0222 vs. 0.0207), it seems that our approximation is indeed reasonable. In general when the value of $\pi = \frac{1}{2}$, as is always the case in the sign test, the normal-curve approximation to the binomial distribution is quite adequate whenever $n \geq 10$.

Example 2

A panel of 12 wine tasters is asked to test two varieties of wine: Cabernet Sauvignon and Pinot Noir. Each member of the panel is instructed to rank the two wines according to his personal preference. Rank No. 2 is to be assigned to the most preferred wine and rank No. 1 to the least preferred one. Table 13–2 shows the ranks assigned to the two wines by members of the panel. In addition, the rightmost column of the table shows the sign of the difference of the ranks received by the two wines.

Table 13–2

Ranks received by two types of wine

Panel member	Cabernet Sauvignon	Pinot Noir	Sign of difference
A	2	1	+
B	2	1	+
C	1	2	−
D	2	1	+
E	2	1	+
F	2	1	+
G	1	2	−
H	2	1	+
I	2	1	+
J	1	2	−
K	1	2	−
L	2	1	+

Using the information presented in Table 13–2, let us now test the null hypothesis that the two wines are equally preferred by consumers, against the alternative hypothesis that preference for the two wines is not the same. We will use a level of significance of 0.05.

If the null hypothesis is true (if consumers have equal preference for the two types of wines), we would expect an equal number of plus signs and minus signs in the population; we would consider the probability of getting a plus sign $\frac{1}{2}$ and the probability of getting a minus sign $\frac{1}{2}$.

Now, if the probability of getting a plus sign is $\frac{1}{2}$, then the number of plus signs in a random sample of size 12 is described by a binomial distribution with $\pi = \frac{1}{2}$ and $n = 12$. The mean and the standard deviation of this binomial distribution are

$$\mu = n\pi = 12(\tfrac{1}{2}) = 6,$$

and

$$\sigma = \sqrt{n\pi(1 - \pi)} = \sqrt{12(\tfrac{1}{2})(\tfrac{1}{2})} = \sqrt{3} = 1.73.$$

Thus, if the null hypothesis is true, the number of plus signs in a random sample of size 12 is expected to be 6. By contrast, Table 13–2 shows eight plus signs.

In order to test whether the disparity between the observed number of plus signs and the expected number of plus signs is significant at the 0.05 level, we will use the normal-curve approximation to the binomial distribution and determine z as

$$z = \frac{x - \mu}{\sigma} = \frac{7.5 - 6}{1.73} = \frac{1.5}{1.73} = 0.87.$$

Since the actual value of z is 0.87, which is less than 1.96, the null hypothesis cannot be rejected using the 0.05 level of significance (the test is two-sided). Thus we can conclude that the ranks assigned to the two types of wine by members of the board is not enough evidence to reject the hypothesis that consumers have equal preference for the two types of wine.

The Mann–Whitney *U*-Test

We will once again state that the use of the *t*-distribution to test the significance of the difference between the means of two small samples requires the following assumptions:

1. that the two samples selected are independent random samples, and

2. that the parent populations are normally distributed with equal variances.

We have also learned that the use of the sign test requires none of these stringent assumptions.

However, in situations in which the first assumption is satisfied (the two samples are independent), and only the assumptions concerning the parent populations are in doubt, we can use a better nonparametric test called the Mann–Whitney *U*-test.   Compared to the sign test, the Mann–Whitney *U*-test utilizes more of the information contained in the sample, and hence, is more efficient.   We will illustrate the Mann–Whitney *U*-test with the following example.

Example 3

An arithmetic test is administered to a random sample of 10 sixth-graders from the Los Angeles School District.   The same test is also administered to a random sample of 11 sixth-graders from the Pasadena School District. The scores achieved by the two groups are

Los Angeles	Pasadena
70	72
68	67
73	74
81	65
66	63
56	77
62	71
75	60
83	76
48	61
	64

Let us now use the Mann–Whitney *U* test to test the null hypothesis that the performance of all sixth-graders in the Los Angeles School District in arithmetic is the same as those in the Pasadena School District.   This hypothesis is tested against the alternative hypothesis that their performance is not the same.   We will use a level of significance of 0.05.*

Solution: The Mann–Whitney *U*-Test consists of the following steps:

Step 1.   We assign a rank to every score.   Combining the scores of both groups, we assign rank 1 to the lowest score in the combined data, rank 2 to the second lowest, rank 3 to the third lowest, etc.   The ranks assigned to the various scores are shown in Table 13–3.

---

*   The *t*-test is not applicable here because the Los Angeles scores show greater variability than the Pasadena scores.

Table 13–3

Ranking of scores (Mann–Whitney U-Test)

Los Angeles		Pasadena	
Score	Rank	Score	Rank
70	12	72	14
68	11	67	10
73	15	74	16
81	20	65	8
66	9	63	6
56	2	77	19
62	5	71	13
75	17	60	3
83	21	76	18
48	1	61	4
		64	7
Rank sum	$R_1 = 113$		$R_2 = 118$

Step 2. We sum the ranks received by each group. The rank sum is denoted by R. Table 13–3 shows that the sum of the ranks in group 1 (Los Angeles) is 113 and the sum of the ranks in group 2 (Pasadena) is 118. Hence, $R_1 = 113$ and $R_2 = 118$.

Step 3. We determine the value of the Mann–Whitney U-statistic as

$$U = n_1 n_2 + \left(\frac{n_1(n_1 + 1)}{2} \right) - R_1$$

or

$$U = n_1 n_2 + \left(\frac{n_2(n_2 + 1)}{2} \right) - R_2.$$

We will choose the first formula to determine the value of U in our problem. Hence with $n_1 = 10$, $n_2 = 11$, and $R_1 = 113$, the value of U is

$$U = n_1 n_2 + \left(\frac{n_1(n_1 + 1)}{2} \right) - R_1$$

$$= (10)(11) + \left(\frac{10(10 + 1)}{2} \right) - 113 = 52.$$

Step 4. We determine the mean and the standard deviation of the statistic U. The mean of the statistic U, denoted by $E(U)$, is

$$E(U) = \frac{n_1 n_2}{2} = \frac{(10)(11)}{2} = 55.$$

The standard deviation of U, denoted by σ_U, is

$$\sigma_U = \sqrt{\frac{n_1 n_2 (n_1 + n_2 + 1)}{12}} = \sqrt{\frac{(10)(11)(10 + 11 + 1)}{12}} = 14.2.$$

Step 5. If both n_1 and $n_2 \geq 8$ (as is the case in our problem), the statistic U is approximately normally distributed. Hence, we calculate the value of z as

$$z \doteq \frac{U - E(U)}{\sigma_U} = \frac{52 - 55}{14.2} = -0.21.$$

Now since the absolute value of z is 0.21, which is less than 1.96, we cannot reject the null hypothesis using the 0.05 level of significance. In other words, based on the information contained in the two samples, there is not enough evidence to reject the hypothesis that the arithmetic performance of sixth-graders in both school districts is the same.

We will conclude our discussion now with two important remarks.

1. When there are ties between observations, the identical values are assigned the mean of their tied ranks. For example, if the 5th and 6th values are the same, the rank of 5.5 is assigned to each of the two observations, and the next observation in the sequence is assigned the rank of 7. Similarly, if there is a tie between the 9th, 10th, and 11th observations, the rank of 10 is assigned to each of these three observations and the rank 12 is assigned to the next observation in the sequence.

2. The statistic U is approximately normally distributed only when both n_1 and n_2 are 8 or more. If this condition is not met, the use of the normal distribution is no longer appropriate and special tables are needed to test the significance of U.

The Kruskal–Wallis H-Test

The Kruskal–Wallis H-test is an extended version of the Mann–Whitney test. It is used to test the null hypothesis that several independent samples belong to identical populations. As is the case in the Mann–Whitney test, we assign a rank to every observation. Considering the observations of all samples together, we assign the rank of 1 to the lowest value, the rank of 2 to the second lowest value, etc.   After summing the ranks which have been assigned to each sample, we compute the statistic H as

$$H = \frac{12}{n(n+1)} \left(\frac{R_1^2}{n_1} + \frac{R_2^2}{n_2} + \cdots + \frac{R_k^2}{n_k} \right) - 3(n+1),$$

where

k = number of samples,
R_1 = sum of ranks assigned to the n_1 observations of the first sample,
R_2 = sum of ranks assigned to the n_2 observations of the second sample,
R_k = sum of ranks assigned to the n_k observations of the kth sample,
n = $n_1 + n_2 + \cdots + n_k$ (total number of observations in all samples).

Now, assuming that the null hypothesis is true, and assuming that each sample consists of at least five observations, then the statistic H has a probability distribution that can be approximated by a chi-square distribution with $(k-1)$ degrees of freedom.   Hence, the null hypothesis is rejected at the 0.05 level of significance if the value of $H \geq \chi_{0.05}^2$; and it is rejected at the 0.01 level of significance if the value of $H \geq \chi_{0.01}^2$.   We will illustrate the Kruskal–Wallis H-test with the following example.

Example 4

To compare the effectiveness of three types of weight-reducing diets, a homogeneous group of 22 coeds was divided into three subgroups, and each subgroup followed one of these diet plans for a period of two weeks.   The weight losses, in pounds, during these two weeks were as follows:

First Plan	Second Plan	Third Plan
5.3	6.3	2.4
4.2	8.4	3.1
3.7	9.3	3.7
7.2	6.5	4.1
6.0	7.7	2.5
4.8	8.2	1.7
	9.5	5.3
		4.5
		1.3

Use the Kruskal–Wallis H-test to test the null hypothesis that the effectiveness of the three reducing plans is the same, against the alternative hypothesis that their effectiveness is not the same.   Use a level of significance of 0.01.

Table 13–4

Assignment of ranks (Kruskal–Wallis Test)

First Plan		Second Plan		Third Plan	
Loss	Rank	Loss	Rank	Loss	Rank
5.3	12.5	6.3	15	2.4	3
4.2	9	8.4	20	3.1	5
3.7	6.5	9.3	21	3.7	6.5
7.2	17	6.5	16	4.1	8
6.0	14	7.7	18	2.5	4
4.9	11	8.2	19	1.7	2
	$R_1 = \overline{70}$	9.5	22	5.3	12.5
			$R_2 = \overline{131}$	4.5	10
				1.3	1
					$R_3 = \overline{52}$

Solution: The first step in our solution is to assign a rank to every observation in the combined data.   The ranks assigned to these observations and the sum of the ranks received by each sample are shown in Table 13–4.   With $n_1 = 6$, $n_2 = 7$, $n_3 = 9$, and $R_1 = 70$, $R_2 = 131$, $R_3 = 52$, we can now determine the value of the statistic H as

$$H = \frac{12}{n(n+1)} \left(\frac{R_1^2}{n_1} + \frac{R_2^2}{n_2} + \frac{R_3^2}{n_3} \right) - 3(n+1)$$

$$= \frac{12}{22(22+1)} \left(\frac{70^2}{6} + \frac{131^2}{7} + \frac{52^2}{9} \right) - 3(22+1)$$

$$= 15.633.$$

To test the null hypothesis at the 0.01 level of significance, we now compare this value of H against $\chi^2_{0.01}$ with $(k-1)$ degrees of freedom.   According to Table D, the value of $\chi^2_{0.01}$ (d.f. $= 2$) is 9.210.   Hence, since the value of H is greater than $\chi^2_{0.01}$, the null hypothesis is rejected at the 0.01 level of significance.   In other words, based on the information provided by the three samples, we conclude that the effectiveness of the three diet plans is not the same.

Rank Correlation

The coefficient of rank correlation (also called Spearman's rank-correlation coefficient) is used to measure the degree of association between two sets of ranks observations.   For example, if the teaching performances of a group of college professors are ranked independently by their dean and by their department chairman, the coefficient of rank correlation can be used to measure the consistency of the two sets of ranks assigned by the two administrators.

Some observations, such as the teaching abilities of the group of college professors, cannot be measured but can only be ranked; and the coefficient of rank correlations is the best means for measuring the degree of relationship between two sets of ranks.   However, the coefficient of rank correlation can also be used to determine the degree of relationship between two sets of *measurements* once the measurements are converted into ranks.   We will illustrate the use of the rank-correlation coefficient with two examples.   In the first example, the coefficient of rank correlation is used to determine the consistency of two sets of ratings.   In the second example, in addition to computing the coefficient of correlation, we will describe the method used to convert two sets of measurements into two sets of ranks.

Example 5

The teaching abilities of a group of 10 college professors are ranked independently by their dean and by their department chairman.   Rank 1 is given to the best professor, Rank 2 to the second best, ..., and Rank 10 to the professor with the worst teaching performance.   The two sets of ranks are as follows:

Professor	Dean's ratings	Chairman's ratings
A	5	6
B	4	4
C	3	5
D	1	2
E	2	1
F	6	3
G	7	10
H	10	9
I	9	7
J	8	8

Calculate the coefficient of rank correlation as a measure of the consistency of the two ratings.

Solution: The Spearman's rank-correlation coefficient, denoted by r_s, is determined as

$$r_s = 1 - \frac{6 \sum d^2}{n(n^2 - 1)},$$

where n is the number of paired observations and d is the difference between each pair of ranks. The preliminary computations for finding r_s are shown in Table 13–5.

Table 13–5

Computations for coefficient of rank correlation

Professor	Dean's ratings	Chairman's ratings	d	d^2
A	5	6	−1	1
B	4	4	0	0
C	3	5	−2	4
D	1	2	−1	1
E	2	1	+1	1
F	6	3	+3	9
G	7	10	−3	9
H	10	9	+1	1
I	9	7	+2	4
J	8	8	0	0
				$\sum d^2 = 30$

Substituting $\sum d^2 = 30$ (Table 13–5), and $n = 10$ into the formula for r_s, we get

$$r_s = 1 - \frac{6 \sum d^2}{n(n^2 - 1)}$$

$$= 1 - \frac{6(30)}{10(10^2 - 1)} = 0.82.$$

Example 6

A large automobile agency wishes to determine the relationship between a salesman's aptitude test score and the number of cars sold by the salesman during his first year of employment. A random sample of 15 salesmen's files reveals the following information:

Salesman	Test score x	Number of cars y
A	72	341
B	88	422
C	70	322
D	87	440
E	71	287
F	85	415
G	89	463
H	93	497
I	98	510
J	96	512
K	86	432
L	82	390
M	88	453
N	83	374
O	80	385

Calculate the coefficient of rank correlation to measure the degree of relationship between test scores and the number of cars sold.

Table 13–6

Computations for coefficient of rank correlation

Score x	No. of Cars y	Rank of x	Rank of y	d	d^2
72	341	13	13	0	0
88	422	5.5	8	−2.5	6.25
70	322	15	14	+1	1
87	440	7	6	+1	1
71	287	14	15	−1	1
85	415	9	9	0	0
89	463	4	4	0	0
93	497	3	3	0	0
98	510	1	2	−1	1
96	512	2	1	+1	1
86	432	8	7	+1	1
82	390	11	10	+1	1
88	453	5.5	5	+0.5	0.25
83	374	10	12	−2	4
80	385	12	11	+1	1
					$\sum d^2 = 18.5$

Solution: The first step in our solution is to replace the two sets of measurements by two sets of ranks. Beginning with the test scores, we assign Rank 1 to the highest score, Rank 2 to the second highest score, ..., and Rank 15 to the lowest score. The numbers of cars sold are ranked in a similar fashion, giving Rank 1 to the greatest number of cars sold and Rank 15 to the smallest number.

Table 13–6 shows the two sets of measurements and their corresponding sets of ranks, together with the preliminary computations needed to compute r_s.

Substituting $\sum d^2 = 18.5$ (Table 13–6), and $n = 15$ into the formula for r_s, we get

$$r_s = 1 - \frac{6 \sum d^2}{n(n^2 - 1)} = 1 - \frac{(6)(18.5)}{15(15^2 - 1)} = 0.97.$$

The Significance of Rank Correlation

To test the significance of the rank-correlation coefficient, r_s, we begin with the null hypothesis that there *is* no correlation between the two sets of ranks in the population. Now, assuming that the null hypothesis is true, and assuming that $n \geq 25$, then the statistic r_s is approximately normally distributed, with a mean of 0 and a standard deviation of $1/\sqrt{n-1}$. Hence, to test the significance of r_s, we compute the value of z as

$$z = \frac{r_s - 0}{1/\sqrt{n-1}} = r_s \sqrt{n-1}.$$

The coefficient of rank correlation is significant at the 0.05 level when the absolute value of z is greater than 1.96. It is significant at the 0.01 level when the value of z is greater than 2.58. Special tables are used to test the significance of r_s when the sample size is small.

Example 7

Given that $n = 26$ and $r_s = 0.4$, test the significance of r_s using the 0.05 level of significance.

Solution: Substituting $r_s = 0.4$ and $n = 26$ into the formula

$$z = r_s \sqrt{n-1},$$

we get

$$z = 0.4 \sqrt{26 - 1} = 2.$$

Since the value of z is greater than 1.96, the coefficient of rank correlation is significant at the 0.05 level of significance.

378   *Nonparametric Statistics*

EXERCISES

1. Ten viewers selected at random are asked to rate two comedy television programs, X and Y. A rank of 2 is to be assigned to the most preferred program and
a rank of 1 to the least preferred one. The ratings received by the two programs
are shown below.

Viewer	Program X	Program Y
A	2	1
B	2	1
C	1	2
D	2	1
E	1	2
F	2	1
G	2	1
H	1	2
I	1	2
J	2	1

Perform a sign test to test the hypothesis that programs X and Y are rated the
same by television viewers. Use a level of significance of 0.05.

2. An engineering consulting firm is considering moving its main offices to a new
location. To assess employees' preferences for the proposed location, a random
sample of 16 employees are asked whether they prefer the present or the new
location. Employees' responses are shown below.

Employee	Location preferred	Employee	Location preferred
A	New	I	New
B	Present	J	New
C	New	K	New
D	New	L	Present
E	New	M	New
F	Present	N	Present
G	Present	O	New
H	New	P	New

Use the sign test to test the hypothesis that the two locations are equally
preferred by the company employees. Use a level of significance of 0.05.

3. The play constructiveness of 12 school children was measured during a free-play
period and while they played in a frustrating situation. The two ratings are

shown below.

Child	Constructiveness index during free play	Constructiveness index during frustration
A	3.7	3.1
B	4.2	3.8
C	5.1	4.7
D	6.3	6.1
E	5.3	5.4
F	4.7	3.2
G	6.2	5.3
H	6.3	5.7
I	6.7	4.6
J	7.1	7.6
K	4.2	3.2
L	4.8	3.5

Perform a sign test to test the hypothesis that frustration has no effect on play constructiveness. Use a level of significance of 0.01.

4. An anxiety test was given to a group of 14 male college students while sober. The same test was administered to the same group two hours after each student drank 3 ounces of vodka. The results of the two tests are shown below.

Subject	Anxiety ratings before drinking	Anxiety ratings after drinking
1	97	82
2	81	71
3	72	63
4	63	54
5	92	77
6	63	72
7	81	66
8	73	54
9	63	51
10	82	88
11	54	50
12	88	73
13	93	82
14	68	52

Perform a sign test to test the null hypothesis that alcohol has no effect on anxiety against the alternative hypothesis that alcohol reduces anxiety. Use a level of significance of 0.010.

5. The Army Alpha Test, an intelligence test, was administered to two age groups, a group of 20-year-olds and a group of 40-year-olds. The test scores of the two groups are shown below.

20-year-old group	40-year-old group
142	122
130	125
132	132
137	127
145	130
143	135
147	124
138	140
141	126
148	133
152	
140	

Perform the Mann–Whitney test to test the hypothesis that the average intelligence of both age groups are the same.   Use a level of significance of 0.05.

6. In an experiment concerning dream reporting, 26 healthy female college students were observed for a period of two weeks.   According to the number and the detail of the dreams reported in these two weeks, the students are categorized into two groups: Recallers and nonrecallers.   Among the various personality characteristics of the students observed during the two weeks was time asleep each day.   The data below represents the average number of minutes of sleep per day for each student in the two categories.

Dream recallers	Non-recallers
477	430
432	453
445	418
460	392
480	427
457	441
422	411
430	397
475	424
463	360
452	372
447	438
431	
472	

Perform the Mann–Whitney test on the hypothesis that the length of sleep for dream recallers and nonrecallers is the same.   Use a level of significance of 0.05.

7. In an investigation concerning the physiological status of schizophrenic patients, 19 patients were classified into three groups according to the severity of their psychosis: 1) remission; 2) moderate; 3) severe.   The respiratory rates for individual patients in the three subgroups are shown below.

Remission	Moderate	Severe
16.6	17.8	18.6
16.3	17.2	18.4
17.3	17.7	18.3
16.5	17.5	18.8
17.1	18.2	17.6
16.9	17.4	18.5
	16.8	

Perform the Kruskal–Wallis test of the hypothesis that the respiratory rates of patients in the three stages of schizophrenia are the same.   Use a level of significance of 0.01.

8. The social maturity scores received by 40 persons belonging to five occupational groups are shown below.

Dentists	Bankers	Machinists	Military officers	Research scientists
56.3	54.3	43.4	50.3	55.7
54.7	56.7	44.3	51.7	55.7
58.2	55.4	47.3	52.6	57.5
53.3	57.4	41.2	49.8	57.3
57.6	56.4	42.7	52.3	58.9
55.1	58.5	45.2	53.7	54.5
56.2		45.7	46.3	
58.4		42.5	54.6	
59.2		46.2		
		40.6		

Use the Kruskal–Wallis test to test the hypothesis that social maturity of the five occupational groups is the same, against the alternative hypothesis that their social maturity is *not* the same.   Use a level of significance of 0.01.

9. Two judges rated the participants of a beauty contest as follows:

Contestant	Rating by first judge	Rating by second judge
A	5	3
B	1	2
C	6	4
D	2	1
E	7	7
F	4	5
G	3	8
H	8	6

Compute the coefficient of rank correlation as a measure of the consistency of ratings assigned by the two judges.

10. The following are the scores achieved by a group of sixth-graders on tests in arithmetic and reading:

Student	Arithmetic test score	Reading test score
1	72	75
2	63	68
3	82	85
4	75	82
5	93	95
6	77	79
7	66	63
8	57	60
9	83	88
10	74	86

Compute the coefficient of rank correlation to measure the degree of relationship between the reading ability and the arithmetic performance of a sixth-grader.

11. A consumer panel rated 10 different makes of color television sets according to their overall quality. The quality rank and the average retail price of each set is shown below.

Set	Quality rank	Average price
A	10	$450
B	1	600
C	2	525
D	5	515
E	4	495
F	3	475
G	6	550
H	7	480
I	9	560
J	8	530

Compute the coefficient of rank correlation to measure the degree of relationship between the quality and the price of a color television set.

12. Test the significance of the rank-correlation coefficient in each of the following samples:

a) $n = 37$, $r_s = 0.4$ ($\alpha = 0.01$)
b) $n = 50$, $r_s = 0.3$ ($\alpha = 0.05$)
c) $n = 65$, $r_s = 0.2$ ($\alpha = 0.05$)

13. Sixteen families are randomly selected in a given community. The sample shows the following information concerning annual family income and monthly

housing expenditures:

Family	Annual income (thousands of dollars)	Housing expenditures
A	22	306
B	30	332
C	15	350
D	40	740
E	7	143
F	24	320
G	15	451
H	14	250
I	21	314
J	17	265
K	19	503
L	50	350
M	25	400
N	19	362
O	7	245
P	13	400

Compute the coefficient of rank correlation to measure the degree of relationship between family annual income and housing expenditures.

Tables

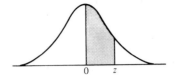

Table A
Areas under the normal curve

z	.00	.01	.02	.03	.04	.05	.06	.07	.08	.09
0.0	.0000	.0040	.0080	.0120	.0160	.0199	.0239	.0279	.0319	.0359
0.1	.0398	.0438	.0478	.0517	.0557	.0596	.0636	.0675	.0714	.0753
0.2	.0793	.0832	.0871	.0910	.0948	.0987	.1026	.1064	.1103	.1141
0.3	.1179	.1217	.1255	.1293	.1331	.1368	.1406	.1443	.1480	.1517
0.4	.1554	.1591	.1628	.1664	.1700	.1736	.1772	.1808	.1844	.1879
0.5	.1915	.1950	.1985	.2019	.2054	.2088	.2123	.2157	.2190	.2224
0.6	.2257	.2291	.2324	.2357	.2389	.2422	.2454	.2486	.2517	.2549
0.7	.2580	.2611	.2642	.2673	.2704	.2734	.2764	.2794	.2823	.2852
0.8	.2881	.2910	.2939	.2967	.2995	.3023	.3051	.3078	.3106	.3133
0.9	.3159	.3186	.3212	.3238	.3264	.3289	.3315	.3340	.3365	.3389
1.0	.3413	.3438	.3461	.3485	.3508	.3531	.3554	.3577	.3599	.3621
1.1	.3643	.3665	.3686	.3708	.3729	.3749	.3770	.3790	.3810	.3830
1.2	.3849	.3869	.3888	.3907	.3925	.3944	.3962	.3980	.3997	.4015
1.3	.4032	.4049	.4066	.4082	.4099	.4115	.4131	.4147	.4162	.4177
1.4	.4192	.4207	.4222	.4236	.4251	.4265	.4279	.4292	.4306	.4319
1.5	.4332	.4345	.4357	.4370	.4382	.4394	.4406	.4418	.4429	.4441
1.6	.4452	.4463	.4474	.4484	.4495	.4505	.4515	.4525	.4535	.4545
1.7	.4554	.4564	.4573	.4582	.4591	.4599	.4608	.4616	.4625	.4633
1.8	.4641	.4649	.4656	.4664	.4671	.4678	.4686	.4693	.4699	.4706
1.9	.4713	.4719	.4726	.4732	.4738	.4744	.4750	.4756	.4761	.4767
2.0	.4772	.4778	.4783	.4788	.4793	.4798	.4803	.4808	.4812	.4817
2.1	.4821	.4826	.4830	.4834	.4838	.4842	.4846	.4850	.4854	.4857
2.2	.4861	.4864	.4868	.4871	.4875	.4878	.4881	.4884	.4887	.4890
2.3	.4893	.4896	.4898	.4901	.4904	.4906	.4909	.4911	.4913	.4916
2.4	.4918	.4920	.4922	.4925	.4927	.4929	.4931	.4932	.4934	.4936
2.5	.4938	.4940	.4941	.4943	.4945	.4946	.4948	.4949	.4951	.4952
2.6	.4953	.4955	.4956	.4957	.4959	.4960	.4961	.4962	.4963	.4964
2.7	.4965	.4966	.4967	.4968	.4969	.4970	.4971	.4972	.4973	.4974
2.8	.4974	.4975	.4976	.4977	.4977	.4978	.4979	.4979	.4980	.4981
2.9	.4981	.4982	.4982	.4983	.4984	.4984	.4985	.4985	.4986	.4986
3.0	.4987	.4987	.4987	.4988	.4988	.4989	.4989	.4989	.4990	.4990

Table B
Values of $F_{.05}$

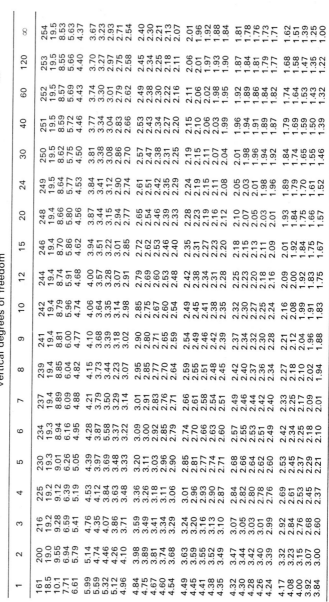

$F_{.05}$

Vertical degrees of freedom

	1	2	3	4	5	6	7	8	9	10	12	15	20	24	30	40	60	120	∞
1	161	200	216	225	230	234	237	239	241	242	244	246	248	249	250	251	252	253	254
2	18.5	19.0	19.2	19.2	19.3	19.3	19.4	19.4	19.4	19.4	19.4	19.4	19.4	19.5	19.5	19.5	19.5	19.5	19.5
3	10.1	9.55	9.28	9.12	9.01	8.94	8.89	8.85	8.81	8.79	8.74	8.70	8.66	8.64	8.62	8.59	8.57	8.55	8.53
4	7.71	6.94	6.59	6.39	6.26	6.16	6.09	6.04	6.00	5.96	5.91	5.86	5.80	5.77	5.75	5.72	5.69	5.66	5.63
5	6.61	5.79	5.41	5.19	5.05	4.95	4.88	4.82	4.77	4.74	4.68	4.62	4.56	4.53	4.50	4.46	4.43	4.40	4.37
6	5.99	5.14	4.76	4.53	4.39	4.28	4.21	4.15	4.10	4.06	4.00	3.94	3.87	3.84	3.81	3.77	3.74	3.70	3.67
7	5.59	4.74	4.35	4.12	3.97	3.87	3.79	3.73	3.68	3.64	3.57	3.51	3.44	3.41	3.38	3.34	3.30	3.27	3.23
8	5.32	4.46	4.07	3.84	3.69	3.58	3.50	3.44	3.39	3.35	3.28	3.22	3.15	3.12	3.08	3.04	3.01	2.97	2.93
9	5.12	4.26	3.86	3.63	3.48	3.37	3.29	3.23	3.18	3.14	3.07	3.01	2.94	2.90	2.86	2.83	2.79	2.75	2.71
10	4.96	4.10	3.71	3.48	3.33	3.22	3.14	3.07	3.02	2.98	2.91	2.85	2.77	2.74	2.70	2.66	2.62	2.58	2.54
11	4.84	3.98	3.59	3.36	3.20	3.09	3.01	2.95	2.90	2.85	2.79	2.72	2.65	2.61	2.57	2.53	2.49	2.45	2.40
12	4.75	3.89	3.49	3.26	3.11	3.00	2.91	2.85	2.80	2.75	2.69	2.62	2.54	2.51	2.47	2.43	2.38	2.34	2.30
13	4.67	3.81	3.41	3.18	3.03	2.92	2.83	2.77	2.71	2.67	2.60	2.53	2.46	2.42	2.38	2.34	2.30	2.25	2.21
14	4.60	3.74	3.34	3.11	2.96	2.85	2.76	2.70	2.65	2.60	2.53	2.46	2.39	2.35	2.31	2.27	2.22	2.18	2.13
15	4.54	3.68	3.29	3.06	2.90	2.79	2.71	2.64	2.59	2.54	2.48	2.40	2.33	2.29	2.25	2.20	2.16	2.11	2.07
16	4.49	3.63	3.24	3.01	2.85	2.74	2.66	2.59	2.54	2.49	2.42	2.35	2.28	2.24	2.19	2.15	2.11	2.06	2.01
17	4.45	3.59	3.20	2.96	2.81	2.70	2.61	2.55	2.49	2.45	2.38	2.31	2.23	2.19	2.15	2.10	2.06	2.01	1.96
18	4.41	3.55	3.16	2.93	2.77	2.66	2.58	2.51	2.46	2.41	2.34	2.27	2.19	2.15	2.11	2.06	2.02	1.97	1.92
19	4.38	3.52	3.13	2.90	2.74	2.63	2.54	2.48	2.42	2.38	2.31	2.23	2.16	2.11	2.07	2.03	1.98	1.93	1.88
20	4.35	3.49	3.10	2.87	2.71	2.60	2.51	2.45	2.39	2.35	2.28	2.20	2.12	2.08	2.04	1.99	1.95	1.90	1.84
21	4.32	3.47	3.07	2.84	2.68	2.57	2.49	2.42	2.37	2.32	2.25	2.18	2.10	2.05	2.01	1.96	1.92	1.87	1.81
22	4.30	3.44	3.05	2.82	2.66	2.55	2.46	2.40	2.34	2.30	2.23	2.15	2.07	2.03	1.98	1.94	1.89	1.84	1.78
23	4.28	3.42	3.03	2.80	2.64	2.53	2.44	2.37	2.32	2.27	2.20	2.13	2.05	2.01	1.96	1.91	1.86	1.81	1.76
24	4.26	3.40	3.01	2.78	2.62	2.51	2.42	2.36	2.30	2.25	2.18	2.11	2.03	1.98	1.94	1.89	1.84	1.79	1.73
25	4.24	3.39	2.99	2.76	2.60	2.49	2.40	2.34	2.28	2.24	2.16	2.09	2.01	1.96	1.92	1.87	1.82	1.77	1.71
30	4.17	3.32	2.92	2.69	2.53	2.42	2.33	2.27	2.21	2.16	2.09	2.01	1.93	1.89	1.84	1.79	1.74	1.68	1.62
40	4.08	3.23	2.84	2.61	2.45	2.34	2.25	2.18	2.12	2.08	2.00	1.92	1.84	1.79	1.74	1.69	1.64	1.58	1.51
60	4.00	3.15	2.76	2.53	2.37	2.25	2.17	2.10	2.04	1.99	1.92	1.84	1.75	1.70	1.65	1.59	1.53	1.47	1.39
120	3.92	3.07	2.68	2.45	2.29	2.18	2.09	2.02	1.96	1.91	1.83	1.75	1.66	1.61	1.55	1.50	1.43	1.35	1.25
∞	3.84	3.00	2.60	2.37	2.21	2.10	2.01	1.94	1.88	1.83	1.75	1.67	1.57	1.52	1.46	1.39	1.32	1.22	1.00

Horizontal degrees of freedom

Table C
Values of $F_{.01}$

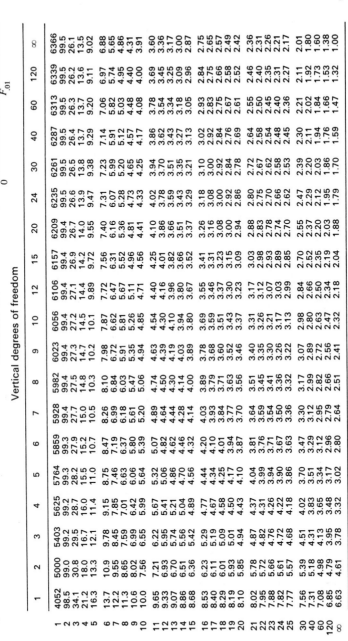

0.01

$F_{.01}$

0

Vertical degrees of freedom

Horizontal degrees of freedom

	1	2	3	4	5	6	7	8	9	10	12	15	20	24	30	40	60	120	∞
1	4052	5000	5403	5625	5764	5859	5928	5982	6023	6056	6106	6157	6209	6235	6261	6287	6313	6339	6366
2	98.5	99.0	99.2	99.2	99.3	99.3	99.4	99.4	99.4	99.4	99.4	99.4	99.4	99.5	99.5	99.5	99.5	99.5	99.5
3	34.1	30.8	29.5	28.7	28.2	27.9	27.7	27.5	27.3	27.2	27.1	26.9	26.7	26.6	26.5	26.4	26.3	26.2	26.1
4	21.2	18.0	16.7	16.0	15.5	15.2	15.0	14.8	14.7	14.5	14.4	14.2	14.0	13.9	13.8	13.7	13.7	13.6	13.5
5	16.3	13.3	12.1	11.4	11.0	10.7	10.5	10.3	10.2	10.1	9.89	9.72	9.55	9.47	9.38	9.29	9.20	9.11	9.02
6	13.7	10.9	9.78	9.15	8.75	8.47	8.26	8.10	7.98	7.87	7.72	7.56	7.40	7.31	7.23	7.14	7.06	6.97	6.88
7	12.2	9.55	8.45	7.85	7.46	7.19	6.99	6.84	6.72	6.62	6.47	6.31	6.16	6.07	5.99	5.91	5.82	5.74	5.65
8	11.3	8.65	7.59	7.01	6.63	6.37	6.18	6.03	5.91	5.81	5.67	5.52	5.36	5.28	5.20	5.12	5.03	4.95	4.86
9	10.6	8.02	6.99	6.42	6.06	5.80	5.61	5.47	5.35	5.26	5.11	4.96	4.81	4.73	4.65	4.57	4.48	4.40	4.31
10	10.0	7.56	6.55	5.99	5.64	5.39	5.20	5.06	4.94	4.85	4.71	4.56	4.41	4.33	4.25	4.17	4.08	4.00	3.91
11	9.65	7.21	6.22	5.67	5.32	5.07	4.89	4.74	4.63	4.54	4.40	4.25	4.10	4.02	3.94	3.86	3.78	3.69	3.60
12	9.33	6.93	5.95	5.41	5.06	4.82	4.64	4.50	4.39	4.30	4.16	4.01	3.86	3.78	3.70	3.62	3.54	3.45	3.36
13	9.07	6.70	5.74	5.21	4.86	4.62	4.44	4.30	4.19	4.10	3.96	3.82	3.66	3.59	3.51	3.43	3.34	3.25	3.17
14	8.86	6.51	5.56	5.04	4.70	4.46	4.28	4.14	4.03	3.94	3.80	3.66	3.51	3.43	3.35	3.27	3.18	3.09	3.00
15	8.68	6.36	5.42	4.89	4.56	4.32	4.14	4.00	3.89	3.80	3.67	3.52	3.37	3.29	3.21	3.13	3.05	2.96	2.87
16	8.53	6.23	5.29	4.77	4.44	4.20	4.03	3.89	3.78	3.69	3.55	3.41	3.26	3.18	3.10	3.02	2.93	2.84	2.75
17	8.40	6.11	5.19	4.67	4.34	4.10	3.93	3.79	3.68	3.59	3.46	3.31	3.16	3.08	3.00	2.92	2.83	2.75	2.65
18	8.29	6.01	5.09	4.58	4.25	4.01	3.84	3.71	3.60	3.51	3.37	3.23	3.08	3.00	2.92	2.84	2.75	2.66	2.57
19	8.19	5.93	5.01	4.50	4.17	3.94	3.77	3.63	3.52	3.43	3.30	3.15	3.00	2.92	2.84	2.76	2.67	2.58	2.49
20	8.10	5.85	4.94	4.43	4.10	3.87	3.70	3.56	3.46	3.37	3.23	3.09	2.94	2.86	2.78	2.69	2.61	2.52	2.42
21	8.02	5.78	4.87	4.37	4.04	3.81	3.64	3.51	3.40	3.31	3.17	3.03	2.88	2.80	2.72	2.64	2.55	2.46	2.36
22	7.95	5.72	4.82	4.31	3.99	3.76	3.59	3.45	3.35	3.26	3.12	2.98	2.83	2.75	2.67	2.58	2.50	2.40	2.31
23	7.88	5.66	4.76	4.26	3.94	3.71	3.54	3.41	3.30	3.21	3.07	2.93	2.78	2.70	2.62	2.54	2.45	2.35	2.26
24	7.82	5.61	4.72	4.22	3.90	3.67	3.50	3.36	3.26	3.17	3.03	2.89	2.74	2.66	2.58	2.49	2.40	2.31	2.21
25	7.77	5.57	4.68	4.18	3.86	3.63	3.36	3.32	3.22	3.13	2.99	2.85	2.70	2.62	2.53	2.45	2.36	2.27	2.17
30	7.56	5.39	4.51	4.02	3.70	3.47	3.30	3.17	3.07	2.98	2.84	2.70	2.55	2.47	2.39	2.30	2.21	2.11	2.01
40	7.31	5.18	4.31	3.83	3.51	3.29	3.12	2.99	2.89	2.80	2.66	2.52	2.37	2.29	2.20	2.11	2.02	1.92	1.80
60	7.08	4.98	4.13	3.65	3.34	3.12	2.95	2.82	2.72	2.63	2.50	2.35	2.20	2.12	2.03	1.94	1.84	1.73	1.60
120	6.85	4.79	3.95	3.48	3.17	2.96	2.79	2.66	2.56	2.47	2.34	2.19	2.03	1.95	1.86	1.76	1.66	1.53	1.38
∞	6.63	4.61	3.78	3.32	3.02	2.80	2.64	2.51	2.41	2.32	2.18	2.04	1.88	1.79	1.70	1.59	1.47	1.32	1.00

Tables B and C are reproduced from Table 18 of *Biometrika Tables for Statistics*, Vol. 1, 1966, by permission of the Biometrica Trustees.

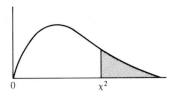

Table D
X^2-Distribution

df	.10	.05	.025	.01	.005
1	2.706	3.841	5.024	6.635	7.879
2	4.605	5.991	7.378	9.210	10.597
3	6.251	7.815	9.348	11.345	12.838
4	7.779	9.488	11.143	13.277	14.860
5	9.236	11.070	12.832	15.086	16.750
6	10.645	12.592	14.449	16.812	18.548
7	12.017	14.067	16.013	18.475	20.278
8	13.362	15.507	17.535	20.090	21.955
9	14.684	16.919	19.023	21.666	23.589
10	15.987	18.307	20.483	23.209	25.188
11	17.275	19.675	21.920	24.725	26.757
12	18.549	21.026	23.337	26.217	28.300
13	19.812	22.362	24.736	27.688	29.819
14	21.064	23.685	26.119	29.141	31.319
15	22.307	24.996	27.488	30.578	32.801
16	23.542	26.296	28.845	32.000	34.267
17	24.769	27.587	30.191	33.409	35.718
18	25.989	28.869	31.526	34.805	37.156
19	27.204	30.144	32.852	36.191	38.582
20	28.412	31.410	34.170	37.566	39.997
21	29.615	32.671	35.479	38.932	41.401
22	30.813	33.924	36.781	40.289	42.796
23	32.007	35.172	38.076	41.638	44.181
24	33.196	36.415	39.364	42.980	45.558
25	34.382	37.652	40.646	44.314	46.928
26	35.563	38.885	41.923	45.642	48.290
27	36.741	40.113	43.194	46.963	49.645
28	37.916	41.337	44.461	48.278	50.993
29	39.087	42.557	45.722	49.588	52.336
30	40.256	43.773	46.979	50.892	53.672

This table is an abridged version of Table III from R. A. Fisher, *Statistical Methods for Research Workers*, published by Oliver and Boyd, Ltd., Edinburgh. Reproduced by permission of the author and publishers.

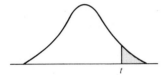

Table E

t-Distribution

df	$t_{.100}$	$t_{.050}$	$t_{.025}$	$t_{.010}$	$t_{.005}$
1	3.078	6.314	12.706	31.821	63.657
2	1.886	2.920	4.303	6.965	9.925
3	1.638	2.353	3.182	4.541	5.841
4	1.533	2.132	2.776	3.747	4.604
5	1.476	2.015	2.571	3.365	4.032
6	1.440	1.943	2.447	3.143	3.707
7	1.415	1.895	2.365	2.998	3.499
8	1.397	1.860	2.306	2.896	3.355
9	1.383	1.833	2.262	2.821	3.250
10	1.372	1.812	2.228	2.764	3.169
11	1.363	1.796	2.201	2.718	3.106
12	1.356	1.782	2.179	2.681	3.055
13	1.350	1.771	2.160	2.650	3.012
14	1.345	1.761	2.145	2.624	2.977
15	1.341	1.753	2.131	2.602	2.947
16	1.337	1.746	2.120	2.583	2.921
17	1.333	1.740	2.110	2.567	2.898
18	1.330	1.734	2.101	2.552	2.878
19	1.328	1.729	2.093	2.539	2.861
20	1.325	1.725	2.086	2.528	2.845
21	1.323	1.721	2.080	2.518	2.831
22	1.321	1.717	2.074	2.508	2.819
23	1.319	1.714	2.069	2.500	2.807
24	1.318	1.711	2.064	2.492	2.797
25	1.316	1.708	2.060	2.485	2.787
26	1.315	1.706	2.056	2.479	2.779
27	1.314	1.703	2.052	2.473	2.771
28	1.313	1.701	2.048	2.467	2.763
29	1.311	1.699	2.045	2.462	2.756
∞	1.282	1.645	1.960	2.326	2.576

This table is an abridged version of Table IV from R. A. Fisher, *Statistical Methods for Research Workers*, published by Oliver and Boyd, Ltd., Edinburgh. Reproduced by permission of the author and publishers.

Table G

Critical levels of r at 5% and 1% levels of significance

df	$r_{.05}$*	$r_{.01}$*	df	$r_{.05}$*	$r_{.01}$*
1	0.997	1.000	24	0.388	0.496
2	0.950	0.990	25	0.381	0.487
3	0.878	0.959	26	0.374	0.478
4	0.811	0.917	27	0.367	0.470
5	0.754	0.874	28	0.361	0.463
6	0.707	0.834	29	0.355	0.456
7	0.666	0.798	30	0.349	0.449
8	0.632	0.765	35	0.325	0.418
9	0.602	0.735	40	0.304	0.393
10	0.576	0.708	45	0.288	0.372
11	0.553	0.684	50	0.273	0.354
12	0.532	0.661	60	0.250	0.325
13	0.514	0.641	70	0.232	0.302
14	0.497	0.623	80	0.217	0.283
15	0.482	0.606	90	0.205	0.267
16	0.468	0.590	100	0.195	0.254
17	0.456	0.575	125	0.174	0.228
18	0.444	0.561	150	0.159	0.208
19	0.433	0.549	200	0.138	0.181
20	0.423	0.537	300	0.113	0.148
21	0.413	0.526	400	0.098	0.128
22	0.404	0.515	500	0.088	0.115
23	0.396	0.505	1000	0.062	0.081

* Reproduced by the courtesy of the author and of the publisher from G. W. Snedecor and W. G. Cochran, *Statistical Methods*, The Iowa State University Press, Ames, Iowa, 1967, Table A11, p. 557.

Table H

Squares and Square Roots

N	N^2	$\sqrt{N}$	$\sqrt{10N}$	N	N^2	$\sqrt{N}$	$\sqrt{10N}$
1.00	1.0000	1.00000	3.16228	**1.40**	1.9600	1.18322	3.74166
1.01	1.0201	1.00499	3.17805	1.41	1.9881	1.18743	3.75500
1.02	1.0404	1.00995	3.19374	1.42	2.0164	1.19164	3.76829
1.03	1.0609	1.01489	3.20936	1.43	2.0449	1.19583	3.78153
1.04	1.0816	1.01980	3.22490	1.44	2.0736	1.20000	3.79473
1.05	1.1025	1.02470	3.24037	1.45	2.1025	1.20416	3.80789
1.06	1.1236	1.02956	3.25576	1.46	2.1316	1.20830	3.82099
1.07	1.1449	1.03441	3.27109	1.47	2.1609	1.21244	3.83406
1.08	1.1664	1.03923	3.28634	1.48	2.1904	1.21655	3.84708
1.09	1.1881	1.04403	3.30151	1.49	2.2201	1.22066	3.86005
1.10	1.2100	1.04881	3.31662	**1.50**	2.2500	1.22474	3.87298
1.11	1.2321	1.05357	3.33167	1.51	2.2801	1.22882	3.88587
1.12	1.2544	1.05830	3.34664	1.52	2.3104	1.23288	3.89872
1.13	1.2769	1.06301	3.36155	1.53	2.3409	1.23693	3.91152
1.14	1.2996	1.06771	3.37639	1.54	2.3716	1.24097	3.92428
1.15	1.3225	1.07238	3.39116	1.55	2.4025	1.24499	3.93700
1.16	1.3456	1.07703	3.40588	1.56	2.4336	1.24900	3.94968
1.17	1.3689	1.08167	3.42053	1.57	2.4649	1.25300	3.96232
1.18	1.3924	1.08628	3.43511	1.58	2.4964	1.25698	3.97492
1.19	1.4161	1.09087	3.44964	1.59	2.5281	1.26095	3.98748
1.20	1.4400	1.09545	3.46410	**1.60**	2.5600	1.26491	4.00000
1.21	1.4641	1.10000	3.47851	1.61	2.5921	1.26886	4.01248
1.22	1.4884	1.10454	3.49285	1.62	2.6244	1.27279	4.02492
1.23	1.5129	1.10905	3.50714	1.63	2.6569	1.27671	4.03733
1.24	1.5376	1.11355	3.52136	1.64	2.6896	1.28062	4.04969
1.25	1.5625	1.11803	3.53553	1.65	2.7225	1.28452	4.06202
1.26	1.5876	1.12250	3.54965	1.66	2.7556	1.28841	4.07431
1.27	1.6129	1.12694	3.56371	1.67	2.7889	1.29228	4.08656
1.28	1.6384	1.13137	3.57771	1.68	2.8224	1.29615	4.09878
1.29	1.6641	1.13578	3.59166	1.69	2.8561	1.30000	4.11096
1.30	1.6900	1.14018	3.60555	**1.70**	2.8900	1.30384	4.12311
1.31	1.7161	1.14455	3.61939	1.71	2.9241	1.30767	4.13521
1.32	1.7424	1.14891	3.63318	1.72	2.9584	1.31149	4.14729
1.33	1.7689	1.15326	3.64692	1.73	2.9929	1.31529	4.15933
1.34	1.7956	1.15758	3.66060	1.74	3.0276	1.31909	4.17133
1.35	1.8225	1.16190	3.67423	1.75	3.0625	1.32288	4.18330
1.36	1.8496	1.16619	3.68782	1.76	3.0976	1.32665	4.19524
1.37	1.8769	1.17047	3.70135	1.77	3.1329	1.33041	4.20714
1.38	1.9044	1.17473	3.71484	1.78	3.1684	1.33417	4.21900
1.39	1.9321	1.17898	3.72827	1.79	3.2041	1.33791	4.23804

(Continued)

T–8

Squares and Square Roots *(Continued)*

N	N²	√N	√10N	N	N²	√N	√10N
1.80	3.2400	1.34164	4.24264	**2.20**	4.8400	1.48324	4.69042
1.81	3.2761	1.34536	4.25441	2.21	4.8841	1.48661	4.70106
1.82	3.3124	1.34907	4.26615	2.22	4.9284	1.48997	4.77169
1.83	3.3489	1.35277	4.27785	2.23	4.9729	1.49332	4.72229
1.84	3.3856	1.35647	4.28952	2.24	5.0176	1.49666	4.73286
1.85	3.4225	1.36015	4.30116	2.25	5.0625	1.50000	4.74342
1.86	3.4596	1.36382	4.31277	2.26	5.1076	1.50333	4.75395
1.87	3.4969	1.36748	4.32435	2.27	5.1529	1.50665	4.76445
1.88	3.5344	1.37113	4.33590	2.38	5.1984	1.50997	4.77493
1.89	3.5721	1.37477	4.34741	2.29	5.2441	1.51327	4.78539
1.90	3.6100	1.37840	4.35890	**2.30**	5.2900	1.51658	4.79583
1.91	3.6481	1.38203	4.37035	2.31	5.3361	1.51987	4.80625
1.92	3.6864	1.38564	4.38178	2.32	5.3824	1.52315	4.81664
1.93	3.7249	1.38924	4.39318	2.33	5.4289	1.52643	4.82701
1.94	3.7636	1.39284	4.40454	2.34	5.4756	1.52971	4.83735
1.95	3.8025	1.39642	4.41588	2.35	5.5225	1.53297	4.84768
1.96	3.8416	1.40000	4.42719	2.36	5.5696	1.53623	4.85798
1.97	3.8809	1.40357	4.43847	2.37	5.6169	1.53948	4.86826
1.98	3.9204	1.40712	4.44972	2.38	5.6644	1.54272	4.87852
1.99	3.9601	1.41067	4.46094	2.39	5.7121	1.54596	4.88876
2.00	4.0000	1.41421	4.47214	**2.40**	5.7600	1.54919	4.89898
2.01	4.0401	1.41774	4.48330	2.41	5.8081	1.55242	4.90918
2.02	4.0804	1.42127	4.49444	2.42	5.8564	1.55563	4.91935
2.03	4.1209	1.42478	4.50555	2.43	5.9049	1.55885	4.92950
2.04	4.1616	1.42829	4.51664	2.44	5.9536	1.56205	4.93964
2.05	4.2025	1.43178	4.52769	2.45	6.0025	1.56525	4.94975
2.06	4.2436	1.43527	4.53872	2.46	6.0516	1.56844	5.95984
2.07	4.2849	1.43875	4.54973	2.47	6.1009	1.57162	4.96991
2.08	4.3264	1.44222	4.56070	2.48	6.1504	1.57480	4.97996
2.09	4.3681	1.44568	4.57165	2.49	6.2001	1.57797	4.98999
2.10	4.4100	1.44914	4.58258	**2.50**	6.2500	1.58114	5.00000
2.11	4.4521	1.45258	4.59347	2.51	6.3001	1.58430	5.00999
2.12	4.4944	1.45602	4.60435	2.52	6.3504	1.58745	5.01996
2.13	4.5369	1.45945	4.61519	2.53	6.4009	1.59060	5.02991
2.14	4.5796	1.46287	4.62601	2.54	6.4516	1.59374	5.03984
2.16	4.6225	1.46629	4.63681	2.55	6.5025	1.59687	5.04975
2.16	4.6656	1.46969	4.64758	2.56	6.5536	1.60000	5.05964
2.17	4.7089	1.47309	4.65833	2.57	6.6049	1.60312	5.06952
2.18	4.7524	1.47648	4.66905	2.58	6.6564	1.60624	5.07937
2.19	4.7961	1.47986	4.67974	2.59	6.7081	1.60935	5.08920

Squares and Square Roots *(Continued)*

N	N²	√N	√10N	N	N²	√N	√10N
2.60	6.7600	1.61245	5.09902	**3.00**	9.0000	1.73205	5.47723
2.61	6.8121	1.61555	5.10882	3.01	9.0601	1.73494	5.48635
2.62	6.8644	1.61864	5.11859	3.02	9.1204	1.73781	5.49545
2.63	6.9169	1.62173	5.12835	3.03	9.1809	1.74069	5.50454
2.64	6.9696	1.62481	5.13809	3.04	9.2416	1.74356	5.51362
2.65	7.0225	1.62788	5.14782	3.05	9.3025	1.74642	5.52268
2.66	7.0756	1.63095	5.15752	3.06	9.3636	1.74929	5.53173
2.67	7.1289	1.63401	5.16720	3.07	9.4249	1.75214	5.54076
2.68	7.1824	1.63707	5.17687	3.08	9.4864	1.75499	5.54977
2.69	7.2361	1.64012	5.18652	3.09	9.5481	1.75784	5.55878
2.70	7.2900	1.64317	5.19615	**3.10**	9.6100	1.76068	5.56776
2.71	7.3441	1.64621	5.20577	3.11	9.6721	1.76352	5.57674
2.72	7.3984	1.64924	5.21536	3.12	9.7344	1.76636	5.58570
2.73	7.4529	1.65227	5.22494	3.13	9.7969	1.76918	5.59464
2.74	7.5076	1.65529	5.23450	3.14	9.8596	1.77200	5.60357
2.75	7.5625	1.65831	5.24404	3.15	9.9225	1.77482	5.61249
2.76	7.6176	1.66132	5.25357	3.16	9.9856	1.77764	5.62139
2.77	7.6729	1.66433	5.26308	3.17	10.0489	1.78045	5.63028
2.78	7.7284	1.66733	5.27257	3.18	10.1124	1.78326	5.63915
2.79	7.7841	1.67033	5.28205	3.19	10.1761	1.78606	5.64801
2.80	7.8400	1.67332	5.29150	**3.20**	10.2400	1.78885	5.65685
2.81	7.8961	1.67631	5.30094	3.21	10.3041	1.79165	5.66569
2.82	7.9524	1.67929	5.31037	3.22	10.3684	1.79444	5.67450
2.83	8.0089	1.68226	5.31977	3.23	10.4329	1.79722	5.68331
2.84	8.0656	1.68523	5.32917	3.24	10.4976	1.80000	5.69210
2.85	8.1225	1.68819	5.33854	3.25	10.5625	1.80278	5.70088
2.86	8.1796	1.69115	5.34790	3.26	10.6276	1.80555	5.70964
2.87	8.2369	1.69411	5.35724	3.27	10.6929	1.80831	5.71839
2.88	8.2944	1.69706	5.36656	3.28	10.7584	1.81108	5.72713
2.89	8.3521	1.70000	5.37587	3.29	10.8241	1.81384	5.73585
2.90	8.4100	1.70294	5.38516	**3.30**	10.8900	1.81659	5.74456
2.91	8.4681	1.70587	5.39444	3.31	10.9561	1.81934	5.75326
2.92	8.5264	1.70880	5.40370	3.32	11.0224	1.82209	5.76194
2.93	8.5849	1.71172	5.41295	3.33	11.0889	1.82483	5.77062
2.94	8.6436	1.71464	5.42218	3.34	11.1556	1.82757	5.77927
2.95	8.7025	1.71756	5.43139	3.35	11.2225	1.83030	5.78792
2.96	8.7616	1.72047	5.44059	3.36	11.2896	1.83303	5.79655
2.97	8.8209	1.72337	5.44977	3.37	11.3569	1.83576	5.80517
2.98	8.8804	1.72627	5.45894	3.38	11.4244	1.83848	5.81378
2.99	8.9401	1.72916	5.46809	3.39	11.4921	1.84120	5.82237

(Continued)

Squares and Square Roots *(Continued)*

N	N²	√N̄	√1̄0̄N̄	N	N²	√N̄	√1̄0̄N̄
3.40	11.5600	1.84391	5.83095	**3.80**	14.4400	1.94936	6.16441
3.41	11.6281	1.84662	5.83952	3.81	14.5161	1.95192	6.17252
3.42	11.6964	1.84932	5.84808	3.82	14.5924	1.95448	6.18061
3.43	11.7649	1.85203	5.85662	3.83	14.6689	1.95704	6.18870
3.44	11.8336	1.85472	5.86515	3.84	14.7456	1.95959	6.19677
3.45	11.9025	1.85742	5.87367	3.85	14.8225	1.96214	6.20484
3.46	11.9716	1.86011	5.88218	3.86	14.8996	1.96469	6.21289
3.47	12.0409	1.86279	5.89067	3.87	14.9769	1.96732	6.22093
3.48	12.1104	1.86548	5.89915	3.88	15.0544	1.96977	6.22896
3.49	12.1801	1.86815	5.90762	3.89	15.1321	1.97231	6.23699
3.50	12.2500	1.87083	5.91608	**3.90**	15.2100	1.97484	6.24500
3.51	12.3201	1.87350	5.92453	3.91	15.2881	1.97737	6.25300
3.52	12.3904	1.87617	5.93296	3.92	15.3664	1.97990	6.26099
3.53	12.4609	1.87883	5.94138	3.93	15.4449	1.98242	6.26897
3.54	12.5316	1.88149	5.94979	3.94	15.5236	1.98494	6.27694
3.55	12.6025	1.88414	5.95819	3.95	15.6025	1.98746	6.28490
3.56	12.6736	1.88680	5.96657	3.96	15.6816	1.98997	6.29285
3.57	12.7449	1.88944	5.97495	3.97	15.7609	1.99249	6.30079
3.58	12.8164	1.89209	5.98331	3.98	15.8404	1.99499	6.30872
3.59	12.8881	1.89473	5.99166	3.99	15.9201	1.99750	6.31664
3.60	12.9600	1.89737	6.00000	**4.00**	16.0000	2.00000	6.32456
3.61	13.0321	1.90000	6.00833	4.01	16.0801	2.00250	6.33246
3.62	13.1044	1.90263	6.01664	4.02	16.1604	2.00499	6.34035
3.63	13.1769	1.90526	6.02495	4.03	16.2409	2.00749	6.34823
3.64	13.2496	1.90788	6.03324	4.04	16.3216	2.00998	6.35610
3.65	13.3225	1.91050	6.04152	4.05	16.4025	2.01246	6.36396
3.66	13.3956	1.91311	6.04979	4.06	16.4836	2.01494	6.37181
3.67	13.4689	1.91572	6.05805	4.07	16.5649	2.01742	6.37966
3.68	13.5424	1.91833	6.06630	4.08	16.6464	2.01990	6.38749
3.69	13.6161	1.92094	6.07454	4.09	16.7281	2.02237	6.39531
3.70	13.6900	1.92354	6.08276	**4.10**	16.8100	2.02485	6.40312
3.71	13.7641	1.92614	6.09098	4.11	16.8921	2.02731	6.41093
3.72	13.8384	1.92873	6.09918	4.12	16.9744	2.02978	6.41872
3.73	13.9129	1.93132	6.10737	4.13	17.0569	2.03224	6.42651
3.74	13.9876	1.93391	6.11555	4.14	17.1396	2.03470	6.43428
3.75	14.0625	1.93649	6.12372	4.15	17.2225	2.03715	6.44205
3.76	14.1376	1.93907	6.13188	4.16	17.3056	2.03961	6.44981
3.77	14.2129	1.94165	6.14003	4.17	17.3889	2.04206	6.45755
3.78	14.2884	1.94422	6.14817	4.18	17.4724	2.04450	6.46529
3.79	14.3641	1.94679	6.15630	4.19	17.5561	2.04695	6.47302

Squares and Square Roots *(Continued)*

N	N²	√N	√10N	N	N²	√N	√10N
4.20	17.6400	2.04939	6.48074	**4.60**	21.1600	2.14476	6.78233
4.21	17.7241	2.05183	6.48845	4.61	21.2521	2.14709	6.78970
4.22	17.8084	2.05426	6.49615	4.62	21.3444	2.14942	6.79706
4.23	17.8929	2.05670	6.50384	4.63	21.4369	2.15174	6.80441
4.24	17.9776	2.05913	6.51153	4.64	21.5296	2.15407	6.81175
4.25	18.0625	2.06155	6.51920	4.65	21.6225	2.15639	6.81909
4.26	18.1476	2.06398	6.52687	4.66	21.7156	2.15870	6.82642
4.27	18.2329	2.06640	6.53452	4.67	21.8089	2.16102	6.83374
4.28	18.3184	2.06882	6.54217	4.68	21.9024	2.16333	6.84105
4.29	18.4041	2.07123	6.54981	4.69	21.9961	2.16564	6.84836
4.30	18.4900	2.07364	6.55744	**4.70**	22.0900	2.16795	6.85565
4.31	18.5761	2.07605	6.56506	4.71	22.1841	2.17025	6.86294
4.32	18.6624	2.07846	6.57267	4.72	22.2784	2.17256	6.87023
4.33	18.7489	2.08087	6.58027	4.73	22.3729	2.17486	6.87750
4.34	18.8356	2.08327	6.58787	4.74	22.4676	2.17715	6.88477
4.35	18.9225	2.08567	6.59545	4.75	22.5625	2.17945	6.89202
4.36	19.0096	2.08806	6.60303	4.76	22.6576	2.18174	6.89928
4.27	19.0969	2.09045	6.61060	4.77	22.7529	2.18403	6.90652
4.38	19.1844	2.09284	6.61816	4.78	22.8484	2.18632	6.91375
4.39	19.2721	2.09523	6.62571	4.79	22.9441	2.18861	6.92098
4.40	19.3600	2.09762	6.63325	**4.80**	23.0400	2.19089	6.92820
4.41	19.4481	2.10000	6.64078	4.81	23.1361	2.19317	6.93542
4.42	19.5364	2.10238	6.64831	4.82	23.2324	2.19545	6.94262
4.43	19.6249	2.10476	6.65582	4.83	23.3289	2.19773	6.94982
4.44	19.7136	2.10713	6.66333	4.84	23.4256	2.20000	6.95701
4.45	19.8025	2.10950	6.67083	4.85	23.5225	2.20227	6.96419
4.46	19.8916	2.11187	6.67832	4.86	23.6196	2.20454	6.97137
4.47	19.9809	2.11424	6.68581	4.87	23.7169	2.20681	6.97854
4.48	20.0704	2.11660	6.69328	4.88	23.8144	2.20907	6.98570
4.49	20.1601	2.11896	6.70075	4.89	23.9121	2.21133	6.99285
4.50	20.2500	2.12132	6.70820	**4.90**	24.0100	2.21359	7.00000
4.51	20.3401	2.12368	6.71565	4.91	24.1081	2.21585	7.00714
4.52	20.4304	2.12603	6.72309	4.92	24.2064	2.21811	7.01427
4.53	20.5209	2.12838	6.73053	4.93	24.3049	2.22036	7.02140
4.54	20.6116	2.13073	6.73795	4.94	24.4036	2.22261	7.02851
4.55	20.7025	2.13307	6.74537	4.95	24.5025	2.22486	7.03562
4.56	20.7936	2.13542	6.75278	4.96	24.6016	2.22711	7.04273
4.57	20.8849	2.13776	6.76018	4.97	24.7009	2.22935	7.04982
5.48	20.9764	2.14009	6.76757	4.98	24.8004	2.23159	7.05691
4.59	21.0681	2.14243	6.77495	4.99	24.9901	2.23383	7.06399

(Continued)

Squares and Square Roots *(Continued)*

N	N²	√N	√10N	N	N²	√N	√10N
5.00	25.0000	2.23607	7.07107	**5.40**	29.1600	2.32379	7.34847
5.01	25.1001	2.23830	7.07814	5.41	29.2681	2.32594	7.35527
5.02	25.2004	2.24054	7.08520	5.42	29.3764	2.32809	7.36206
5.03	25.3009	2.24277	7.09225	5.43	29.4849	2.33024	7.36885
5.04	25.4016	2.24499	7.09930	5.44	29.5936	2.33238	7.37564
5.05	25.5025	2.24722	7.10634	5.45	29.7025	2.33452	7.38241
5.06	25.6036	2.24944	7.11337	5.46	29.8116	2.33666	7.38918
5.07	25.7049	2.25167	7.12039	5.47	29.9209	2.33880	7.39594
5.08	25.8064	2.25389	7.12741	5.48	30.0304	2.34094	7.40270
5.09	25.9081	2.25610	7.13442	5.49	30.1401	2.34307	7.40945
5.10	26.0100	2.25832	7.14143	**5.50**	30.2500	2.34521	7.41620
5.11	26.1121	2.26053	7.14843	5.51	30.3601	2.34734	7.42294
5.12	26.2144	2.26274	7.15542	5.52	30.4704	2.34947	7.42967
5.13	26.3169	2.26495	7.16240	5.53	30.5809	2.35160	7.43640
5.14	26.4196	2.26716	7.16938	5.54	30.6916	2.35372	7.44312
5.15	26.5225	2.26936	7.17635	5.55	30.8025	2.35584	7.44983
5.16	26.6256	2.27156	7.18331	5.56	30.9136	2.35797	7.45654
5.17	26.7289	2.27376	7.19027	5.57	31.0249	2.36008	7.46324
5.18	26.8324	2.27596	7.19722	5.58	21.1364	2.36220	7.46994
5.19	26.9361	2.27816	7.20417	5.59	31.2481	2.36432	7.47663
5.20	27.0400	2.28035	7.21110	**5.60**	31.3600	2.36643	7.48331
5.21	27.1441	2.28254	7.21803	5.61	31.4721	2.26854	7.48999
5.22	27.2484	2.28473	7.22496	5.62	31.5844	2.37065	7.49667
5.23	27.3529	2.28692	7.23187	5.63	31.6969	2.37276	7.50333
5.24	27.4576	2.28910	7.23878	5.64	31.8096	2.37487	7.50999
5.25	27.5625	2.29129	7.24569	5.65	31.9225	2.37697	7.51665
5.26	27.6676	2.29347	7.25259	5.66	32.0356	2.37908	7.52330
5.27	27.7729	2.29565	7.25948	5.67	32.1489	2.38118	7.52994
5.28	27.8784	2.29783	7.26636	5.68	32.2624	2.38328	7.53658
5.29	27.9841	2.30000	7.27324	5.69	32.3761	2.38537	7.54321
5.30	28.0900	2.30217	7.28011	**5.70**	32.4900	2.38747	7.54983
5.31	28.1961	2.30434	7.28697	5.71	32.6041	2.38956	7.55645
5.32	28.3024	2.30651	7.29383	5.72	32.7184	2.39165	7.56307
5.33	28.4089	2.30868	7.30068	5.73	32.8329	2.39374	7.56968
5.34	28.5156	2.31084	7.30753	5.74	32.9476	2.39583	7.57628
5.35	28.6225	2.31301	7.31437	5.75	33.0625	2.39792	7.58288
5.36	28.7296	2.31517	7.32120	5.76	33.1776	2.40000	7.58947
5.37	28.8369	2.31733	7.32803	5.77	33.2929	2.40208	7.59605
5.38	28.9444	2.31948	7.33485	5.78	33.4048	2.40416	7.60263
5.39	29.0521	2.32164	7.34166	5.79	33.5241	2.40624	7.60920

Squares and Square Roots *(Continued)*

N	N²	√N	√10N	N	N²	√N	√10N
5.80	33.6400	2.40832	7.61577	**6.20**	38.4400	2.48998	7.87401
5.81	33.7561	2.41039	7.62234	6.21	38.5641	2.49199	7.88036
5.82	33.8724	2.41247	7.62889	6.22	38.6884	2.49399	7.88670
5.83	33.9889	2.41454	7.63544	6.23	38.8129	2.49600	7.89303
5.84	34.1056	2.41661	7.64199	6.24	38.9376	2.49800	7.89937
5.85	34.2226	2.41868	7.64853	6.25	39.0625	2.50000	7.90569
5.86	34.3396	2.42074	7.65506	6.26	39.1876	2.50200	7.91202
5.87	34.4569	2.42281	7.66159	6.27	39.3129	2.50400	7.91833
5.88	34.5744	2.42487	7.66812	6.28	39.4384	2.50599	7.92465
5.89	34.6921	2.42693	7.67463	6.29	39.5641	2.50799	7.93095
5.90	34.8100	2.42899	7.68115	**6.30**	39.6900	2.50998	7.93725
5.91	34.9281	2.43105	7.68765	6.31	39.8161	2.51197	7.94355
5.92	35.0464	2.43311	7.69415	6.32	39.9424	2.51396	7.94984
5.93	35.1649	2.43516	7.70065	6.33	40.0689	2.51595	7.95613
5.94	35.2836	2.43721	7.70714	6.34	40.1956	2.51794	7.96241
5.95	35.4025	2.43926	7.71362	6.35	40.3225	2.51992	7.96869
5.96	35.5216	2.44131	7.72010	6.36	40.4496	2.52190	7.97496
5.97	35.6409	2.44336	7.72658	6.37	40.5769	2.52389	7.98123
5.98	35.7604	2.44540	7.73305	6.38	40.7044	2.52587	7.98749
5.99	35.8801	2.44745	7.73951	6.39	40.8321	2.52784	7.99375
6.00	36.0000	2.44949	7.74597	**6.40**	40.9600	2.52982	8.00000
6.01	36.1201	2.45153	7.75242	6.41	41.0881	2.53180	8.00625
6.02	36.2404	2.45357	7.75887	6.42	41.2164	2.53377	8.01249
6.03	36.3609	2.45561	7.76531	6.43	41.3449	2.53574	8.01873
6.04	36.4816	2.45764	7.77174	6.44	41.4736	2.53772	8.02496
6.05	36.6025	2.45967	7.77817	6.45	41.6025	2.53969	8.03119
6.06	36.7236	2.46171	7.78460	6.46	41.7316	2.54165	8.03741
6.07	36.8449	2.46374	7.79102	6.47	41.8609	2.54362	8.04363
6.08	36.9664	2.46577	7.79744	6.48	41.9904	2.54558	8.04984
6.09	37.0881	2.46779	7.80385	6.49	42.1201	2.54755	8.05605
6.10	37.2100	2.46982	7.81025	**6.50**	42.2500	2.54951	8.06226
6.11	37.3321	2.47184	7.81665	6.51	42.3801	2.55147	8.06846
6.12	37.4544	2.47386	7.82304	6.52	42.5104	2.55343	8.07465
6.13	37.5769	2.47588	7.82943	6.53	42.6409	2.55539	8.08084
6.14	37.6996	2.47790	7.83582	6.54	42.7716	2.55734	8.08703
6.15	37.8225	2.47992	7.84219	6.55	42.9025	2.55930	8.09321
6.16	37.9456	2.48193	7.84857	5.56	43.0336	2.56125	8.09938
6.17	38.0689	2.48395	7.85493	6.57	43.1649	2.56320	8.10555
6.18	38.1924	2.48596	7.86130	6.58	43.2964	2.56515	8.11172
6.19	38.3161	2.48797	7.86766	6.59	43.4281	2.56710	8.11788

(Continued)

T–14

Squares and Square Roots *(Continued)*

N	N²	√N	√10N	N	N²	√N	√10N
6.60	43.5600	2.56905	8.12404	**7.00**	49.0000	2.64575	8.36660
6.61	43.6921	2.57099	8.13019	7.01	49.1401	2.64764	8.37257
6.62	43.8244	2.57294	8.13634	7.02	49.2804	2.64953	8.37854
6.63	43.9569	2.57488	8.14248	7.03	49.4209	2.65141	8.38451
6.64	44.0896	2.57682	8.14862	7.04	49.5616	2.65330	8.39047
6.65	44.2225	2.57876	8.15475	7.05	49.7025	2.65518	8.39643
6.66	44.3556	2.58070	8.16088	7.06	49.8436	2.65707	8.40238
6.67	44.4889	2.58263	8.16701	7.07	49.9849	2.65895	8.40833
6.68	44.6224	2.58457	8.17313	7.08	50.1264	2.66083	8.41427
6.69	44.7561	2.58650	8.17924	7.09	50.2681	2.66271	8.42021
6.70	44.8900	2.58844	8.18535	**7.10**	50.4100	2.66458	8.42615
6.71	45.0241	2.59037	8.19145	7.11	50.5521	2.66646	8.43208
6.72	45.1584	2.59230	8.19755	7.12	50.6944	2.66833	8.43801
6.73	45.2929	2.59422	8.20366	7.13	50.8369	2.67021	8.44393
6.74	45.4276	2.59615	8.20975	7.14	50.9796	2.67208	8.44985
6.75	45.5625	2.59808	8.21584	7.15	51.1225	2.67395	8.45577
6.76	45.6976	2.60000	8.22192	7.16	51.2656	2.67582	8.46168
6.77	45.8329	2.60192	8.22800	7.17	51.4089	2.67769	8.46759
6.78	45.9684	2.60384	8.23408	7.18	51.5524	2.67955	8.47349
6.79	46.1041	2.60576	8.24015	7.19	51.6961	2.68142	8.47939
6.80	46.2400	2.60768	8.24621	**7.20**	51.8400	2.68328	8.48528
6.81	46.3761	2.60960	8.25227	7.21	51.9841	2.68514	8.49117
6.82	46.5124	2.61151	8.25833	7.22	52.1284	2.68701	8.49706
6.83	46.6489	2.61343	8.26438	7.23	52.2729	2.68887	8.50294
6.84	46.7856	2.61534	8.27043	7.24	52.4176	2.69072	8.50882
6.85	46.9225	2.61725	8.27647	7.25	52.5625	2.69258	8.51469
6.86	47.0596	2.61916	8.28251	7.26	52.7076	2.69444	8.52056
6.87	47.1969	2.62107	8.28855	7.27	52.8529	2.69629	8.52643
6.88	47.3344	2.62298	8.29458	7.28	52.9984	2.69815	8.53229
6.89	47.4721	2.62488	8.30060	7.29	53.1441	2.70000	8.53815
6.90	47.6100	2.62679	8.30662	**7.30**	53.2900	2.70185	8.54400
6.91	47.7481	2.62869	8.31264	7.31	53.4361	2.70370	8.54985
6.92	47.8864	2.63059	8.31865	7.32	53.5824	2.70555	8.55570
6.93	48.0249	2.63249	8.32466	7.33	53.7289	2.70740	8.56154
6.94	48.1636	2.63439	8.33067	7.34	53.8756	2.70924	8.56738
6.95	48.3025	2.63629	8.33667	7.35	54.0225	2.71109	8.57321
6.96	48.4416	2.63818	8.34266	7.36	54.1696	2.71293	8.57904
6.97	48.5809	2.64008	8.34865	7.37	54.3169	2.71477	8.58487
6.98	48.7204	2.64197	8.35464	7.38	54.4644	2.71662	8.59069
6.99	48.8601	2.64386	8.36062	7.39	54.6121	2.71846	8.59651

Squares and Square Roots *(Continued)*

N	N²	√N	√10N	N	N²	√N	√10N
7.40	54.7600	2.72029	8.60233	**7.80**	60.8400	2.79285	8.83176
7.41	54.9081	2.72213	8.60814	7.81	60.9961	2.79464	8.83742
7.42	55.0564	2.72397	8.61394	7.82	61.1524	2.79643	8.84308
7.43	55.2049	2.72580	8.61974	7.83	61.3089	2.79821	8.84873
7.44	55.3536	2.72764	8.62554	7.84	61.4656	2.80000	8.85438
7.45	55.5025	2.72947	8.63134	7.85	61.6225	2.80179	8.86002
7.46	55.6516	2.73130	8.63713	7.86	61.7796	2.80357	8.86566
7.47	55.8009	2.73313	8.64292	7.87	61.9369	2.80535	8.87130
7.48	55.9504	2.73496	8.64870	7.88	62.0944	2.80713	8.87694
7.49	56.1001	2.73679	8.65448	7.89	62.2521	2.80891	8.88257
7.50	56.2500	2.73861	8.66025	**7.90**	62.4100	2.81069	8.88819
7.51	56.4001	2.74044	8.66603	7.91	62.5681	2.81247	8.98382
7.52	56.5504	2.74226	8.67179	7.92	62.7264	2.81425	8.89944
7.53	56.7009	2.74408	8.67756	7.93	62.8849	2.81603	8.90505
7.54	56.8516	2.74591	8.68332	7.94	63.0436	2.81780	8.91067
7.55	57.0025	2.74773	8.68907	7.95	63.2025	2.81957	8.91628
7.56	57.1536	2.74955	8.69483	7.96	63.3616	2.82135	8.92188
7.57	57.3049	2.75136	8.70057	7.97	63.5209	2.82312	8.92749
7.58	57.4564	2.75318	8.70632	7.98	63.6804	2.82489	8.93308
7.59	57.6081	2.75500	8.71206	7.99	63.8401	2.82666	8.93868
7.60	57.7600	2.75681	8.71780	**8.00**	64.0000	2.82843	8.94427
7.61	57.9121	2.75862	8.72353	8.01	64.1601	2.83019	8.94986
7.62	58.0644	2.76043	8.72926	8.02	64.3204	2.83196	8.95545
7.63	58.2169	2.76225	8.73499	8.03	64.4809	2.83373	8.96103
7.64	58.3696	2.76405	8.74071	8.04	64.6416	2.83549	8.96660
7.65	58.5225	2.76586	8.74643	8.05	64.8025	2.83725	8.97218
7.66	58.6756	2.76767	8.75214	8.06	64.9636	2.83901	8.97775
7.67	58.8289	2.76948	8.75785	8.07	65.1249	2.84077	8.98332
7.68	58.9824	2.77128	8.76356	8.08	65.2864	2.84253	8.98888
7.69	59.1361	2.77308	8.76926	8.09	65.4481	2.84429	8.99444
7.70	59.2900	2.77489	8.77496	**8.10**	65.6100	2.84605	9.00000
7.71	59.4441	2.77669	8.78066	8.11	65.7721	2.84781	9.00555
7.72	59.5984	2.77849	8.78635	8.12	65.9344	2.84956	9.01110
7.73	59.7529	2.78029	8.79204	8.13	66.0969	2.85132	9.01665
7.74	59.9076	2.78209	8.79773	8.14	66.2596	2.85307	9.02219
7.75	60.0625	2.78388	8.80341	8.15	66.4225	2.85482	9.02774
7.76	60.2176	2.78568	8.80909	8.16	66.5856	2.85657	9.03327
7.77	60.3729	2.78747	8.81476	8.17	66.7489	2.85832	9.03881
7.78	60.5284	2.78927	8.82043	8.18	66.9124	2.86007	9.04434
7.79	60.6841	2.79106	8.82610	8.19	67.0761	2.86182	9.04986

(Continued)

Squares and Square Roots *(Continued)*

N	N²	√N	√10N	N	N²	√N	√10N
8.20	67.2400	2.86356	9.05539	**8.60**	73.9600	2.93258	9.27362
8.21	67.4041	2.86531	9.06091	8.61	74.1321	2.93428	9.27901
8.22	67.5684	2.86705	9.06642	8.62	74.3044	2.93598	9.28440
8.23	67.7329	2.86880	9.07193	8.63	74.4769	2.93769	9.28978
8.24	67.8976	2.87054	9.07744	8.64	74.6496	2.93939	9.29516
8.25	68.0625	2.87228	9.08295	8.65	74.8225	2.94109	9.30054
8.26	68.2276	2.87402	9.08845	8.66	74.9956	2.94279	9.30591
8.27	68.3929	2.87576	9.09395	8.67	75.1689	2.94449	9.31128
8.28	68.5584	2.87750	9.09945	8.68	75.3424	2.94618	9.31665
8.29	68.7241	2.87924	9.10494	8.69	75.5161	2.94788	9.32202
8.30	68.8900	2.88097	9.11043	**8.70**	75.6900	2.94958	9.32738
8.31	69.0561	2.88271	9.11592	8.71	75.8641	2.95127	9.33274
8.32	69.2224	2.88444	9.12140	8.72	76.0384	2.95296	9.33809
8.33	69.3889	2.88617	9.12688	8.73	76.2129	2.95466	9.34345
8.34	69.5556	2.88791	9.13236	8.74	76.3876	2.95635	9.34880
8.35	69.7225	2.88964	9.13783	8.75	76.5625	2.95804	9.35414
8.36	69.8896	2.89137	9.14330	8.76	76.7376	2.95973	9.35949
8.37	70.0569	2.89310	9.14877	8.77	76.9129	2.96142	9.36483
8.38	70.2244	2.89482	9.15423	8.78	77.0884	2.96311	9.37017
8.39	70.3921	2.89655	9.15969	8.79	77.2641	2.96479	9.37550
8.40	70.5600	2.89828	9.16515	**8.80**	77.4400	2.96648	9.38083
8.41	70.7281	2.90000	9.17061	8.81	77.6161	2.96816	9.38616
8.42	70.8964	2.90172	9.17606	8.82	77.7924	2.96985	9.39149
8.43	71.0649	2.90345	9.18150	8.83	77.9689	2.97153	9.39681
8.44	71.2336	2.90517	9.18695	8.84	78.1456	2.97321	9.40213
8.45	71.4025	2.90689	9.19239	8.85	78.3225	2.97489	9.40744
8.46	71.5716	2.90861	9.19783	8.86	78.4996	2.97658	9.41276
8.47	71.7490	2.91033	9.20326	8.87	78.6769	2.97825	9.41807
8.48	71.9104	2.91204	9.20869	8.88	78.8544	2.97993	9.42338
8.49	72.0801	2.91376	9.21412	8.89	79.0321	2.98161	9.42868
8.50	72.2500	2.91548	9.21954	**8.90**	79.2100	2.98329	9.43398
8.51	72.4201	2.91719	9.22497	8.91	79.3881	2.98496	9.43928
8.52	72.5904	2.91890	9.23038	8.92	79.5664	2.98664	9.44458
8.53	72.7609	2.92062	9.23580	8.93	79.7449	2.98831	9.44987
8.54	72.9316	2.92233	9.24121	8.94	79.9236	2.98998	9.45516
8.55	73.1025	2.92404	9.24662	8.95	80.1025	2.99166	9.46044
8.56	73.2736	2.92575	9.25203	8.96	80.2816	2.99333	9.46573
8.57	73.4449	2.92746	9.25743	8.97	80.4609	2.99500	9.47101
8.58	73.6164	2.92916	9.26283	8.98	80.6404	2.99666	9.47629
8.59	73.7881	2.93087	9.26823	8.99	80.8201	2.99833	9.48156

Squares and Square Roots *(Continued)*

N	N²	√N	√10N	N	N²	√N	√10N
9.00	81.0000	3.00000	9.48683	**9.40**	88.3600	3.06594	9.69536
9.01	81.1801	3.00167	9.49210	9.41	88.5481	3.06757	9.70052
9.02	81.3604	3.00333	9.49737	9.42	88.7364	3.06920	9.70567
9.03	81.5409	3.00500	9.50263	9.43	88.9249	3.07083	9.71082
9.04	81.7216	3.00666	9.50789	9.44	89.1136	3.07246	9.71597
9.05	81.9025	3.00832	9.51315	9.45	89.3025	3.07409	9.72111
9.06	82.0836	3.00998	9.51840	9.46	89.4916	3.07571	9.72625
9.07	82.2649	3.01164	9.52365	9.47	89.6809	3.07734	9.73139
9.08	82.4464	3.01330	9.52890	9.48	89.8704	3.07896	9.73653
9.09	82.6281	3.01496	9.53415	9.49	90.0601	3.08058	9.74166
9.10	82.8100	3.01662	9.53939	**9.50**	90.2500	3.08221	9.74679
9.11	82.9921	3.01828	9.54463	9.51	90.4401	3.08383	9.75192
9.12	83.1744	3.01993	9.54987	9.52	90.6304	3.08545	9.75705
9.13	83.3569	3.02159	9.55510	9.53	90.8209	3.08707	9.76217
9.14	83.5396	3.02324	9.56033	9.54	91.0116	3.08869	9.76729
9.15	83.7225	3.02490	9.56556	9.55	91.2025	3.09031	9.77241
9.16	83.9056	3.02655	9.57079	9.56	91.3936	3.09192	9.77753
9.17	84.0889	3.02820	9.57601	9.57	91.5849	3.09354	9.78264
9.18	84.2724	3.02985	9.58123	9.58	91.7764	3.09516	9.78775
9.19	84.4561	3.03150	9.58645	9.59	91.9681	3.09677	9.79285
9.20	84.6400	3.03315	9.59166	**9.60**	92.1600	3.09839	9.79796
9.21	84.8241	3.03480	9.59687	9.61	92.3521	3.10000	9.80306
9.22	85.0084	3.03645	9.60208	9.62	92.5444	3.10161	9.80816
9.23	85.1929	3.03809	9.60729	9.63	92.7369	3.10322	9.81326
9.24	85.3776	3.03974	9.61249	9.64	92.9296	3.10483	9.81835
9.25	85.5625	3.04138	9.61769	9.65	93.1225	3.10644	9.82344
9.26	85.7476	3.04302	9.62289	9.66	93.3156	3.10805	9.82853
9.27	85.9329	3.04467	9.62808	9.67	93.5089	3.10966	9.83362
9.28	86.1184	3.04631	9.63328	9.68	93.7024	3.11127	9.83870
9.29	86.3041	3.04795	9.63846	9.69	93.8961	3.11288	9.84378
9.30	86.4900	3.04959	9.64365	**9.70**	94.0900	3.11448	9.84886
9.31	86.6761	3.05123	9.64883	9.71	94.2841	3.11609	9.85393
9.32	86.8624	3.05287	9.65401	9.72	94.4784	3.11769	9.85901
9.33	87.0489	3.05450	9.65919	9.73	94.6729	3.11929	9.86408
9.34	87.2356	3.05614	9.66437	9.74	94.8676	3.12090	9.86914
9.35	87.4225	3.05778	9.66954	9.75	95.0625	3.12250	9.87421
9.36	87.6096	3.05941	9.67471	9.76	95.2576	3.12410	9.87927
9.37	87.7969	3.06105	9.67988	9.77	95.4529	3.12570	9.88433
9.38	87.9844	3.06268	9.68504	9.78	95.6484	3.12730	9.88939
9.39	88.1721	3.06431	9.69020	9.79	95.8441	3.12890	9.89444

(Continued)

Squares and Square Roots *(Continued)*

N	N^2	$\sqrt{N}$	$\sqrt{10N}$	N	N^2	$\sqrt{N}$	$\sqrt{10N}$
9.80	96.0400	3.13050	9.89949	**9.90**	98.0100	3.14643	9.94987
9.81	96.2361	3.13209	9.90454	9.91	98.2081	3.14802	9.95490
9.82	96.4324	3.13369	9.90959	9.92	98.4064	3.14960	9.95992
9.83	96.6289	3.13528	9.91464	9.93	98.6049	3.15119	9.96494
9.84	96.8256	3.13688	9.91968	9.94	98.8036	3.15278	9.96995
9.85	97.0225	3.13847	9.92472	9.95	99.0025	3.15436	9.97497
9.86	97.2196	3.14006	9.92975	9.96	99.2016	3.15595	9.97998
9.87	97.4169	3.14166	9.93479	9.97	99.4009	3.15753	9.98499
9.88	97.6144	3.14325	9.93982	9.98	99.6004	3.15911	9.98999
9.89	97.8121	3.14484	9.94485	9.99	99.8001	3.16070	9.99500
9.90	98.0100	3.14643	9.94987	**10.00**	100.0000	3.16228	10.00000

Table I

Logarithms

N	0	1	2	3	4	5	6	7	8	9
10	0000	0043	0086	0128	0170	0212	0253	0294	0334	0374
11	0414	0453	0492	0531	0569	0607	0645	0682	0719	0755
12	0792	0828	0864	0899	0934	0969	1004	1038	1072	1106
13	1139	1173	1206	1239	1271	1303	1335	1367	1399	1430
14	1461	1492	1523	1553	1584	1614	1644	1673	1703	1732
15	1761	1790	1818	1847	1875	1903	1931	1959	1987	2014
16	2041	2068	2095	2122	2148	2175	2201	2227	2253	2279
17	2304	2330	2355	2380	2405	2430	2455	2480	2504	2529
18	2553	2577	2601	2625	2648	2672	2695	2718	2742	2765
19	2788	2810	2833	2856	2878	2900	2923	2945	2967	2989
20	3010	3032	3054	3075	3096	3118	3139	3160	3181	3201
21	3222	3243	3263	3284	3304	3324	3345	3365	3385	3404
22	3424	3444	3464	3483	3502	3522	3541	3560	3579	3598
23	3617	3636	3655	3674	3692	3711	3729	3747	3766	3784
24	3802	3820	3838	3856	3874	3892	3909	3927	3945	3962
25	3979	3997	4014	4031	4048	4065	4082	4099	4116	4133
26	4150	4166	4183	4200	4216	4232	4249	4265	4281	4298
27	4314	4330	4346	4362	4378	4393	4409	4425	4440	4456
28	4472	4487	4502	4518	4533	4548	4564	4579	4594	4609
29	4624	4639	4654	4669	4683	4698	4713	4728	4742	4757
30	4771	4786	4800	4814	4829	4843	4857	4871	4886	4900
31	4914	4928	4942	4955	4969	4983	4997	5011	5024	5038
32	5051	5065	5079	5092	5105	5119	5132	5145	5159	5172
33	5185	5198	5211	5224	5237	5250	5263	5276	5289	5302
34	5315	5328	5340	5353	5366	5378	5391	5403	5416	5428
35	5441	5453	5465	5478	5490	5502	5514	5527	5539	5551
36	5563	5575	5587	5599	5611	5623	5635	5647	5658	5670
37	5682	5694	5705	5717	5729	5740	5752	5763	5775	5786
38	5798	5809	5821	5832	5843	5855	5866	5877	5888	5899
39	5911	5922	5933	5944	5955	5966	5977	5988	5999	6010
40	6021	6031	6042	6053	6064	6075	6085	6096	6107	6117
41	6128	6138	6149	6160	6170	6180	6191	6201	6212	6222
42	6232	6243	6253	6263	6274	6284	6294	6304	6314	6325
43	6335	6345	6355	6365	6375	6385	6395	6405	6415	6425
44	6435	6444	6454	6464	6474	6484	6493	6503	6513	6522
45	6532	6542	6551	6561	6571	6580	6590	6599	6609	6618
46	6628	6637	6646	6656	6665	6675	6684	6693	6702	6712
47	6721	6730	6739	6749	6758	6767	6776	6785	6794	6803
48	6812	6821	6830	6839	6848	6857	6866	6875	6884	6893
49	6902	6911	6920	6928	6937	6946	6955	6964	6972	6981
50	6990	6998	7007	7016	7024	7033	7042	7050	7059	7067
51	7076	7084	7093	7101	7110	7118	7126	7135	7143	7152
52	7160	7168	7177	7185	7193	7202	7210	7218	7226	7235
53	7243	7251	7259	7267	7275	7284	7292	7300	7308	7316
54	7324	7332	7340	7348	7356	7364	7372	7380	7388	7396

(Continued)

Logarithms *(Continued)*

N	0	1	2	3	4	5	6	7	8	9
55	7404	7412	7419	7427	7435	7443	7451	7459	7466	7474
56	7482	7490	7497	7505	7513	7520	7528	7536	7543	7551
57	7559	7566	7574	7582	7589	7597	7604	7612	7619	7627
58	7634	7642	7649	7657	7664	7672	7679	7686	7694	7701
59	7709	7716	7723	7731	7738	7745	7752	7760	7767	7774
60	7782	7789	7796	7803	7810	7818	7825	7832	7839	7846
61	7853	7860	7868	7875	7882	7889	7896	7903	7910	7917
62	7924	7931	7938	7945	7952	7959	7966	7973	7980	7987
63	7993	8000	8007	8014	8021	8028	8035	8041	8048	8055
64	8062	8069	8075	8082	8089	8096	8102	8109	8116	8122
65	8129	8136	8142	8149	8156	8162	8169	8176	8182	8189
66	8195	8202	8209	8215	8222	8228	8235	8241	8248	8254
67	8261	8267	8274	8280	8287	8293	8299	8306	8312	8319
68	8325	8331	8338	8344	8351	8357	8363	8370	8376	8382
69	8388	8395	8401	8407	8414	8420	8426	8432	8439	8445
70	8451	8457	8463	8470	8476	8482	8488	8494	8500	8506
71	8513	8519	8525	8531	8537	8543	8549	8555	8561	8567
72	8573	8579	8585	8591	8597	8603	8609	8615	8621	8627
73	8633	8639	8645	8651	8657	8663	8669	8675	8681	8686
74	8692	8698	8704	8710	8716	8722	8727	8733	8739	8745
75	8751	8756	8762	8768	8774	8779	8785	8791	8797	8802
76	8808	8814	8820	8825	8831	8837	8842	8848	8854	8859
77	8865	8871	8876	8882	8887	8893	8899	8904	8910	8915
78	8921	8927	8932	8938	8943	8949	8954	8960	8965	8971
79	8976	8982	8987	8993	8998	9004	9009	9015	9020	9025
80	9031	9036	9042	9047	9053	9058	9063	9069	9074	9079
81	9085	9090	9096	9101	9106	9112	9117	9122	9128	9133
82	9138	9143	9149	9154	9159	9165	9170	9175	9180	9186
83	9191	9196	9201	9206	9212	9217	9222	9227	9232	9238
84	9243	9248	9253	9258	9263	9269	9274	9279	9284	9289
85	9294	9299	9304	9309	9315	9320	9325	9330	9335	9340
86	9345	9350	9355	9360	9365	9370	9375	9380	9385	9390
87	9395	9400	9405	9410	9415	9420	9425	9430	9435	9440
88	9445	9450	9455	9460	9465	9469	9474	9479	9484	9489
89	9494	9499	9504	9509	9513	9518	9523	9528	9533	9538
90	9542	9547	9552	9557	9562	9566	9571	9576	9581	9586
91	9590	9595	9600	9605	9609	9614	9619	9624	9628	9633
92	9638	9643	9647	9652	9657	9661	9666	9671	9675	9680
93	9685	9689	9694	9699	9703	9708	9713	9717	9722	9727
94	9731	9736	9741	9745	9750	9754	9759	9763	9768	9773
95	9777	9782	9786	9791	9795	9800	9805	9809	9814	9818
96	9823	9827	9832	9836	9841	9845	9850	9854	9859	9863
97	9868	9872	9877	9881	9886	9890	9894	9899	9903	9908
98	9912	9917	9921	9926	9930	9934	9939	9943	9948	9952
99	9956	9961	9965	9969	9974	9978	9983	9987	9991	9996

Table J

Cumulative binomial probabilities

Left

Right

$P(X \geq x)$ $\qquad n = 1 \qquad$ $P(X \leq x)$

x	$\pi =$	01	02	03	04	05	06	07	08	09	10		
1		0100	0200	0300	0400	0500	0600	0700	0800	0900	1000		0
		99	98	97	96	95	94	93	92	91	90	$= \pi$	x

$n = 1$

x	$\pi =$	11	12	13	14	15	16	17	18	19	20		
1		1100	1200	1300	1400	1500	1600	1700	1800	1900	2000		0
		89	88	87	86	85	84	83	82	81	80	$= \pi$	x

$n = 1$

x	$\pi =$	21	22	23	24	25	26	27	28	29	30		
1		2100	2200	2300	2400	2500	2600	2700	2800	2900	3000		0
		79	78	77	76	75	74	73	72	71	70	$= \pi$	x

$n = 1$

x	$\pi =$	31	32	33	34	35	36	37	38	39	40		
1		3100	3200	3300	3400	3500	3600	3700	3800	3900	4000		0
		69	68	67	66	65	64	63	62	61	60	$= \pi$	x

$n = 1$

x	$\pi =$	41	42	43	44	45	46	47	48	49	50		
1		4100	4200	4300	4400	4500	4600	4700	4800	4900	5000		0
		59	58	57	56	55	54	53	52	51	50	$= \pi$	x

$n = 2$

x	$\pi =$	01	02	03	04	05	06	07	08	09	10		
1		0199	0396	0591	0784	0975	1164	1351	1536	1719	1900		1
2		0001	0004	0009	0016	0025	0036	0049	0064	0081	0100		0
		99	98	97	96	95	94	93	92	91	90	$= \pi$	x

$n = 2$

x	$\pi =$	11	12	13	14	15	16	17	18	19	20		
1		2079	2256	2431	2604	2775	2944	3111	3276	3439	3600		1
2		0121	0144	0169	0196	0225	0256	0289	0324	0361	0400		0
		89	88	87	86	85	84	83	82	81	80	$= \pi$	x

(Continued)

Left   *Right*

$P(X \geq x)$ $n = 2$ $P(X \leq x)$

x	$\pi =$	21	22	23	24	25	26	27	28	29	30		
1		3759	3916	4071	4224	4375	4524	4671	4816	4959	5100		1
2		0441	0484	0529	0576	0625	0676	0729	0784	0841	0900		0
		79	78	77	76	75	74	73	72	71	70	$= \pi$	x

$n = 2$

x	$\pi =$	31	32	33	34	35	36	37	38	39	40		
1		5239	5376	5511	5644	5775	5904	6031	6156	6279	6400		1
2		0961	1024	1089	1156	1225	1296	1369	1444	1521	1600		0
		69	68	67	66	65	64	63	62	61	60	$= \pi$	x

$n = 2$

x	$\pi =$	41	42	43	44	45	46	47	48	49	50		
1		6519	6636	6751	6864	6975	7084	7191	7296	7399	7500		1
2		1681	1764	1849	1936	2025	2116	2209	2304	2401	2500		0
		59	58	57	56	55	54	53	52	51	50	$= \pi$	x

$n = 3$

x	$\pi =$	01	02	03	04	05	06	07	08	09	10		
1		0297	0588	0873	1153	1426	1694	1956	2213	2464	2710		2
2		0003	0012	0026	0047	0073	0104	0140	0182	0228	0280		1
3					0001	0001	0002	0003	0005	0007	0010		0
		99	98	97	96	95	94	93	92	91	90	$= \pi$	x

$n = 3$

x	$\pi =$	11	12	13	14	15	16	17	18	·19	20		
1		2950	3185	3415	3639	3859	4073	4282	4486	4686	4880		2
2		0336	0397	0463	0533	0608	0686	0769	0855	0946	1040		1
3		0013	0017	0022	0027	0034	0041	0049	0058	0069	0080		0
		89	88	87	86	85	84	83	82	81	80	$= \pi$	x

$n = 3$

x	$\pi =$	21	22	23	24	25	26	27	28	29	30		
1		5070	5254	5435	5610	5781	5948	6110	6268	6421	6570		2
2		1138	1239	1344	1452	1563	1676	1793	1913	2035	2160		1
3		0093	0106	0122	0138	0156	0176	0197	0220	0244	0270		0
		79	78	77	76	75	74	73	72	71	70	$= \pi$	x

(*Continued*)

$P(X \geq x)$ $n = 3$ $P(X \leq x)$

x	$\pi =$	31	32	33	34	35	36	37	38	39	40		
1		6715	6856	6992	7125	7254	7379	7500	7617	7730	7840		2
2		2287	2417	2548	2682	2818	2955	3094	3235	3377	3520		1
3		0298	0328	0359	0393	0429	0467	0507	0549	0593	0640		0
		69	68	67	66	65	64	63	62	61	60	$= \pi$	x

$n = 3$

x	$\pi =$	41	42	43	44	45	46	47	48	49	50		
1		7946	8049	8148	8244	8336	8425	8511	8594	8673	8750		2
2		3665	3810	3957	4104	4253	4401	4551	4700	4850	5000		1
3		0689	0741	0795	0852	0911	0973	1038	1106	1176	1250		0
		59	58	57	56	55	54	53	52	51	50	$= \pi$	x

$n = 4$

x	$\pi =$	01	02	03	04	05	06	07	08	09	10		
1		0394	0776	1147	1507	1855	2193	2519	2836	3143	3439		3
2		0006	0023	0052	0091	0140	0199	0267	0344	0430	0523		2
3				0001	0002	0005	0008	0013	0019	0027	0037		1
4										0001	0001		0
		99	98	97	96	95	94	93	92	91	90	$= \pi$	x

$n = 4$

x	$\pi =$	11	12	13	14	15	16	17	18	19	20		
1		3726	4003	4271	4530	4780	5021	5254	5479	5695	5904		3
2		0624	0732	0847	0968	1095	1228	1366	1509	1656	1808		2
3		0049	0063	0079	0098	0120	0144	0171	0202	0235	0272		1
4		0001	0002	0003	0004	0005	0007	0008	0010	0013	0016		0
		89	88	87	86	85	84	83	82	81	80	$= \pi$	x

$n = 4$

x	$\pi =$	21	22	23	24	25	26	27	28	29	30		
1		6105	6298	6485	6664	6836	7001	7160	7313	7459	7599		3
2		1963	2122	2285	2450	2617	2787	2959	3132	3307	3483		2
3		0312	0356	0403	0453	0508	0566	0628	0694	0763	0837		1
4		0019	0023	0028	0033	0039	0046	0053	0061	0071	0081		0
		79	78	77	76	75	74	73	72	71	70	$= \pi$	x

(*Continued*)

Left   Right

$P(X \geq x)$ $n = 4$ $P(X \leq x)$

x	$\pi =$	31	32	33	34	35	36	37	38	39	40		
1		7733	7862	7985	8103	8215	8322	8425	8522	8615	8704		3
2		3660	3837	4015	4193	4370	4547	4724	4900	5075	5248		2
3		0915	0996	1082	1171	1265	1362	1464	1569	1679	1792		1
4		0092	0105	0119	0134	0150	0168	0187	0209	0231	0256		0
x		69	68	67	66	65	64	63	62	61	60	$= \pi$	x

$n = 4$

x	$\pi =$	41	42	43	44	45	46	47	48	49	50		
1		8788	8868	8944	9017	9085	9150	9211	9269	9323	9375		3
2		5420	5590	5759	5926	6090	6252	6412	6569	6724	6875		2
3		1909	2030	2155	2283	2415	2550	2689	2831	2977	3125		1
4		0283	0311	0342	0375	0410	0448	0488	0531	0576	0625		0
		59	58	57	56	55	54	53	52	51	50	$= \pi$	x

$n = 5$

x	$\pi =$	01	02	03	04	05	06	07	08	09	10		
1		0490	0961	1413	1846	2262	2661	3043	3409	3760	4095		4
2		0010	0038	0085	0148	0226	0319	0425	0544	0674	0815		3
3			0001	0003	0006	0012	0020	0031	0045	0063	0086		2
4							0001	0001	0002	0003	0005		1
		99	98	97	96	95	94	93	92	91	90	$= \pi$	x

$n = 5$

x	$\pi =$	11	12	13	14	15	16	17	18	19	20		
1		4416	4723	5016	5296	5563	5818	6061	6293	6513	6723		4
2		0965	1125	1292	1467	1648	1835	2027	2224	2424	2627		3
3		0112	0143	0179	0220	0266	0318	0375	0437	0505	0579		2
4		0007	0009	0013	0017	0022	0029	0036	0045	0055	0069		1
5					0001	0001	0001	0001	0002	0002	0003		0
		89	88	87	86	85	84	83	82	81	80	$= \pi$	x

$n = 5$

x	$\pi =$	21	22	23	24	25	26	27	28	29	30		
1		6923	7113	7293	7464	7627	7781	7927	8065	8196	8319		4
2		2833	3041	3251	3461	3672	3883	4093	4303	4511	4718		3
3		0659	0744	0836	0933	1035	1143	1257	1376	1501	1631		2
4		0081	0097	0114	0134	0156	0181	0208	0238	0272	0308		1
5		0004	0005	0006	0008	0010	0012	0014	0017	0021	0024		0
		79	78	77	76	75	74	73	72	71	70	$= \pi$	x

(*Continued*)

$P(X \geq x)$ $n = 5$ $P(X \leq x)$

x	$\pi =$	31	32	33	34	35	36	37	38	39	40		
1		8436	8546	8650	8748	8840	8926	9008	9084	9155	9222		4
2		4923	5125	5325	5522	5716	5906	6093	6276	6455	6630		3
3		1766	1905	2050	2199	2352	2509	2670	2835	3003	3174		2
4		0347	0390	0436	0486	0540	0598	0660	0726	0796	0780		1
5		0029	0034	0039	0045	0053	0060	0069	0079	0090	0102		0
		69	68	67	66	65	64	63	62	61	60	$= \pi$	x

$n = 5$

x	$\pi =$	41	42	43	44	45	46	47	48	49	50		
1		9285	9344	9398	9449	9497	9541	9582	9620	9655	9688		4
2		6801	6967	7129	7286	7438	7585	7728	7865	7998	8125		3
3		3349	3525	3705	3886	4069	4253	4439	4625	4813	5000		2
4		0949	1033	1121	1214	1312	1415	1522	1635	1753	1875		1
5		0116	0131	0147	0165	0185	0206	0229	0255	0282	0313		0
		59	58	57	56	55	54	53	52	51	50	$= \pi$	x

$n = 6$

x	$\pi =$	01	02	03	04	05	06	07	08	09	10		
1		0585	1142	1670	2172	2649	3101	3530	3936	4321	4686		5
2		0015	0057	0125	0216	0328	0459	0608	0773	0952	1143		4
3			0002	0005	0012	0022	0038	0058	0085	0118	0159		3
4						0001	0002	0003	0005	0008	0013		2
5											0001		1
		99	98	97	96	95	94	93	92	91	90	$= \pi$	x

$n = 6$

x	$\pi =$	11	12	13	14	15	16	17	18	19	20		
1		5030	5356	5664	5954	6229	6487	6731	6960	7176	7379		5
2		1345	1556	1776	2003	2235	2472	2713	2956	3201	3446		4
3		0206	0261	0324	0395	0473	0560	0655	0759	0870	0989		3
4		0018	0025	0034	0045	0059	0075	0094	0116	0141	0170		2
5		0001	0001	0002	0003	0004	0005	0007	0010	0013	0016		1
6											0001		0
		89	88	87	86	85	84	83	82	81	80	$= \pi$	x

(*Continued*)

Left   **Right**

$P(X \geq x)$ $n = 6$ $P(X \leq x)$

x	$\pi =$	21	22	23	24	25	26	27	28	29	30	
1		7569	7748	7916	8073	8220	8358	8487	8607	8719	8824	5
2		3692	3937	4180	4422	4661	4896	5128	5356	5580	5798	4
3		1115	1250	1391	1539	1694	1856	2023	2196	2374	2557	3
4		0202	0239	0280	0326	0376	0431	0492	0557	0628	0705	2
5		0020	0025	0031	0038	0046	0056	0067	0079	0093	0109	1
6		0001	0001	0001	0002	0002	0003	0004	0005	0006	0007	0
		79	78	77	76	75	74	73	72	71	70	$= \pi$

$n = 6$

x	$\pi =$	31	32	33	34	35	36	37	38	39	40	
1		8921	9011	9095	9173	9246	9313	9375	9432	9485	9533	5
2		6012	6220	6422	6619	6809	6994	7172	7343	7508	7667	4
3		2744	2936	3130	3328	3529	3732	3937	4143	4350	4557	3
4		0787	0875	0969	1069	1174	1286	1404	1527	1657	1792	2
5		0127	0148	0170	0195	0223	0254	0288	0325	0365	0410	1
6		0009	0011	0013	0015	0018	0022	0026	0030	0035	0041	0
		69	68	67	66	65	64	63	62	61	60	$= \pi$ x

$n = 6$

x	$\pi =$	41	42	43	44	45	46	47	48	49	50	
1		9578	9169	9657	9692	9723	9752	9778	9802	9824	9844	5
2		7819	7965	8105	8238	8364	8485	8599	8707	8810	8906	4
3		4764	4971	5177	5382	5585	5786	5985	6180	6373	6563	3
4		1938	2080	2232	2390	2553	2721	2893	3070	3252	3438	2
5		0458	0510	0566	0627	0692	0762	0837	0917	1003	1094	1
6		0048	0055	0063	0073	0083	0095	0108	0122	0138	0156	0
		59	58	57	56	55	54	53	52	51	50	$= \pi$ x

$n = 7$

x	$\pi =$	01	02	03	04	05	06	07	08	09	10	
1		0679	1319	1920	2486	3017	3515	3983	4422	4832	5217	6
2		0020	0079	0171	0294	0444	0618	0813	1026	1255	1497	5
3			0003	0009	0020	0038	0063	0097	0140	0193	0257	4
4					0001	0002	0004	0007	0012	0018	0027	3
5									0001	0001	0002	2
		99	98	97	96	95	94	93	92	91	90	$= \pi$ x

(Continued)

Left											Right

$P(X \geq x)$ $n = 7$ $P(X \leq x)$

x	$\pi =$	11	12	13	14	15	16	17	18	19	20	
1		5577	5913	6227	6521	6794	7049	7286	7507	7712	7903	6
2		1750	2012	2281	2556	2834	3115	3396	3677	3956	4233	5
3		0331	0416	0513	0620	0738	0866	1005	1154	1313	1480	4
4		0039	0054	0072	0094	0121	0153	0189	0231	0279	0333	3
5		0003	0004	0006	0009	0012	0017	0022	0029	0037	0047	2
6						0001	0001	0001	0002	0003	0004	1
		89	88	87	86	85	84	83	82	81	80	$= \pi$	x

$n = 7$

x	$\pi =$	21	22	23	24	25	26	27	28	29	30	
1		8080	8243	8395	8535	8665	8785	8895	8997	9090	9176	6
2		4506	4775	5040	5298	5551	5796	6035	6266	6490	6706	5
3		1657	1841	2033	2231	2436	2646	2861	3081	3304	3529	4
4		0394	0461	0536	0617	0706	0802	0905	1016	1134	1260	3
5		0058	0072	0088	0107	0129	0153	0181	0213	0248	0288	2
6		0005	0006	0008	0011	0013	0017	0021	0026	0031	0038	1
7						0001	0001	0001	0001	0002	0002	0
		79	78	77	76	75	74	73	72	71	70	$= \pi$	x

$n = 7$

x	$\pi =$	31	32	33	34	35	36	37	38	39	40	
1		9255	9328	9394	9454	9510	9560	9606	9648	9686	9720	6
2		6914	7113	7304	7487	7662	7828	7987	8137	8279	8414	5
3		3757	3987	4217	4447	4677	4906	5134	5359	5581	5801	4
4		1394	1534	1682	1837	1998	2167	2341	2521	2707	2898	3
5		0332	0380	0434	0492	0556	0625	0701	0782	0869	0963	2
6		0046	0055	0065	0077	0090	0105	0123	0142	0164	0188	1
7		0003	0003	0004	0005	0006	0008	0009	0011	0014	0016	0
		69	68	67	66	65	64	63	62	61	60	$= \pi$	x

$n = 7$

x	$\pi =$	41	42	43	44	45	46	47	48	49	50	
1		9751	9779	9805	9827	9848	9866	9883	9897	9910	9922	6
2		8541	8660	8772	8877	8976	9068	9153	9233	9307	9375	5
3		6017	6229	6436	6638	6836	7027	7213	7393	7567	7734	4
4		3094	3294	3498	3706	3917	4131	4346	4563	4781	5000	3
5		1063	1169	1282	1402	1529	1663	1803	1951	2105	2266	2
6		0216	0246	0279	0316	0357	0402	0451	0504	0562	0625	1
7		0019	0023	0027	0032	0037	0044	0051	0059	0068	0078	0
		59	58	57	56	55	54	53	52	51	50	$= \pi$	x

(*Continued*)

Cumulative binomial probabilities (*Continued*)

Left   *Right*

$P(X \geq x)$ $n = 8$ $P(X \leq x)$

x $\pi =$	01	02	03	04	05	06	07	08	09	10	
1	0773	1492	2163	2786	3366	3904	4404	4868	5297	5695	7
2	0027	0103	0223	0381	0572	0792	1035	1298	1577	1869	6
3	0001	0004	0013	0031	0058	0096	0147	0211	0289	0381	5
4			0001	0002	0004	0007	0013	0022	0034	0050	4
5							0001	0001	0003	0004	3
	99	98	97	96	95	94	93	92	91	90 $= \pi$	x

$n = 8$

x $\pi =$	11	12	13	14	15	16	17	18	19	20	
1	6063	6404	6718	7008	7275	7521	7748	7956	8147	8322	7
2	2171	2480	2794	3111	3428	3744	4057	4366	4670	4967	6
3	0487	0608	0743	0891	1052	1226	1412	1608	1815	2031	5
4	0071	0097	0129	0168	0214	0267	0328	0397	0476	0563	4
5	0007	0010	0015	0021	0029	0038	0050	0065	0083	0104	3
6		0001	0001	0002	0002	0003	0005	0007	0009	0012	2
7									0001	0001	1
	89	88	87	86	85	84	83	82	81	80 $= \pi$	x

$n = 8$

x $\pi =$	21	22	23	24	25	26	27	28	29	30	
1	8483	8630	8764	8887	8999	9101	9194	9278	9354	9424	7
2	5257	5538	5811	6075	6329	6573	6807	7031	7244	7447	6
3	2255	2486	2724	2967	3215	3465	3718	3973	4228	4482	5
4	0659	0765	0880	1004	1138	1281	1433	1594	1763	1941	4
5	0129	0158	0191	0230	0273	0322	0377	0438	0505	0580	3
6	0016	0021	0027	0034	0042	0052	0064	0078	0094	0113	2
7	0001	0002	0002	0003	0004	0005	0006	0008	0010	0013	1
8									0001	0001	0
	79	78	77	76	75	74	73	72	71	70 $= \pi$	x

$n = 8$

x $\pi =$	31	32	33	34	35	36	37	38	39	40	
1	9486	9543	9594	9640	9681	9719	9752	9782	9808	9832	7
2	7640	7822	7994	8156	8309	8452	8586	8711	8828	8936	6
3	4736	4987	5236	5481	5722	5958	6189	6415	6634	6846	5
4	2126	2319	2519	2724	2936	3153	3374	3599	3828	4059	4
5	0661	0750	0846	0949	1061	1180	1307	1443	1586	1737	3
6	0134	0159	0187	0218	0253	0293	0336	0385	0439	0498	2
7	0016	0020	0024	0030	0036	0043	0051	0061	0072	0085	1
8	0001	0001	0001	0002	0002	0003	0004	0004	0005	0007	0
	69	68	67	66	65	64	63	62	61	60 $= \pi$	x

(*Continued*)

Left   Right

$P(X \geq x)$ $n = 8$ $P(X \leq x)$

x	$\pi =$	41	42	43	44	45	46	47	48	49	50	
1		9853	9872	9889	9903	9916	9928	9938	9947	9954	9961	7
2		9037	9130	9216	9295	9368	9435	9496	9552	9602	9648	6
3		7052	7250	7440	7624	7799	7966	8125	8276	8419	8555	5
4		4292	4527	4762	4996	5230	5463	5694	5922	6146	6367	4
5		1895	2062	2235	2416	2604	2798	2999	3205	3416	3633	3
6		0563	0634	0711	0794	0885	0982	1086	1198	1318	1445	2
7		0100	0117	0136	0157	0181	0208	0239	0272	0310	0352	1
8		0008	0010	0012	0014	0017	0020	0024	0028	0033	0039	0
		59	58	57	56	55	54	53	52	51	50	$= \pi$	x

$n = 9$

x	$\pi =$	01	02	03	04	05	06	07	08	09	10	
1		0865	1663	2398	3075	3698	4270	4796	5278	5721	6126	8
2		0034	0131	0282	0478	0712	0978	1271	1583	1912	2252	7
3		0001	0006	0020	0045	0084	0138	0209	0298	0405	0530	6
4				0001	0003	0006	0013	0023	0037	0057	0083	5
5							0001	0002	0003	0005	0009	4
6											0001	3
		99	98	97	96	95	94	93	92	91	90	$= \pi$	x

$n = 9$

z	$\pi =$	11	12	13	14	15	16	17	18	19	20	
1		6496	6835	7145	7427	7684	7918	8131	8324	8499	8658	8
2		2599	2951	3304	3657	4005	4348	4685	5012	5330	5638	7
3		0672	0833	1009	1202	1409	1629	1861	2105	2357	2618	6
4		0117	0158	0209	0269	0339	0420	0512	0615	0730	0856	5
5		0014	0021	0030	0041	0056	0075	0098	0125	0158	0196	4
6		0001	0002	0003	0004	0006	0009	0013	0017	0023	0031	3
7							0001	0001	0002	0002	0003	2
		89	88	87	86	85	84	83	82	81	80	$= \pi$	x

$n = 9$

x	$\pi =$	21	22	23	24	25	26	27	28	29	30	
1		8801	8931	9048	9154	9249	9335	9411	9480	9542	9596	8
2		5934	6218	6491	6750	6997	7230	7452	7660	7856	8040	7
3		2885	3158	3434	3713	3993	4273	4552	4829	5102	5372	6
4		0994	1144	1304	1475	1657	1849	2050	2260	2478	2703	5
5		0240	0291	0350	0416	0489	0571	0662	0762	0870	0988	4
6		0040	0051	0065	0081	0100	0122	0149	0179	0213	0253	3
7		0004	0006	0008	0010	0013	0017	0022	0028	0035	0043	2
8				0001	0001	0001	0001	0002	0003	0003	0004	1
		79	78	77	76	75	74	73	72	71	70	$= \pi$	x

(*Continued*)

$P(X \geq x)$ $n = 9$ $P(X \leq x)$

x $\pi =$	31	32	33	34	35	36	37	38	39	40	
1	9645	9689	9728	9762	9793	9820	9844	9865	9883	9899	8
2	8212	8372	8522	8661	8789	8908	9017	9118	9210	9295	7
3	5636	5894	6146	6390	6627	6856	7076	7287	7489	7682	6
4	2935	3173	3415	3662	3911	4163	4416	4669	4922	5174	5
5	1115	1252	1398	1553	1717	1890	2072	2262	2460	2666	4
6	0298	0348	0404	0467	0536	0612	0696	0787	0886	0994	3
7	0053	0064	0078	0094	0112	0133	0157	0184	0215	0250	2
8	0006	0007	0009	0011	0014	0017	0021	0026	0031	0038	1
9				0001	0001	0001	0001	0002	0002	0003	0
x	69	68	67	66	65	64	63	62	61	60 $= \pi$	x

$n = 9$

x $\pi =$	41	42	43	44	45	46	47	48	49	50	
1	9913	9926	9936	9946	9954	9961	9967	9972	9977	9980	8
2	9372	9442	9505	9563	9615	9662	9704	9741	9775	9805	7
3	7866	8039	8204	8359	8505	8642	8769	8889	8999	9102	6
4	5424	5670	5913	6152	6386	6614	6836	7052	7260	7461	5
5	2878	3097	3322	3551	3786	4024	4265	4509	4754	5000	4
6	1109	1233	1366	1508	1658	1817	1985	2161	2346	2539	3
7	0290	0334	0383	0437	0498	0564	0637	0717	0804	0898	2
8	0046	0055	0065	0077	0091	0107	0125	0145	0169	0195	1
9	0003	0004	0005	0006	0008	0009	0011	0014	0016	0020	0
x	59	58	57	56	55	54	53	52	51	50 $= \pi$	x

$n = 10$

x $\pi =$	01	02	03	04	05	06	07	08	09	10	
1	0956	1829	2626	3352	4013	4614	5160	5656	6106	6513	9
2	0043	0162	0345	0582	0861	1176	1517	1879	2254	2639	8
3	0001	0009	0028	0062	0115	0188	0283	0401	0540	0702	7
4			0001	0004	0010	0020	0036	0058	0088	0128	6
5					0001	0002	0003	0006	0010	0016	5
6									0001	0001	4
x	99	98	97	96	95	94	93	92	91	90 $= \pi$	x

$n = 10$

x $\pi =$	11	12	13	14	15	16	17	18	19	20	
1	6882	7215	7516	7787	8031	8251	8448	8626	8784	8926	9
2	3028	3417	3804	4184	4557	4920	5270	5608	5932	6242	8
3	0884	1087	1308	1545	1798	2064	2341	2628	2922	3222	7
4	0178	0239	0313	0400	0500	0614	0741	0883	1039	1209	6
5	0025	0037	0053	0073	0099	0130	0168	0213	0266	0328	5
6	0003	0004	0006	0010	0014	0020	0027	0037	0049	0064	4
7			0001	0001	0001	0002	0003	0004	0006	0009	3
8									0001	0001	2
x	89	88	87	86	85	84	83	82	81	80 $= \pi$	x

(*Continued*)

Left   Right

$P(X \geq x)$ $n = 10$ $P(X \leq x)$

x	$\pi =$	21	22	23	24	25	26	27	28	29	30	
1		9053	9166	9267	9357	9437	9508	9570	9626	9674	9718	9
2		6536	6815	7079	7327	7560	7778	7981	8170	8345	8507	8
3		3526	3831	4137	4442	4744	5042	5335	5622	5901	6172	7
4		1391	1587	1794	2012	2241	2479	2726	2979	3239	3504	6
5		0399	0479	0569	0670	0781	0904	1037	1181	1337	1503	5
6		0082	0104	0130	0161	0197	0239	0287	0342	0404	0473	4
7		0012	0016	0021	0027	0035	0045	0056	0070	0087	0106	3
8		0001	0002	0002	0003	0004	0006	0007	0010	0012	0016	2
9								0001	0001	0001	0001	1
		79	78	77	76	75	74	73	72	71	70 $= \pi$	x

$n = 10$

x	$\pi =$	31	32	33	34	35	36	37	38	39	40	
1		9755	9789	9818	9843	9865	9885	9902	9916	9929	9940	9
2		8656	8794	8920	9035	9140	9236	9323	9402	9473	9536	8
3		6434	6687	6930	7162	7384	7595	7794	7983	8160	8327	7
4		3772	4044	4316	4589	4862	5132	5400	5664	5923	6177	6
5		1679	1867	2064	2270	2485	2708	2939	3177	3420	3669	5
6		0551	0637	0732	0836	0949	1072	1205	1348	1500	1662	4
7		0129	0155	0185	0220	0260	0305	0356	0413	0477	0548	3
8		0020	0025	0032	0039	0048	0059	0071	0086	0103	0123	2
9		0002	0003	0003	0004	0005	0007	0009	0011	0014	0017	1
10									0001	0001	0001	0
		69	68	67	66	65	64	63	62	61	60 $= \pi$	x

$n = 10$

x	$\pi =$	41	42	43	44	45	46	47	48	49	50	
1		9949	9957	9964	9970	9975	9979	9983	9986	9988	9990	9
2		9594	9645	9691	9731	9767	9799	9827	9852	9874	9893	8
3		8483	8628	8764	8889	9004	9111	9209	9298	9379	9453	7
4		6425	6665	6898	7123	7340	7547	7745	7933	8112	8281	6
5		3922	4178	4436	4696	4956	5216	5474	5730	5982	6230	5
6		1834	2016	2207	2407	2616	2832	3057	3288	3526	3770	4
7		0626	0712	0806	0908	1020	1141	1271	1410	1560	1719	3
8		0146	0172	0202	0236	0274	0317	0366	0420	0480	0547	2
9		0021	0025	0031	0037	0045	0054	0065	0077	0091	0107	1
10		0001	0002	0002	0003	0003	0004	0005	0006	0008	0010	0
		59	58	57	56	55	54	53	52	51	50 $= \pi$	x

(*Continued*)

Cumulative binomial probabilities (*Continued*)

Left   *Right*

$P(X \geq x)$ $n = 20$ $P(X \leq x)$

x	$\pi =$	01	02	03	04	05	06	07	08	09	10	
1		1821	3324	4562	5580	6415	7099	7658	8113	8484	8784	19
2		0169	0599	1198	1897	2642	3395	4131	4831	5484	6083	18
3		0010	0071	0210	0439	0755	1150	1610	2121	2666	3231	17
4			0006	0027	0074	0159	0290	0471	0706	0993	1330	16
5				0003	0010	0026	0056	0107	0183	0290	0432	15
6					0001	0003	0009	0019	0038	0068	0013	14
7							0001	0003	0006	0013	0024	13
8									0001	0002	0004	12
9											0001	11
		99	98	97	96	95	94	93	92	91	90	$= \pi$ x

$n = 20$

x	$\pi =$	11	12	13	14	15	16	17	18	19	20	
1		9028	9224	9383	9510	9612	9694	9759	9811	9852	9885	19
2		6624	7109	7539	7916	8244	8529	8773	8982	9159	9308	18
3		3802	4369	4920	5450	5951	6420	6854	7252	7614	7939	17
4		1710	2127	2573	3041	3523	4010	4496	4974	5439	5886	16
5		0610	0827	1083	1375	1702	2059	2443	2849	3271	3704	15
6		0175	0260	0370	0507	0673	0870	1098	1356	1643	1958	14
7		0041	0067	0103	0153	0219	0304	0409	0537	0689	0867	13
8		0008	0014	0024	0038	0059	0088	0127	0177	0241	0321	12
9		0001	0002	0005	0008	0013	0021	0033	0049	0071	0100	11
10				0001	0001	0002	0004	0007	0011	0017	0026	10
11							0001	0001	0002	0004	0006	9
12										0001	0001	8
		89	88	87	86	85	84	83	82	81	80	$= \pi$ x

$n = 20$

x	$\pi =$	21	22	23	24	25	26	27	28	29	30	
1		9910	9931	9946	9959	9968	9976	9982	9986	9989	9992	19
2		9434	9539	9626	9698	9757	9805	9845	9877	9903	9924	18
3		8230	8488	8716	8915	9087	9237	9365	9474	9567	9654	17
4		6310	6711	7085	7431	7748	8038	8300	8534	8744	8929	16
5		4142	4580	5014	5439	5852	6248	6625	6981	7315	7625	15
6		2297	2657	3035	3427	3828	4235	4643	5048	5447	5836	14
7		1071	1301	1557	1838	2142	2467	2810	3169	3540	3920	13
8		0419	0536	0675	0835	1018	1225	1455	1707	1982	2277	12
9		0138	0186	0246	0320	0409	0515	0640	0784	0948	1133	11
10		0038	0054	0075	0103	0139	0183	0238	0305	0385	0480	10
11		0009	0013	0019	0028	0039	0055	0074	0100	0132	0171	9
12		0002	0003	0004	0006	0009	0014	0019	0027	0038	0051	8
13				0001	0001	0002	0003	0004	0006	0009	0013	7
14								0001	0001	0002	0003	6
		79	78	77	76	75	74	73	72	71	70	$= \pi$ x

(*Continued*)

Left   **Right**

$P(X \geq x)$ $n = 20$ $P(X \leq x)$

x	$\pi=$ 31	32	33	34	35	36	37	38	39	40	
1	9994	9996	9997	9998	9998	9999	9999	9999	9999	10000	19
2	9940	9953	9964	9972	9979	9984	9988	9991	9993	9995	18
3	9711	9765	9811	9848	9879	9904	9924	9940	9953	9964	17
4	9092	9235	9358	9465	9556	9634	9700	9755	9802	9840	16
5	7911	8173	8411	8626	8818	8989	9141	9274	9390	9490	15
6	6213	6574	6917	7242	7546	7829	8090	8329	8547	8744	14
7	4305	4693	5079	5460	5834	6197	6547	6882	7200	7500	13
8	2591	2922	3268	3624	3990	4361	4735	5108	5478	5841	12
9	1340	1568	1818	2087	2376	2683	3005	3341	3688	4044	11
10	0591	0719	0866	1032	1218	1424	1650	1897	2163	2447	10
11	0220	0279	0350	0434	0532	0645	0775	0923	1090	1275	9
12	0069	0091	0119	0154	0196	0247	0308	0381	0466	0565	8
13	0018	0025	0034	0045	0060	0079	0102	0132	0167	0210	7
14	0004	0006	0008	0011	0015	0021	0028	0037	0049	0065	6
15	0001	0001	0001	0002	0003	0004	0006	0009	0012	0016	5
16						0001	0001	0002	0002	0003	4
	69	68	67	66	65	64	63	62	61	60	x

$n = 20$

x	$\pi=$ 41	42	43	44	45	46	47	48	49	50	
1	10000	10000	10000	10000	10000	10000	10000	10000	10000	10000	19
2	9996	9997	9998	9998	9999	9999	9999	10000	10000	10000	18
3	9972	9979	9984	9988	9991	9993	9995	9996	9997	9998	17
4	9872	9898	9920	9937	9951	9962	9971	9977	9983	9987	16
5	9577	9651	9714	9767	9811	9848	9879	9904	9924	9941	15
6	8921	9078	9217	9340	9447	9539	9619	9687	9745	9793	14
7	7780	8041	8281	8501	8701	8881	9042	9186	9312	9423	13
8	6196	6539	6868	7183	7480	7759	8020	8261	8482	8684	12
9	4406	4771	5136	5499	5857	6207	6546	6873	7186	7483	11
10	2748	3064	3394	3736	4086	4443	4804	5166	5525	5881	10
11	1480	1705	1949	2212	2493	2791	3104	3432	3771	4119	9
12	0679	0810	0958	1123	1308	1511	1734	1977	2238	2517	8
13	0262	0324	0397	0482	0580	0694	0823	0969	1133	1316	7
14	0084	0107	0136	0172	0214	0265	0326	0397	0480	0577	6
15	0022	0029	0038	0050	0064	0083	0105	0133	0166	0207	5
16	0004	0006	0008	0011	0015	0020	0027	0035	0046	0059	4
17	0001	0001	0001	0002	0003	0004	0005	0007	0010	0013	3
18						0001	0001	0001	0001	0002	2
	59	58	57	56	55	54	53	52	51	50 $=\pi$	x

Table K*

Cumulative Poisson probabilities

$P(X \leq x)$

x	0.1	0.2	0.3	0.4	0.5	0.6	0.7	0.8	0.9
					μ				
0	0.9048	0.8187	0.7408	0.6730	0.6065	0.5488	0.4966	0.4493	0.4066
1	0.9953	0.9825	0.9631	0.9384	0.9098	0.8781	0.8442	0.8088	0.7725
2	0.9998	0.9989	0.9964	0.9921	0.9856	0.9769	0.9659	0.9526	0.9371
3	1.0000	0.9999	0.9997	0.9992	0.9982	0.9966	0.9942	0.9909	0.9865
4		1.0000	1.0000	0.9999	0.9998	0.9996	0.9992	0.9986	0.9977
5				1.0000	1.0000	1.0000	0.9999	0.9998	0.9997
6							1.0000	1.0000	1.0000

x	1.0	1.5	2.0	2.5	3.0	3.5	4.0	4.5	5.0
					μ				
0	0.3679	0.2231	0.1353	0.0821	0.0498	0.0302	0.0183	0.0111	0.0067
1	0.7358	0.5578	0.4060	0.2873	0.1991	0.1359	0.0916	0.0611	0.0404
2	0.9197	0.8088	0.6767	0.5438	0.4232	0.3208	0.2381	0.1736	0.1247
3	0.9810	0.9344	0.8571	0.7576	0.6472	0.5366	0.4335	0.3423	0.2650
4	0.9963	0.9814	0.9473	0.8912	0.8153	0.7254	0.6288	0.5321	0.4405
5	0.9994	0.9955	0.9834	0.9580	0.9161	0.8576	0.7851	0.7029	0.6160
6	0.9999	0.9991	0.9955	0.9858	0.9665	0.9347	0.8893	0.8311	0.7622
7	1.0000	0.9998	0.9989	0.9958	0.9881	0.9733	0.9489	0.9134	0.8666
8		1.0000	0.9998	0.9989	0.9962	0.9901	0.9786	0.9597	0.9319
9			1.0000	0.9997	0.9989	0.9967	0.9919	0.9829	0.9682
10				0.9999	0.9997	0.9990	0.9972	0.9933	0.9863
11				1.0000	0.9999	0.9997	0.9991	0.9976	0.9945
12					1.0000	0.9999	0.9997	0.9992	0.9980
13						1.0000	0.9999	0.9997	0.9993
14							1.0000	0.9999	0.9998
15								1.0000	0.9999
16									1.0000

Cumulative Poisson probabilities (*Continued*)

$$P(X \le x)$$

x	5.5	6.0	6.5	7.0	7.5	8.0	8.5	9.0	9.5	10.0
					μ					
0	0.0041	0.0025	0.0015	0.0009	0.0006	0.0003	0.0002	0.0001	0.0001	0.0000
1	0.0266	0.0174	0.0113	0.0073	0.0047	0.0030	0.0019	0.0012	0.0008	0.0005
2	0.0884	0.0620	0.0430	0.0296	0.0203	0.0138	0.0093	0.0062	0.0042	0.0028
3	0.2017	0.1512	0.1118	0.0818	0.0591	0.0424	0.0301	0.0212	0.0149	0.0103
4	0.3575	0.2851	0.2237	0.1730	0.1321	0.0996	0.0744	0.0550	0.0403	0.0293
5	0.5289	0.4457	0.3690	0.3007	0.2414	0.1912	0.1496	0.1157	0.0885	0.0671
6	0.6860	0.6063	0.5265	0.4497	0.3782	0.3134	0.2562	0.2068	0.1649	0.1301
7	0.8095	0.7440	0.6728	0.5987	0.5246	0.4530	0.3856	0.3239	0.2687	0.2202
8	0.8944	0.8472	0.7916	0.7291	0.6620	0.5925	0.5231	0.4557	0.3918	0.3328
9	0.9462	0.9161	0.8774	0.8305	0.7764	0.7166	0.6530	0.5874	0.5218	0.4579
10	0.9747	0.9574	0.9332	0.9015	0.8622	0.8159	0.7634	0.7060	0.6453	0.5830
11	0.9890	0.9799	0.9661	0.9466	0.9208	0.8881	0.8487	0.8030	0.7520	0.6968
12	0.9955	0.9912	0.9840	0.9730	0.9573	0.9362	0.9091	0.8758	0.8364	0.7916
13	0.9983	0.9964	0.9929	0.9872	0.9784	0.9658	0.9486	0.9261	0.8981	0.8645
14	0.9994	0.9986	0.9970	0.9943	0.9897	0.9827	0.9726	0.9585	0.9400	0.9165
15	0.9998	0.9995	0.9988	0.9976	0.9954	0.9918	0.9862	0.9780	0.9665	0.9513
16	0.9999	0.9998	0.9996	0.9990	0.9980	0.9963	0.9934	0.9889	0.9823	0.9730
17	1.0000	0.9999	0.9998	0.9996	0.9992	0.9984	0.9970	0.9947	0.9911	0.9857
18		1.0000	0.9999	0.9999	0.9997	0.9994	0.9987	0.9976	0.9957	0.9928
19			1.0000	1.0000	0.9999	0.9997	0.9995	0.9989	0.9980	0.9965
20					1.0000	0.9999	0.9998	0.9996	0.9991	0.9984
21						1.0000	0.9999	0.9998	0.9996	0.9993
22							1.0000	0.9999	0.9999	0.9997
23								1.0000	0.9999	0.9999
24									1.0000	1.0000

* Reproduced from E. C. Molina, *Poisson's Exponential Binomial Limit*. Copyright 1942, Princeton, N.J.: D. Van Nostrand Company, Inc., by permission of the publisher.

Answers
to Exercises

Answers
to Exercises

Chapter 1

1.

Class interval	Lower boundary	Upper boundary	Class width	Class mark
10–14	9.5	14.5	5	12
15–19	14.5	19.5	5	17
20–24	19.5	24.5	5	22
25–29	24.5	29.5	5	27

Chapter 2

1. a) Mean = 8, mode = 11, median = 9
   b) Range = 7, average deviation = 2.8, σ = 2.97

2. Mean = 8, median = 8, mode = 8   **3.** It is possible

4. Mean = 1, σ = 5.69

5. Median = $7950, mean = $8600. The median is a better average

6. Median = $15,000, mode = $15,000

7. μ and σ must be measured in the same units

8. Annual income: median = $12,750
   IQ: median = 104
   Weight: mean = 163.4
   Hat size: mode = $6\frac{7}{8}$

10. Population A: $\mu = 11$, $\sigma = 3.16$, median = 10
   Population B: $\mu = 9$, $\sigma = 3.16$, median = 10

11. Population A: mean = 5, $R = 8$, $\sigma = 2.61$
   Population B: mean = 5, $R = 8$, $\sigma = 3.16$

12. a) $\mu = 12$, $\sigma = 3$ b) $\mu = 20$, $\sigma = 6$

13. $\mu = 65$, $\sigma = 12.19$;   a) 70%   b) 100%   c) 100%

15. a) Mean = 32, median = 33, mode = 33
  b) Range = 24, average deviation = 5.84, standard deviation = 6.87
  c) (1) 64%   (2) 100%   (3) 100%

Chapter 3

Set 1

1. a) {10, 11, 12}   b) {1, 2, 3, 4, 5, 6, 7, 8}   c) {3, 6, 9, 12, 15, 18}

2. a) {9, 10, 11, 12}   b) {10, 11, 12, 13}   c) $C = \{-2\}$ d) $D = \{-7, 4\}$

3. $\{x, y, 3\}, \{x\}, \{x, y\}, \{y\}, \{x, 3\}, \{3\}, \{y, 3\}, \{\varnothing\}$

4. {man, woman, baby, home}, {man, woman, baby}, {man, woman, home}, {man, baby, home}, {woman, baby, home}, {man, woman}, {man, baby}, {man, home} {woman, baby}, {woman, home}, {baby, home}, {man}, {woman}, {baby}, {home}, $\{\varnothing\}$

5. a) $A \cup B = \{10, 11, 12, 13, 14\}$ b) $(A \cap B) = \{12, 13\}$ c) $A' = \{14, 15\}$

6. a) $A \cup B = \{20, 22, 24\}$ b) $(A \cup B)' = \{26\}$ c) $(A \cap B) = \{22\}$
  d) $(A \cap B)' = \{20, 24, 26\}$

7. a) $(A \cup B) = \{v, w, x, y, z\}$ b) $(A \cup B)' = \{\varnothing\}$ c) $(A \cap B) = \{\varnothing\}$
  d) $(A \cap B)' = \{v, w, x, y, z\}$

8. $A = \{13, 14, 15, 16, 17, 18, 19\}$
  $B = \{18, 19, \ldots, \infty\}$
  $(A \cap B) = \{18, 19\}$

9. $A \cap B = \{8\}$

10. a) $(X \cup Y) = \{-4, 2, -2\}$ b) $(X \cap Y) = \{2\}$

11. a) $(A \cup B) = U$ b) $(A \cup B)' = \{\varnothing\}$ c) $(A \cap B) = \{\varnothing\}$

Set 2

1. a) {1, 2, 3, 4, 5, 6}   b) {2, 4, 6}   c) {5, 6}

2. a) $\{HH, HT, TH, TT\}$ b) $\{HT, TH\}$ c) $\{HH, HT, TH\}$

3. a) $S = \{(1, 1), (1, 2), (1, 3), (1, 4), (1, 5), (1, 6),$
        $(2, 1), (2, 2), (2, 3), (2, 4), (2, 5), (2, 6),$
        $(3, 1), (3, 2), (3, 3), (3, 4), (3, 5), (3, 6),$
        $(4, 1), (4, 2), (4, 3), (4, 4), (4, 5), (4, 6),$
        $(5, 1), (5, 2), (5, 3), (5, 4), (5, 5), (5, 6),$
        $(6, 1), (6, 2), (6, 3), (6, 4), (6, 5), (6, 6)\}$
  b) {(6, 3), (5, 4), (4, 5), (3, 6)}
  c) {(3, 1), (2, 2), (1, 3), (4, 1), (3, 2), (2, 3), (1, 4)}

4. a) $S = \{DDD, DDD', DD'D, DD'D', D'DD, D'DD', D'D'D, D'D'D'\}$
  b) $\{D'D'D'\}$
  c) There are exactly two defective parts

5. $S = \{(H, 1), (H, 2), (H, 3), (H, 4), (H, 5), (H, 6),$
$(T, 1), (T, 2), (T, 3), (T, 4), (T, 5), (T, 6)\}$

Set 3

1. a) $\frac{1}{2}$ b) $\frac{1}{2}$

2. a) $\frac{1}{2}$ b) $\frac{3}{4}$

3. a) $\frac{1}{6}$ b) $\frac{1}{18}$ c) $\frac{2}{9}$ d) $\frac{1}{3}$

4. a) $\frac{4}{52}$ b) $\frac{4}{52}$ c) $\frac{8}{52}$ d) $\frac{28}{52}$ e) $\frac{12}{52}$

5. .12   **6.** .85   **7.** .80   **8.** .30   **9.** a) .95 b) .04 c) .99

Set 4

1. .60   **2.** .10   **3.** a) $\frac{12}{42}$ b) $\frac{16}{49}$

4. $\frac{8}{49}$ **5.** $\frac{114}{441}$

6. a) $\frac{16}{2652}$ b) $\frac{32}{2652}$ c) $\frac{2256}{2652}$ d) $\frac{1892}{2652}$

7. $\frac{32}{1560}$

8. a) $\dfrac{288}{311,875,200}$ b) $\dfrac{54,144}{311,875,200}$ c) $\dfrac{56,448}{311,875,200}$ or $\dfrac{24}{132,600}$

9. a) $\dfrac{9,024}{132,600}$ b) $\dfrac{10,200}{132,600}$ or $\dfrac{1}{13}$

10. $\dfrac{28,824}{132,600}$ **11.** a) $\dfrac{215}{216}$ b) $\dfrac{6}{216}$

12. a) $\frac{2}{1296}$
b) $\frac{5}{1296}$
c) $\frac{1}{216}$

13. a) $\frac{4}{216}$ b) $\frac{212}{216}$

14. .9983616   **15.** .001999   **16.** $\frac{4}{9900}$

17. a) $\dfrac{21984}{24024}$ b) $\dfrac{576}{2184}$ **18.** a) 0.064 b) 0.104

19. a) $\dfrac{2}{12}$ b) $\dfrac{10}{12}$ **20.** .002

21. a) $\dfrac{210}{380}$ b) $\dfrac{30}{380}$ c) $\dfrac{230}{380}$ **22.** a) $\dfrac{5}{24}$ b) $\dfrac{1}{24}$

Set 5

1. a) $\frac{6}{14}$ b) $\frac{2}{3}$ **2.** a) $\frac{32}{252}$ b) $\frac{3}{4}$

3. a) $\frac{30}{210}$ b) $\frac{24}{30}$ **4.** a) .18 b) $\frac{10}{18}$

5. a) .56 b) $\frac{16}{56}$ **6.** a) .44 b) $\frac{6}{44}$

7. a) 0.62 b) $\frac{54}{62}$ **8.** a) 0.34 b) $\frac{16}{34}$

9. a) 0.509 b) $\frac{192}{509}$ **10.** $\frac{2}{65}$

11. $\frac{20}{31}$ **12.** $\frac{2}{3}, \frac{1}{3}$ **13.** $\frac{8}{14}$ **14.** 0.483   **15.** 0.191

Set 6

1. −1 **2.** 0 **3.** $4.20 **4.** $4.20 **5.** 37.45
6. Expected life: $A = 2500$, $B = 2550$
7. Expected cost: $A = \$137.50$, $B = \$142.50$
8. a) $600,000 > $500,000, wait for rezoning b) $\frac{3}{8}$
9. Order 5: the expected profit is $1.05
10. Expected profit is $1,100
11. Expected profit is $3,000
12. a) 1.05 b) $120 c) $126 d) 0.0225
13. a) 0.001 b) 10.8

Chapter 4

Set 1

1. 20 **2.** 16 **3.** 5040 **4.** 151,200 **5.** 35
6. a) 17,576 b) 15,600 **7.** a) 16 b) 0
8. a) 10 b) 7 **9.** a) 126 b) 60 **10.** 24

Set 2

1. a) .2036 b) .2262
2. $\frac{64}{243}$ **3.** a) .3456 b) .0576
4. .05853 **5.** $\frac{640}{3125}$ **6.** .008333928 **7.** a) .1880025 b) .6732
8. a) $\frac{3}{4}$ b) $\frac{11}{16}$

Set 3

1. a) $x: 0\ 1\ 2; P(x): \frac{4}{9}\ \frac{4}{9}\ \frac{1}{9}$ b) $\mu = \frac{2}{3}$, $\sigma = \frac{2}{3}$
2. a) $\mu = 800$, $\sigma = 20$ b) $\mu = 90$, $\sigma = 9$ c) $\mu = 256$, $\sigma = 9.6$
3. $\mu = 20$, $\sigma = 4.216$ **4.** $\mu = 100$

Set 4

1. a) 0.5796 b) 0.3150 c) 0.0802 d) 0.0649
2. a) 0.4348 b) 0.2719 c) 0.0420 d) 0.0345
3. a) 0.0248 b) 0.2815 c) 0.3811 d) 0.6626
4. a) Zero b) 0.1144 c) 0.7723 d) 0.8867

Set 5

1. a) 0.8571 b) 0.1804
2. 0.2231 **3.** 0.1493
4. 0.1755 **5.** 0.2424
6. 0.1438 **7.** 0.3032

Set 6

1. a) .0668 b) .4772 c) 70.68 inches   **2.** 456
3. a) .1151 b) 51.6   **4.** a) 12.616 b) 12.072
5. $\sigma = .78$ **6.** $\mu = 53, \sigma = 2$ **7.** 8:31.8 A.M.   **8.** .4159
9. 0.4782   **10.** $\sigma = 23,810$ **11.** 7,463
12. a) 0.1587 or 159 students b) 0.2119 or 170 students
13. a) 0.1151 b) 0.5328 c) 0.2417
14. 73,840 miles
15. a) 0.0548 b) 418 hours   **16.** 0.1587

Set 7

1. a) .0803 b) .6915   **2.** .8531   **3.** .7517   **4.** $-.5745$
5. a) .1056 b) .1251   **6.** $\$-.37$
7. $z = -4.50$. The claim is untrue

Review Exercises I

1. $\mu = 130, \sigma = 25$
2. $A \cap B = \{-4\}$
3. a) $S = \{RRR, RRB, RBR, RBB, BRR, BRB, BBR, BBB\}$
   b) $A = \{RBB, BRB, BBR\}$
   c) There are exactly three black cards.
4. a) $\mu = 20, \sigma = 4.2$ b) $\frac{7}{10}$
5. a) $\frac{4}{10}$ b) $\frac{12}{90}$ c) $\frac{24}{90}$
6. 0.006
7. 3 engines
8. a) 0.1296 b) 0.8704 c) 6 shots
9. 0.960498
10. $\frac{285}{355}$
11. a) 0.22 b) $\frac{14}{22}$
12. $\$-0.27$
13. a) $\frac{1}{32}$ b) $\frac{10}{32}$
14. a) 0.16 b) 0.488
15. a) 0.4096 b) 0.8
16. a) 0.3456 b) 0.9744 c) 0.0576
17. a) 0.3125 b) 0.2373
18. a) $\frac{3}{8}$ b) $\frac{7}{8}$ c) 4 trials
19. $\sigma = 2, \mu = 10$
20. 0.9772

21. a) 0.0668 b) 10.4 c) 10.4

22. 63.6 hours

23. a) 0.0228 b) $x_1 = 46{,}650$, $x_2 = 53{,}350$
c) 38,350

24. a) 1097.2 units b) 1524.4 units
c) $x_1 = 46.65$, $x_2 = 53.35$

25. a) 0.2327 b) 0.0179

26. a) 0.0087 b) 0.0885

27. a) 0.1151 b) 69.2 hours

28. a) $\frac{2}{12}$ b) $\frac{3}{12}$

29. 88

30. a) 0.25 b) 6.1

31. a) 0.15625 b) 0.96875

32. a) 3.8 b) $5000 c) $19,000

33. a) 0.3085 b) 0.140394

34. a) $\frac{1}{16}$ b) $\frac{270}{1024}$ c) 40

Chapter 5

2. $\bar{x} = \$120$, $s = 30$ **3.** $\bar{x} = 21$, $s = 2$

7. a) 10 b) 4 c) 0.56

Chapter 6

Set 1

1. a) $\mu = 5$, $\sigma = 2.24$ b) 3, 4, 5, 5, 6, 7 c) $E(\bar{x}) = 5$, $\sigma_{\bar{x}} = \sqrt{\frac{5}{3}}$

2. a) Between 68.672 and 69.328 b) Between 68.836 and 69.164

3. a) .0228 b) .1587 c) .50

4. a) .4522 b) .1587 c) .0228

5. 16.280

6. a) 0.1587 b) 0.0228 c) $L_1 = 49.18$, $L_2 = 50.82$

Set 2

1. Between 3.34 and 4.66

2. a) 97 b) 246

3. Between 20.02 and 21.98

4. a) 1537 b) 385

5. a) Between 16.452 and 19.548 b) 1.176

6. 239   **7.** 54.68%

8. a) Between 34.208 and 45.792 b) Between 18.699 and 21.301
c) Between 65.016 and 74.984

9. a) Between \$111.80 and \$128.20 b) \$9.80

10. Between 10.30 and 10.70

11. Between \$5,845,200 and \$6,154,800

12. Between \$73,360,000 and \$76,640,000

13. a) Between \$10,968 and \$13,032 b) 0.8664

14. Between 44.448 and 55.552

Chapter 7

Set 1

1. $z = .5833$, $.5833 < 1.64$; reserve judgment

2. $z = -2.25$, $2.25 < 2.33$; reserve judgment

3. $z = 3$, $3 > 1.64$; reject

4. $z = 1.5$, $1.5 < 1.96$; reserve judgment

5. $z = 3$, $3 > 2.58$; reject

6. $z = 6$, $6 > 2.33$; reject

7. $t = -2$, $2 < 2.492$; reserve judgment

8. $t = -2$, $2 < 5.841$; reserve judgment

9. $z = -1$, $1 < 1.96$; reserve judgment

10. $t = -3.47$, $3.47 > 2.896$; reject

11. $z = -3$, $3 > 2.33$; reject

12. a) $t = 3$, $3 > 1.753$; reject b) .95

13. $z = 2.2$, $2.2 > 1.64$; reject

14. $t = 1.42$, $1.42 < 2.132$; reserve judgment

15. $z = -2.14$, $2.14 > 1.64$; reject

Set 2

1. a) 0.1142 b) 0.0571 c) 0.0287

2. $n = 63.0436$ or 64, $\overset{*}{c} = \$135.33$

3. a) H_0: $\mu = 2$; H_1: $\mu = 1.5$ b) 0.0228 c) 0.0013

4. a) H_0: $\mu = 502$; H_1: $\mu > 502$ b) 0.0314 c) 0.0262

5. $n = 35.462$ or 36, $\overset{*}{c} = \$882.62$

6. Decision Rule: Select a random sample of 64 observations and determine $\bar{x}$. If $\bar{x} > 153.28$, reject H_0.

7. Decision Rule: Select a random sample of 49 observations and determine $\bar{x}$. If $\bar{x}$ is not between 59.84 and 70.16, reject H_0.

8. Decision Rule: Select a random sample of 40 observations and determine $\bar{x}$. If $\bar{x} < 86.04$, reject H_0.

Chapter 8

Set 1

1. a) .0548   b) .2119
2. a) .0548   b) .6554
3. $z = 1.85$, $1.85 < 1.96$; not significant
4. $z = -2.4$, $2.4 > 1.96$; significant
5. $t = .98$, $.98 < 3.169$; not significant
6. $t = -.641$, $.641 < 2.365$; not significant

Set 2

1. $(F = 3.443) < (F_{.05} = 3.89)$; not significant
2. $(F = 4.8) > (F_{.05} = 4.07)$; significant
3. $(F = 45.02) > (F_{.01} = 6.36)$; significant
4. $(F = 4) > (F_{.05} = 3.10)$; significant

Review Exercises II

1. a) Between 1.9608 and 2.0392
   b) Between 1.99804 and 2.00196   c) 0.0668
2. a) Between 702 and 898   b) Between 787.1 and 812.9
   c) 0.3446   d) Zero
3. a) 0.6892   b) 0.2881
4. a) 0.8185 a   b) 107
5. Zero
6. Between \$2.68 and \$3.32
7. a) Between 9.51 and 10.49
   b) Between 8.9345 and 11.0655
8. 0.90 confidence
9. Between 64,672.5 and 75,327.5
10. Between \$1,045.200 and \$1,354,800
11. $z = 4$, $4 > 1.64$; reject H_0
12. $t = 2.5$, $2.5 < 3.747$; reserve judgment
13. a) Between 38.04 and 41.96   b) 0.90 confidence
    c) $z = -3$, $3 > 2.58$; claim is not valid
14. $z = -0.75$, $0.75 < 1.64$; reserve judgment
15. $z = -3$, $3 > 2.58$; performance is not the same
16. $z = 3$, $3 > 2.58$; blonds do have more fun
17. $F = 5$, $5 > 3.10$; brands are different
18. a) Between \$78.04 and \$81.96   b) \$2.33   c) $Z = 5$; reject H_0
19. Between 4,602,000 and 4,798,000

Chapter 9

Set 1

1. a) 40% b) .5, 1, .5, .5, .5, 0, 0, .5, .5, 0 c) $E(p) = .40$, $\sigma_p = .3$
   d) $\sigma_p = .3$

2. a) .5
   b) $\frac{2}{3}, \frac{1}{3}, \frac{2}{3}, \frac{1}{3}, \frac{2}{3}, \frac{3}{3}, \frac{2}{3}, \frac{2}{3}, \frac{1}{3}, \frac{2}{3}, \frac{1}{3}, \frac{2}{3}, \frac{1}{3}, \frac{1}{3}, 0, \frac{1}{3}, \frac{2}{3}, \frac{1}{3}, \frac{2}{3}, \frac{1}{3}$
   c) .50 d) .90

3. a) .0228 b) .6915 c) .1574

4. a) .0228 b) .0228

5. .018   6. $p^* = .47$, $n = 1568.16$

7. a) 0.0228 b) $P_1 = 0.45$, $P_2 = 0.55$

8. a) 0.0918 b) 0.0021 c) 0.5

9. 0.0828   10. a) 0.0427 b) 0.1814

Set 2

1. Between .1534 and .2466

2. a) 0.0508–0.1492 b) 0.0588

3. 0.8904

4. a) 246 b) 6147   5. 97

6. Between 2,296 and 3,204

7. Between 10,918 and 14,682

8. a) Between 0.1608 and 0.2392 b) 0.9544

9. Between 5,962 and 8,438 accounts

Chapter 10

Set 1

1. $z = 1.67$, $1.67 > 1.64$; reject

2. $z = -.824$, $.824 < 2.33$; reserve judgment

3. $z = -.7$, $.7 < 2.58$; reserve judgment

4. $z = -1.67$, $1.67 < 2.33$; reserve judgment

5. $z = -1$, $1 < 1.96$; reserve judgment

6. $z = 2.5$, $2.5 > 1.28$; reject

7. a) $z = 1.25$; reserve judgment
   b) $z = 5$; reject H_0

8. $z = -0.75$, $0.75 < 2.33$; reserve judgment

9. a) Between 0.7488 and 0.8512
   b) $z = 6$, $6 > 1.64$; reject claim

10. $z = -2.5$, $2.5 > 1.64$; reject H_0

A-10

Set 2

1. a) .6331   b) .0869
2. $z = -1.74$, $1.74 < 2.58$; not significant
3. $z = 2.198$, $2.198 > 1.96$; significant
4. $z = -.77$, $.77 < 1.64$; not significant
5. $z = -3.62$, $3.62 > 1.96$, we reject H_0

Set 3

1. $x^2 = 5.1875$, $5.1875 < 9.488$; not significant
2. $x^2 = 17.28$, $17.28 > 9.210$; significant
3. $x^2 = 26.666$, $26.666 > 9.210$; not independent
4. $27.777 > 9.210$; not independent

Review Exercises III

1. 0.9525
2. a) 0.1587   b) 0.1359
3. a) 0.0427   b) $p_1 = 0.418$, $p_2 = 0.582$
4. a) 0.1335   b) $p_1 = 0.1933$, $p_2 = 0.2067$
   c) Between 0.1488 and 0.2512
5. a) 0.8664   b) Between 0.0608 and 0.1392
6. Between 0.2935 and 0.4265
7. a) Between 0.459 and 0.541   b) 0.9544 confidence
8. a) 0.9544 confidence   b) several possible answers
9. $z = -3.57$, $3.57 > 2.33$; claim is rejected
10. a) $z = -0.5$; claim is not rejected
    b) $z = 3.25$; claim is rejected
11. $z = 1.33$, $1.33 < 1.64$; no change in quality
12. a) Between 0.1752 and 0.2248
    b) $z = 3.08$; $3.08 > 2.58$; claim is not valid
13. a) Between 0.1649 and 0.2351
    b) $z = 3.13$, $3.13 > 2.58$; claim is not valid
    c) Several possible answers
14. a) Between 0.7446 and 0.8554
    b) $z = 2$, $2 < 2.33$; reserve judgment
15. a) Between 0.198 and 0.282
    b) $z = 2$, $2 > 1.64$; sales have increased
16. $z = -3.70$; $3.70 > 1.96$; significant
17. $z = 1.41$, $1.41 < 2.58$; not significant
18. $x^2 = 6.66$, $6.66 > 5.991$; significant

Chapter 11

1. b) $Y' = 36 + 13X$, 81.5 beats
2. a) 13   **b)** 686 beats
3. a) $1750 = 1690 + 60$ **b)** 3.87   **c)** $r^2 = 0.966, r = 0.983$
4. a) $Y' = -0.79 + 0.31X$ **b)** \$293
5. a) $Y' = 5.6 + 1.6X$ **b)** \$21,600
  c) Fixed cost $= \$5,600$ **d)** Variable cost per unit $= \$1,600$
6. a) 0.7   **b)** $0.7 > 0.283$; significant
7. c) For $(66 < X < 72)$, Y' is always positive.
  d) $r = 0.51 > 0.418$; significant
8. a) $Y' = 4.178 + 0.901X$
  b) 1400.728 billions
  c) $r = 0.9997 > 0.487$; significant

Chapter 12

1. a) $Y' = 2.865 + 0.1372X$
  b) $0.0845 < B < 0.1899$
  c) $t = 1.668$, $1.668 < 2.998$; reserve judgment
2. a) $Y' = 0.0731 + 0.5688X$
  b) $0.4187 < B < 0.7189$
  c) $t = 0.693$, $0.693 < 1.860$; reserve judgment
3. a) $Y' = 57.05 + 10.55X$
  b) \$10,000
  c) $t = 0.2329$, $0.2329 < 1.812$; reserve judgment
  d) Between 188.02 and 242.58
4. a) $Y' = 11.824 + 4.696X$
  b) For every additional convention, the hotel expects 470 more customers.
  c) $4.223 < B < 5.169$
  d) Between 46,750 and 52,034 customers
5. a) $Y' = -37.162 + 116.080X$
  b) For $X \geq 1$, Y' is always positive.
  c) \$116.08
  d) $t = -1.738$, $1.738 < 2.650$; reserve judgment
  e) Between 418.40 and 435.80
6. \$321.75; between \$220.41 and \$423.10

Chapter 13

1. $z = 0.32$, $0.32 < 1.96$; reserve judgment
2. $z = 1.25$, $1.25 < 1.96$; reserve judgment
3. $z = 2.02$, $2.02 < 2.33$; reserve judgment
4. $z = 2.41$, $2.41 > 2.33$; reject null hypothesis

5. $z = -3.26$, $3.26 > 1.96$; intelligence is not the same

6. $z = -3.37$, $3.37 > 1.96$; length of sleep is not the same

7. $H = 13.28$, $13.28 > 5.99$; respiratory rate is not the same

8. $H = 30.07$; $30.07 > 13.277$; reject H_0

9. $r_s = 0.524$

10. $r_s = 0.903$

11. $r_s = 0.212$

12. a) $z = 2.4$, $2.4 < 2.58$; not significant
   b) $z = 2.1$, $2.1 > 1.96$; significant
   c) $z = 1.6$, $1.6 < 1.96$; not significant

13. $r_s = 0.421$

Index

Index